Interdisciplinary Statistics

STATISTICAL and PROBABILISTIC METHODS
in
ACTUARIAL SCIENCE

CHAPMAN & HALL/CRC
Interdisciplinary Statistics Series

Series editors: N. Keiding, B. Morgan, T. Speed, P. van der Heijden

Interdisciplinary Statistics

STATISTICAL and PROBABILISTIC METHODS
in
ACTUARIAL SCIENCE

Philip J. Boland
University College Dublin
Ireland

Chapman & Hall/CRC
Taylor & Francis Group
Boca Raton London New York

Chapman & Hall/CRC
Taylor & Francis Group
6000 Broken Sound Parkway NW, Suite 300
Boca Raton, FL 33487-2742

© 2007 by Taylor & Francis Group, LLC
Chapman & Hall/CRC is an imprint of Taylor & Francis Group, an Informa business

No claim to original U.S. Government works
Printed in the United States of America on acid-free paper
10 9 8 7 6 5 4 3 2

International Standard Book Number-10: 1-58488-695-1 (Hardcover)
International Standard Book Number-13: 978-1-58488-695-2 (Hardcover)

Visit the Taylor & Francis Web site at
http://www.taylorandfrancis.com

and the CRC Press Web site at
http://www.crcpress.com

Dedication

To my wife Elizabeth, and my children Daniel and Katherine.

Preface

This book covers many of the diverse methods in applied probability and statistics for students aspiring to careers in insurance, actuarial science and finance. It should also serve as a valuable text and reference for the insurance analyst who commonly uses probabilistic and statistical techniques in practice. The reader will build on an existing basic knowledge of probability and statistics and establish a solid and thorough understanding of these methods, but it should be pointed out that the emphasis here is on the wide variety of practical situations in insurance and actuarial science where these techniques may be used. In particular, applications to many areas of general insurance, including models for losses and collective risk, reserving and experience rating, credibility estimation, and measures of security for risk are emphasized. The text also provides relevant and basic introductions to generalized linear models, decision-making and game theory.

There are eight chapters on a variety of topics in the book. Although there are obvious links between many of the chapters, some of them may be studied quite independently of the others. Chapter 1 stands on its own, but at the same time provides a good introduction to claims reserving via the deterministic chain ladder technique and related methods. Chapters 2, 3 and 4 are closely linked, studying loss distributions, risk models in a fixed period of time, and then a more stochastic approach studying surplus processes and the concept of ruin. Chapter 5 provides a comprehensive introduction to the concept of credibility, where collateral and sample information are brought together to provide reasonable methods of estimation. The Bayesian approach to statistics plays a key role in the establishment of these methods. The final three chapters are quite independent of the previous chapters, but provide solid introductions to methods that any insurance analyst or actuary should know. Experience rating via no claim discount schemes for motor insurance in Chapter 6 provides an interesting application of Markov chain methods. Chapter 7 introduces the powerful techniques of generalized linear models, while Chapter 8 includes a basic introduction to decision and game theory.

There are many worked examples and problems in each of the chapters, with a particular emphasis being placed on those of a more numerical and practical nature. Solutions to selected problems are given in an appendix. There are also appendices on probability distributions, Bayesian statistics and basic tools in probability and statistics. Readers of the text are encouraged (in checking examples and doing problems) to make use of the very versatile and free statistical software package R.

The material for this book has emerged from lecture notes prepared for various courses in actuarial statistics given at University College Dublin (The National University of Ireland – Dublin) over the past 15 years, both at the upper undergraduate and first year postgraduate level. I am grateful to all my colleagues in Statistics and Actuarial Science at UCD for their assistance, but particularly to Marie Doyle, Gareth Colgan, John Connolly and David Williams. The Department of Statistics at Trinity College Dublin kindly provided me with accommodation during a sabbatical year used to prepare this material. I also wish to acknowledge encouragement from the Society of Actuaries in Ireland, which has been supportive of both this venture and our program in Actuarial Science at UCD since its inception in 1991. Patrick Grealy in particular provided very useful advice and examples on the topic of run-off triangles and reserving. John Caslin, Paul Duffy and Shane Whelan were helpful with references and data.

I have been fortunate to have had many excellent students in both statistics and actuarial science over the years, and I thank them for the assistance and inspiration they have given me both in general and in preparing this text. Particular thanks go to John Ferguson, Donal McMahon, Santos Faundez Sekirkin, Adrian O'Hagan and Barry Maher. Many others were helpful in reading drafts and revisions, including Una Scallon, Kevin McDaid and Rob Stapleton. Finally, I wish to thank my family and many friends who along the path to completing this book have been a constant source of support and encouragement.

Introduction

In spite of the stochastic nature of most of this book, the first chapter is rather deterministic in nature, and deals with Claims Reserving and Pricing with Run-off Triangles. In *running-off* a triangle of claims experience, one studies how claims arising from different years have developed, and then makes use of ratios (development factors and/or grossing-up factors) to predict how future claims will evolve. Methods for dealing with past and future inflation in estimating reserves for future claims are considered. The average cost per claim method is a popular tool which takes account of the numbers of claims as well as the amounts. The Bornhuetter–Ferguson method uses additional information such as expected loss ratios (losses relative to premiums) together with the chain ladder technique to estimate necessary reserves. Delay triangles of claims experience can also be useful in pricing new business.

Modeling the size of a claim or loss is of crucial importance for an insurer. In the chapter on Loss Distributions, we study many of the classic probability distributions used to model losses in insurance and finance, such as the exponential, gamma, Weibull, lognormal and Pareto. Particular attention is paid to studying the (right) tail of the distribution, since it is important to not underestimate the size (and frequency) of large losses. Given a data set of claims, there is often a natural desire to *fit* a probability distribution with reasonably tractable mathematical properties to such a data set. Exploratory data analysis can be very useful in searching for a good fit, including basic descriptive statistics (such as the mean, median, mode, standard deviation, skewness, kurtosis and various quantiles) and plots. The method of maximum likelihood is often used to estimate parameters of possible distributions, and various tests may be used to assess the fit of a proposed model (for example, the Kolmogorov–Smirnoff, and χ^2 goodness-of-fit). Often one may find that a mixture of various distributions may be appropriate to model losses due to the varying characteristics of both the policies and policyholders. We also consider the impact of inflation, deductibles, excesses and reinsurance arrangements on the amount of a loss a company is liable for.

Following on from a study of probability distributions for losses and claims, the chapter on Risk Theory investigates various models for the risk consisting of the total or aggregate amount of claims S payable by a company over a relatively short and fixed period of time. Emphasis is placed on two types of models for the aggregate claims S. In the *collective risk model* for S, claims are aggregated as they are reported during the time period under consideration, while in the *individual risk model* there is a term for each individual (or policyholder) irrespective of whether the individual makes a claim or not. Extensive statistical properties of these models are established (including the useful recursion formula of Panjer for the exact distribution of S) as well as methods of approximating the distribution of S. The models can inform analysts about decisions regarding expected profits, premium loadings, reserves necessary to ensure (with high probability) profitability, and the impact of reinsurance and deductibles.

The chapter on Ruin Theory follows the treatment of risk but the emphasis is put on monitoring the surplus (stochastic) process of a portfolio of policies throughout time. The surplus process takes account of initial reserves, net premium income (including, for example, reinsurance payments), and claim payments on a regular basis, and in particular focuses on the possibility of ruin (a *negative* surplus). A precise expression for the probability of ruin does not exist in most situations, but useful surrogates for this measure of security are provided by Lundberg's upper bound and the adjustment coefficient. An emphasis is placed on understanding how one may modify aspects of the process, such as the claim rate, premium loadings, typical claim size and reinsurance arrangements, in order to adjust the security level.

Credibility Theory deals with developing a basis for reviewing and revising premium rates in light of current claims experience (data in hand) and other possibly relevant information from other sources (collateral information). The constant challenge of estimating future claim numbers and/or aggregate claims is done in various ways through a credibility premium formula using a credibility factor $\mathbf{Z}$ for weighting the data in hand. In the classical approach to credibility theory, one addresses the question of how much data is needed for full credibility ($\mathbf{Z} = 1$), and what to do otherwise. In the Bayesian approach the collateral information is summarized by prior information and the credibility estimate is determined from the posterior distribution resulting from incorporating sample (current) claims information. If the posterior estimate is to be linear in the sample information, one uses the greatest accuracy approach to credibility, while if one needs to use the sample information to estimate prior parameters then one uses the Empirical Bayes approach to credibility theory. The chapter on Credibility Theory presents in a unified manner these different approaches to estimating future claims and numbers!

No Claim Discount (NCD) schemes (sometimes called Bonus-Malus systems) are experience rating systems commonly used in European motor insurance. They attempt to create homogeneous groups of policyholders whereby those drivers with bad claims experience pay higher premiums than those who have good records. The theory is that they also reduce the number of small claims, and lead to safer driving because of the penalties associated with making claims. NCD schemes provide a very interesting application of discrete Markov chains, and convergence properties of the limiting distributions for the various (discount) states give interesting insights into the stability of premium income.

Modeling relationships between various observations (responses) and variables is the essence of most statistical research and analysis. Constructing interpretable models for connecting (or linking) such responses to variables can often give one much added insight into the complexity of the relationship which may often be hidden in a huge amount of data. For example, in what way is the size of an employer liability claim related to the personal characteristics of the employee (age, gender, salary) and the working environment (safety standards, hours of work, promotional prospects)? In 1972 Nelder and Wedderburn developed a theory of generalized linear models (GLM) which unified much of the existing theory of linear modeling, and broadened its scope to a wide class of distributions. The chapter on Generalized Linear Models begins with a review of normal linear models. How generalized linear models extend the class of general linear models to a class of distributions known as exponential families and the important concept of a link function are discussed. Several examples are given treating estimation of parameters, the concept of deviance, residual analysis and goodness-of-fit.

All around us, and in all aspects of life, decisions continually need to be made. We are often the decision makers, working as individuals or as part of team. The decisions may be of a personal or business nature, and often enough they may be both! The action or strategy which a decision maker ultimately takes will of course depend on the *criterion* adopted, and in any given situation there may be several possible criteria to consider. In the chapter on Decision and Game Theory, an introduction to the basic elements of zero-sum two-person games is given. Examples are also given of variable-sum games and the concept of a Nash equilibrium. In the treatment of decision theory we concentrate on the *minimax* and *Bayes* criteria for making decisions. A brief introduction to utility theory gives one an insight into the importance of realizing the existence of value systems which are not strictly monetary in nature.

Philip J. Boland

Dublin

September 2006

Contents

1

Claims Reserving and Pricing with Run-Off Triangles

1.1 The evolving nature of claims and reserves

In general insurance, claims due to physical damage (to a vehicle or building) or theft are often reported and settled reasonably quickly. However, in other areas of general insurance, there can be considerable delay between the time of a claim-inducing event, and the determination of the actual amount the company will have to pay in settlement. When an incident leading to a claim occurs, it may not be reported for some time. For example, in employer liability insurance, the exposure of an employee to a dangerous or toxic substance may not be discovered for a considerable amount of time. In medical malpractice insurance, the impact of an erroneous surgical procedure or mistakenly prescribed drug may not be evident for months, or in some cases years. In other situations, a claim may be reported reasonably soon after an incident, but a considerable amount of time may pass before the actual extent of the damage is determined. In the case of an accident the incident may be quickly reported, but it may be some time before it is determined actually who is liable and to what extent. In some situations, one might have to wait for the outcome of legal action before damages can be properly ascertained.

An insurance company needs to know on a regular basis how much it should be setting aside in reserves in order to handle claims arising from incidents that have already occurred, but for which it does not yet know the full extent of its liability. Claims arising from incidents that have already occurred but which have not been reported to the insurer are termed *IBNR (Incurred But Not Reported) claims*, and a reserve set aside for these claims is called an *IBNR reserve*. Claims that have been reported but for which a final settlement has not been determined are called *outstanding claims**. An assessor may make interim payments on a claim (say 5,000 is paid immediately and a further 15,000 at a later stage), thus a claim remains open and outstanding until it has been settled and closed. *Incurred claims* are those which have been

---

*Other terms sometimes used are open claims or IBNER claims (Incurred But Not Enough Reported).

already paid, or which are outstanding. Reserves for those claims that have been reported, but where a final payment has not been paid, are called *case reserves*.

In respect of claims that occur (or originate) in any given accounting or financial year, ultimate losses at any point in time may be estimated as the sum of paid losses, case reserves and IBNR reserves. Estimated incurred claims or losses are the case reserves plus paid claims, while total reserves are the IBNR reserves plus the case reserves. Of course, in practice one usually sets aside combined reserves for claims originating in several different (consecutive) years.

Claims reserving is a challenging exercise in general insurance! One should never underestimate the knowledge and intuition that an experienced claims adjuster makes use of in establishing case reserves and estimating ultimate losses. However, mathematical models and techniques can also be very useful, giving the added advantage of laying a basis for simulation.

In order to give a flavor for the type of problem one is trying to address in claims reserving, consider the triangular representation of cumulative incurred claims given in Table 1.1 for a household contents insurance portfolio. The *origin year* refers to the year in which the incident giving rise to a claim occurred, and the *development year* refers to the delay in reporting relative to the origin year. For example, incremental claims of $(105{,}962 - 50{,}230) = 55{,}732$ were made in 2001 in respect of claims originating in 2000 (hence delayed one year). Of course, one could equivalently present the data using incremental claims instead of cumulative claims. This can be particularly useful when one wants to take account of inflation and standardize payments in some way. Given that the amounts past the first column (development year 0) are indicators of delayed claims, this type of triangular representation of the claims experience is often called a *delay* triangle. Note the significance of the diagonals (those going from the lower left to the upper right) in this delay triangle. For example, the diagonal reading (50,230; 101,093; 108,350) represents cumulative claims for this portfolio at the end of year 2000 for the origin years (2000, 1999, 1998), respectively.

In some cases, the *origin year* might refer to a policy year, or some other accounting or financial year (in this example, it refers to a calendar year). The term *accident year* is commonly used in place of origin year, given the extensive historical use of triangular data of this type in motor insurance. The use of months or quarters may also be used in place of years depending on the reporting procedures of a company. There are perhaps many questions which a company would like answered with respect to information of this type, and certainly one would be to determine what reserves it should set aside at the end of 2002 to handle forthcoming claim payments in respect of incidents originating in the period $1998 - 2002$. In short, how might one *run-off* this triangle?

Of course any good analyst would question the quality of the available data, and make use of any additional information at hand. For example, in

TABLE 1.1

Cumulative incurred claims in a household contents insurance portfolio.

Origin year	Development year 0	1	2	3	4
1998	39,740	85,060	108,350	116,910	124,588
1999	47,597	101,093	128,511	138,537	
2000	50,230	105,962	132,950		
2001	50,542	107,139			
2002	54,567				

this situation is it fair to assume that all claims will be settled by the end of the fourth development year for any origin year? If not, what provisions should be made for this possibility? Can we make the assumption that the way in which claims develop is roughly similar for those originating in different years? Should inflation be taken into account? Is there information at hand with respect to the number of claims reported in each of these years (is there a delay triangle for reported claim numbers)? What other knowledge have we about losses incurred in the past (for example, with respect to premium payments) for this type of business? In this chapter, we will discuss several different ways of addressing the questions posed above. In most cases there is no one definitive answer, and in many situations it is (perhaps best to try several methods to get a reasonable overall estimate of the reserves that should be held.

Certainly one of the most frequently used techniques for estimating reserves is the *chain ladder method*. In this method, one looks at how claims arising from different origin (or cohort) years have developed over subsequent development years, and then use relevant ratios (for example, development factors or grossing-up factors) to predict how future claims from these years will evolve. There are many ways in which one might define a *development factor* for use in projecting into the future. Generally speaking, it will be some ratio (> 1) based on given data which will be used as a multiplier to estimate the progression into the future between consecutive or possibly many years.

The use of *grossing-up* factors to project into the future is similar and in reality dual to the use of development factors. A grossing-up factor is usually (but not necessarily) a proportion (< 1) representing that part of the ultimate (or next year's) estimated cumulative losses which have been incurred or paid to date. Consider, for example, the progression or development of cumulative claims in a portfolio of policies from year 2003 to 2004. We might use a development factor of $d = 11/10$ to estimate how claims will evolve during that one-year period – or, in other words, predict that cumulative claims at the end of 2004 will be 1.1 times those at the end of 2003. Equivalently, we might say that we expect cumulative claims to be grossed-up by a factor of $g = 10/11$ by the end of year 2004. Note of course that $g = 1/d$. Whether one uses development factors or grossing-up factors is often a matter of choice, but in some situations this may be determined by the type of information available.

How the chain ladder method is used to run-off a claims triangle is developed in Section 1.2. In particular, the question of how to deal with past and future inflation in estimating reserves is considered. The average cost per claim method is a popular tool which is dealt with in Section 1.3. This method takes account of the numbers of claims reported (and therefore may be useful for estimating case reserves but not IBNR reserves). The Bornhuetter–Ferguson method [7], which is developed in Section 1.4, uses additional information such as loss ratios (losses relative to premiums) together with the chain ladder technique to estimate necessary reserves. We shall see that there is a Bayesian flavor to the interpretation of the Bornhuetter–Ferguson estimate. Delay triangles of claims experience can also be useful in pricing business, and a detailed practical example of pricing an employer's liability scheme is discussed in Section 1.5. All of the above techniques are rather deterministic in nature, and it seems natural to also consider statistical models which would allow one to evaluate fitness, variability and basic assumptions better. In Section 1.6 we mention very briefly the separation technique/model for claims reserving.

1.2 Chain ladder methods

The chain ladder method for running off a delay triangle of claims is one of the most fundamental tools of a general insurance actuary. The most basic chain ladder method assumes that the pattern of developing claims is reasonably similar over the different origin years. Before explaining this method in detail, we return to the household contents claims data given in Table 1.1. If one felt strongly that the pattern of how cumulative claims developed for the origin year 1998 was representative of how they should develop in other years, then it could be used as a basis for predictions. For example, we note that cumulative claims for origin year 1998 increased by a factor of $(85{,}060/39{,}740) = 2.1404$ in the first development year. We might therefore predict a similar increase in claims for those originating in 2002 – that is, we might estimate the cumulative incurred claims in 2003 for those originating in 2002 to be $54{,}567\,(2.140) = 116{,}796$. In a similar manner, we might estimate ultimate claims (assuming all claims are settled by the end of development year 4) for those originating in 2002 to be

$$54{,}567\left(\frac{124{,}588}{39{,}740}\right) = 54{,}567\left(\frac{85{,}060}{39{,}740}\right)\left(\frac{108{,}350}{85{,}060}\right)\left(\frac{116{,}910}{108{,}350}\right)\left(\frac{124{,}588}{116{,}910}\right)$$

$$= 54{,}567\,(2.1404)\,(1.2738)\,(1.0790)\,(1.0657)$$

$$= 171{,}072.$$

Here the numbers $2.1404, 1.2738, 1.0790, 1.0657$ are the respective *one-year development ratios* for the growth in incurred claims for those which actually originated in 1998.[†] Continuing in this way, we could run-off the claims triangle of Table 1.1 based on these development ratios, and obtain the results (where estimates are in bold) in Table 1.2. An estimate of the reserves which should be set aside for the forthcoming claim settlements in this portfolio would therefore be $(147{,}635 + 152{,}875 + 156{,}927 + 171{,}072) - (138{,}537 + 132{,}950 + 107{,}139 + 54{,}567) = 195{,}316$.

TABLE 1.2
Estimated cumulative incurred claims for Table 1.1 using 1998 development ratios.

		Development year			
Origin year	0	1	2	3	4
1998	39,740	85,060	108,350	116,910	124,588
1999	47,597	101,093	128,511	138,537	**147,635**
2000	50,230	105,962	132,950	**143,453**	**152,875**
2001	50,542	107,139	**136,474**	**147,256**	**156,927**
2002	54,567	**116,796**	**148,775**	**160,529**	**171,072**

1.2.1 Basic chain ladder method

If we feel that claims development is reasonably stable over the years in question, we could probably benefit from making use of the additional information which is available for the years $1999 - 2002$ in making projections. For example, for the evolution of claims from year of origin to development year 1, we have information from the four origin years $1998 - 2001$. The observed one-year development ratios for these years are, respectively, $2.1404, 2.1239, 2.1095$ and 2.1198. Similarly, one may calculate other development ratios for cumulative incurred (or paid if that is the case) claims between development years with the results given in Table 1.3.

In order to predict the evolution of cumulative claims from year of origin to development year 1 in forthcoming years, we might use some average of the ratios $(2.1404, 2.1239, 2.1095, 2.1198)$. One possibility would be to use a straightforward arithmetic average of these ratios, which has some justification in that it puts an equal weight on each of the years $1998 - 2001$. More commonly, however, one uses a weighted average where the weights for a year are proportional to the origin year incurred claims. In other words, more

---

[†]Ratios such as these are also sometimes referred to as *link ratios* or *development factors*.

TABLE 1.3
One-year development ratios for cumulative incurred
household contents claims by year of origin.

| | Development year | | | |
Origin year	$0 \to 1$	$1 \to 2$	$2 \to 3$	$3 \to 4$
1998	2.1404	1.2738	1.0790	1.0657
1999	2.1239	1.2712	1.0780	
2000	2.1095	1.2547		
2001	2.1198			
2002				

weight is put on a factor where larger claim amounts were incurred. This is the technique used in the *basic chain ladder method* for running off a delay triangle. Using the notation $d_{i|j}$ to denote an estimate of the development factor for cumulative claims from development year i to development year j, the *pooled* estimate of the development factor $d_{0|1}$ from year of origin to development year 1 (that is, for the development $0 \to 1$) for our household contents portfolio data would be:

$$
d_{0|1} = \frac{39{,}740(2.1404) + 47{,}597(2.1239) + 50{,}230(2.1095) + 50{,}542(2.1198)}{(39{,}740 + 47{,}597 + 50{,}230 + 50{,}542)}
$$
$$
= \frac{85{,}060 + 101{,}093 + 105{,}962 + 107{,}139}{39{,}740 + 47{,}597 + 50{,}230 + 50{,}542}
$$
$$
= 2.1225.
$$

Note that $d_{0|1}$ is the sum of the four entries in the column for development year 1 divided by the corresponding elements for development year 0 of Table 1.1. In a similar fashion, we obtain estimates for the other development factors of the form $d_{i|i+1}$, the results of which are given in Table 1.4.

TABLE 1.4
Pooled one-year development factors for
cumulative household contents claims.

$d_{0|1}$	$d_{1|2}$	$d_{2|3}$	$d_{3|4}$
2.1225	1.2660	1.0785	1.0657

In general, for $j > i + 1$, one would use $d_{i|j} = \prod_{l=i}^{j-1} d_{l|l+1}$ to estimate development from year i to j. In this basic form of the chain ladder method, one uses these pooled estimates for running off a delay triangle. For example, our estimate of the claims incurred by the end of 2005 which originated in 2001 would be $107{,}139\,(1.2660)(1.0785)(1.0657) = 155{,}885$. Proceeding in this way, one obtains the projected results given in Table 1.5. Estimated total reserves

for the aggregate claims incurred by 2002 using the basic chain ladder method
would therefore be

$$(147,635 + 152,799 + 155,885 + 168,511)$$
$$-(138,537 + 132,950 + 107,139 + 54,567) \quad = \quad 191,637.$$

This is slightly less $(3,679 = 195,316 - 191,637)$ than the estimated reserves
when using the development factors based on the origin year 1998 only! This
represents a difference of less than 2%, which is mainly explained by the fact
that the one-year development factors based only on the origin year 1998 are
in each case slightly larger than those determined by using information over
all of the years.

What is the best estimate in this case? There is no easy answer to a
question like this. However, the estimate using the pooled estimates of the
development factors makes use of all the information available at this stage,
and therefore is probably a safer method to use in general. On the other hand,
this assumes that the development of claims is reasonably stable over the years
being considered, and if for any reason this is in doubt one should modify the
estimate in an appropriate way. So far we have assumed that either inflation
is not a concern, or that the figures have already been appropriately adjusted.
In the next section, we shall consider a chain ladder method that adjusts for
inflation.

TABLE 1.5
Estimated cumulative incurred claims for household contents
claims using the basic chain ladder method.

| | Development year | | | | |
Origin year	0	1	2	3	4
1998	39,740	85,060	108,350	116,910	124,588
1999	47,597	101,093	128,511	138,537	**147,635**
2000	50,230	105,962	132,950	**143,382**	**152,799**
2001	50,542	107,139	**135,636**	**146,279**	**155,885**
2002	54,567	**115,816**	**146,621**	**158,126**	**168,511**

An equivalent way of projecting cumulative claims is through the use of
grossing-up factors. In our household claims data, note that cumulative (incurred) claims of 116,910 at the end of development year 3 for those orig-
inating in 1998 are a (grossing-up) factor $g_{3|4} = 116,910/124,588 = 0.9384$
of the ultimate cumulative claims 124,588 at the end of development year 4.
A grossing-up factor for change from development year 2 to 3 could be determined in several ways, and one possibility (corresponding to the pooling
of development factors presented above) would be a pooled estimate based
on the changes observed from origin years 1998 and 1999, that is, to use

$g_{2|3} = (108{,}350 + 128{,}511)/(116{,}910 + 138{,}537) = 0.9272$. Proceeding in this way by pooling information, one obtains the grossing-up factors given in Table 1.6. Note that for our procedure in this example we have $g_{i|j} = 1/d_{i|j}$. One may then use the grossing-up factors to run-off the cumulative claims table, in this case obtaining the same results as in Table 1.5. Another possibility (and one that is often used and illustrated later) is to use an *arithmetic average* of the grossing-up estimates determined separately from the experience in origin years 1998 and 1999. As mentioned previously, it is often a matter of preference whether to use development or grossing-up factors, although in practice both methods are frequently used. We will see in Section 1.4 that grossing-up factors may be interpreted as credibility factors in the Bornhuetter–Ferguson method for estimating reserves.

TABLE 1.6
Pooled one-year grossing-up factors for
incurred household contents claims.

$g_{0|1}$	$g_{1|2}$	$g_{2|3}$	$g_{3|4}$
0.4712	0.7899	0.9272	0.9384

Although the chain ladder technique is a deterministic method and not necessarily based on a stochastic or statistical model, it is still advisable to investigate how well this technique *fits* the known data. For each entry of cumulative incurred claims in the original Table 1.1, we could use the pooled estimates of the development factors to determine a *fitted* cumulative value and make comparisons. It is, however, perhaps more enlightening to compare *increases in cumulative claims* (or incremental claims) over the various development years, and Table 1.7 gives the calculations for the household contents claims portfolio.

Table 1.7 indicates that there is quite a good fit between the actual incurred incremental claims and those predicted using the pooled development factors from year of origin. Not surprisingly, in view of the comments made above concerning the pooled development factors and those based on the origin year 1998 alone, the fitted values for 1998 are all slightly less than the actual values.

1.2.2 Inflation-adjusted chain ladder method

In the setting of reserves on the basis of information obtained from past years, one should be cognizant of the fact that inflation may have affected the values of claims. Incurred claims of 39,740 in 1998 might be worth considerably more in 2002 prices, hence if we are to set aside reserves in 2002 for future claim payments should we not make projections on the basis of comparable monetary values? One should also bear in mind that inflation which affects claims

TABLE 1.7

Actual and fitted values for increases in household contents incurred claims.

Origin year	Value	Development year				
		0	1	2	3	4
1998	Actual	39,740	45,320	23,290	8,560	7,678
	Fitted	39,740	44,607	22,434	8,379	7,563
	Difference		-713	-856	-181	-115
	% Difference		-1.6	-3.7	-2.1	-1.5
1999	Actual	47,597	53,496	27,418	10,026	
	Fitted	47,597	53,426	26,870	10,035	
	Difference		-70	-548	9	
	% Difference		-0.13	-2.0	0.1	
2000	Actual	50,230	55,732	26,988		
	Fitted	50,230	56,381	28,356		
	Difference		649	1368		
	% Difference		1.2	5.1		
2001	Actual	50,542	56,597			
	Fitted	50,542	56,731			
	Difference		134			
	% Difference		0.2			

might be quite different from inflation as reported in standard consumer price indices. Changes in legislation might affect the way compensation entitlements are determined, and therefore suddenly affect settlements in liability claims. New safety standards might also affect the cost of both repairing and replacing damaged property. In the inflation-adjusted chain ladder method, we adjust the claims incurred in past years for inflation and convert them into equivalent prices (say current prices). We then apply a chain ladder technique (using development factors or grossing-up ratios) to run-off the delay triangle of standardized cumulative claims.

Once again, we use the data on incurred claims in the household contents insurance portfolio of Table 1.1. Suppose that yearly inflation over the four years from mid-1998 until mid-2002 has been, respectively, 2%, 8%, 7% and 3%. As an approximation, we assume that claims in a particular year are on the average incurred in the middle of the year. Table 1.8 gives the *incremental* incurred claims in 2002 prices for this data. For example, in the origin year 1999, claims incurred in the year 2000 were $53,496 = 101,093 - 47,597$, which in mid-2002 money is $58,958 = 53,496(1.07)(1.03)$.

Claims are then accumulated by year of origin and development, then pooled development factors (which more than likely will be different from those determined on the data before inflation was taken into account) are calculated, and the delay triangle can be run-off using these factors. The results are given in Table 1.9, together with the development factors used. For example, estimated cumulative claims from those originating in 2000 by the end of year 2004 (in 2002 money) would be $(55,358+57,404+26,988)(1.0693)(1.0562)$ $= 157,837$. Total reserves that should be set aside for the future would then be 174,950, which is considerably less (by 8.7%) than the necessary reserves 191,637 calculated without taking inflation into account. In this case had we not taken into account inflation, one might easily have set aside too much in 2002 for reserves.

TABLE 1.8
Incremental incurred claims for household contents claims of Table 1.1 in 2002 prices.

Origin year	Development year				
	0	1	2	3	4
1998	48,247	53,943	25,668	8,817	7,678
1999	56,653	58,958	28,241	10,026	
2000	55,358	57,404	26,988		
2001	52,058	56,597			
2002	54,567				

TABLE 1.9
Estimated cumulative incurred claims in 2002 prices using the chain ladder method with inflation.

Origin year	Development year				
	0	1	2	3	4
1998	48,247	102,190	127,858	136,675	144,353
1999	56,653	115,611	143,852	153,878	**162,522**
2000	55,358	112,762	139,750	**149,442**	**157,837**
2001	52,058	108,655	**135,246**	144,625	152,749
2002	54,567	**112,882**	**140,507**	150,251	158,692
$d_{i\mid i+1}$		$d_{0\mid 1} = 2.0687$	$d_{1\mid 2} = 1.2447$	$d_{2\mid 3} = 1.0693$	$d_{3\mid 4} = 1.0562$

1.3 The average cost per claim method

This technique for estimating reserves revolves around analyzing both claim numbers and the average cost per claim as they develop over different origin years. The idea is to study separately frequency as well as severity of claims by using delay triangles for both claim numbers and average costs per claim, then to run-off these triangles to obtain estimates of ultimate claim numbers and mean costs. The results are then multiplied together to give estimates of future payments, thereby determining the necessary reserves. The triangles may be run-off by using the development factor tool introduced already with the basic chain ladder method, but there are also other possibilities. In the following example, we illustrate the use of grossing-up factors.

Example 1.1

Table 1.10 gives cumulative incurred claim amounts (C) in (\$ 000's) and claim numbers (N) over a four-year period for a collection of large vehicle damage claims. We would like to estimate reserves which should be set aside for future payments on this business (i.e., arising out of claims from the given years of origin). Let us assume the figures have already been adjusted for inflation. We will also make the assumption that claims tail off after three years of development for any origin year (e.g., we assume that no further claims arise from 1999 other than the 63 already incurred), although we will return to this point later in the example!

TABLE 1.10
Cumulative incurred claims (C) and numbers (N) for large vehicle damage.

Origin year	\multicolumn{8}{c}{Development year}

	\multicolumn{2}{c}{0}	\multicolumn{2}{c}{1}	\multicolumn{2}{c}{2}	\multicolumn{2}{c}{3}
Origin year	C	N	C	N	C	N	C	N
1999	677	42	792	51	875	57	952	63
2000	752	45	840	54	903	59		
2001	825	52	915	60				
2002	892	59						

Dividing *cumulative* claim amounts by claim numbers in Table 1.10 we obtain Table 1.11, which gives the average size of an incurred claim up to a given development year. For example, the average size or severity of an incurred claim originating in 2000 by the end of 2001 is \$15,560. We now run-off both the triangles for average claim sizes and claim numbers, using

TABLE 1.11
Average cumulative incurred claim size
for large vehicle damage.

Origin year	Development year			
	0	1	2	3
1999	16.12	15.53	15.35	15.11
2000	16.71	15.56	15.31	
2001	15.87	15.25		
2002	15.12			

the grossing-up method.

We begin by running off the claim numbers. For this example, we shall use the notation $g_{i|j}^N$ (respectively, $g_{i|j}^C$) to denote a grossing-up factor from development year i to development year j for claim numbers (average claim size), and again note that there is no unique way to determine such factors. Here we are mainly concerned with ultimate values (for predicting to the ultimate - which in this case is the end of the third development year).

Considering the claim numbers in Table 1.10, we see that we have only one estimate for the grossing-up factor from development year 2 to 3, namely $g_{2|3}^N = 57/63 = 0.90476$, or in other words $57/g_{2|3}^N = 63$. How about an estimate for grossing-up claim numbers from development year 1 to 2? We have the experience of both origin years 1999 and 2000, and we can use this information in several ways. As suggested before, we might weight (proportional to numbers of claims) the two estimates obtained from these two years, but here we will use a straightforward *arithmetic mean*, that is, we use $g_{1|2}^N = [(51/57) + (54/59)]/2 = 0.90500$. In a similar and consistent way we estimate $g_{0|1}^N = [(42/51) + (45/54) + (52/60)]/3 = 0.84118$. Using these grossing-up factors, we could run-off the triangle of claim numbers to obtain estimates for numbers of incurred claims over subsequent development years. In general these can be obtained using $g_{i|l}^N = \prod_{j=i}^{l-1} g_{j|j+1}^N$, where l is the number of the development year. Since here we are interested in predicting to the end of development year 3, we have $g_{1|3}^N = g_{1|2}^N g_{2|3}^N = (0.90500)(0.90476) = 0.81881$ and, similarly, $g_{0|3}^N = 0.68876$. Hence to predict the ultimate number of claims originating in the year 2001 by the end of 2004 (end of development year 3), we would gross-up 60 by $g_{1|3}^N = 0.81881$ to obtain the estimate of $60/(0.81881) = 73.28$. Continuing in this way for other origin years, we obtain the results in Table 1.12.

Next we use grossing-up factors to run-off the triangle of average claim sizes given in Table 1.11. Note, however, that in this example, although both incurred claim amounts and claim numbers increase with development years, the average amount of an incurred claim is decreasing. In some situations claims incurred later on are typically larger, but the opposite is the case here. We proceed to determine grossing-up factors as with the claim numbers, but

TABLE 1.12
Estimated ultimate number of claims: grossing-up
method for large vehicle damage.

| Origin | Development year i | | | | Ultimate |
year	0	1	2	3	# claims
1999	42	51	57	63	63.00
2000	45	54	59		65.21
2001	52	60			73.28
2002	59				85.66
$g_{i\mid3}^{N}$	0.68876	0.81881	0.90476	1	

of course due to the fact that the average incurred claim size is decreasing, our grossing-up factors here will in all cases exceed 1. The grossing-up factor between development years 2 and 3 is $g_{2\mid3}^{C} = 15.35/15.11 = 1.01587$. For the grossing-up factor between development years 1 and 2, we again use an arithmetic average of what we have observed in origin years 1999 and 2000 to obtain $g_{1\mid2}^{C} = [(15.53/15.35) + (15.56/15.31)]/2 = 1.01399$. Similarly, one obtains

$$g_{0\mid1}^{C} = [(16.12/15.53) + (16.71/15.56) + (15.87/15.25)]/3 = 1.05095$$

based on the experience in years 1999 − 2001. We could then use these factors to estimate the average cost of a claim originating in a given year at the end of any development year. For example, the average size of a claim originating in year 2001 at the end of 2003 is estimated to be $15.25/g_{1\mid2}^{C} = (15.25/1.01399) = 15.04$, and the estimate of the ultimate average size would be $15.04/g_{2\mid3}^{C} = 15.04/1.01587 = 14.80$. Estimates of ultimate average claim sizes can, of course, be determined directly by using the ultimate grossing-up factors. Hence, since

$$g_{0\mid3}^{C} = g_{0\mid1}^{C}\, g_{1\mid2}^{C}\, g_{2\mid3}^{C} = (1.05095)\,(1.01399)\,(1.01587) = 1.08256,$$

the estimate of the ultimate average claim size originating in 2002 is $13.97 = 15.12/1.08256$. These results are presented in Table 1.13.

Finally, we can estimate the total amount ultimately payable for incurred claims for each origin year by multiplying the predicted number of claims by the average severity. In origin year 2001, for example, we expect to ultimately pay $(73.28)(14.80) = 1084.84$. Summing over all years, we would expect to pay in total about \$4,215,620 since

$$(63)(15.11) + (65.21)(15.07) + (73.28)(14.80) + (85.66)(13.97) =$$
$$952 + 982.46 + 1084.84 + 1196.31 = 4,215.62.$$

Let us return to reconsider our initial assumption that claims tail off after three years of development. Looking carefully at the development of claim

TABLE 1.13

Estimated ultimate average claim size: grossing-up
method for large vehicle damage.

| Origin | Development year i | | | | Ultimate |
year	0	1	2	3	claim size
1999	16.12	15.53	15.35	15.11	15.11
2000	16.71	15.56	15.31		15.07
2001	15.87	15.25			14.80
2002	15.12				13.97
$g^C_{i\mid i+1}$	1.05095	1.01399	1.01587		
$g^C_{i\mid 3}$	1.08256	1.03008	1.01587	1	

numbers arising from origin year 1999 in Table 1.10, one might have some
misgivings about this assumption. After all, there were 6 claims arising in both
development years 2 and 3 from origin year 1999! In light of this, one might
decide to add a tail factor to the claim numbers of Table 1.10 and reevaluate
the numbers of ultimate claims for each origin year (and, consequently, the
total amount ultimately payable for this business). For example, one might
decide that claims from origin year 1999 have not *finished* developing (i.e.,
there are some IBNR claims) and that 70 is a more reasonable estimate of
the number of ultimate claims arising from this year! Using the grossing-up
factor of 63/70, one would then estimate the number of ultimate claims for
origin years $(2000, 2001, 2002)$ to be, respectively, $(72.46, 81.42, 95.17)$. Using
the previously determined average severity figures, the corresponding estimate
of the total amount ultimately payable on this business would rise to about
$4,684,018.   □

1.4   The Bornhuetter–Ferguson or loss ratio method

In determining future reserves, a good analyst looks at the claims data in
different ways and tries to make the best use of any collateral information
available. In the *Bornhuetter–Ferguson* (B–F) *method*, information on loss
ratios is combined with a standard projection technique like the basic chain
ladder method to estimate necessary reserves. Just as in the average cost per
claim method, this method tries to combine information on how average claim
amounts as well as claim numbers develop over time, and compare them with
changes in losses relative to collected premiums. In their original development
of this method, Bornhuetter and Ferguson [7] applied it to data on incurred
claims, but it can clearly be used for paid claims as well.

There are many ratios that one may consider to evaluate trends in losses in the insurance process (see Chapter 12 in [19]). Here we shall normally use the term *loss ratio*[‡] to mean the ratio of incurred or paid claims to earned premiums over a given period of time. The *combined ratio* is the ratio of incurred claims plus expenses to earned premiums, while the *trading ratio* takes into account the investment return on premiums and reserves. The trading ratio then takes the form of (incurred claims + expenses − investment returns)/earned premiums. Traditionally, the Bornhuetter–Ferguson method takes account of the loss or claim ratio in estimating reserves, but if sufficiently good information is also available about developing costs and investment returns, then the method can be adapted to make use of other relevant ratios.

Normally, one would expect a certain amount of consistency over time in the loss ratios calculated on the basis of business in different origin years. Of course, there could be exceptional years which might arise because of unusual events like floods, hurricanes, market crashes, terrorist activities and other disasters. A very important factor influencing the expected loss ratio is the *market cycle* itself! The market cycle is where premium rates on a class of business rise or fall due to the economic effects of supply and demand in the provision of insurance capacity. The more (less) insurers there are competing for business, the lower (higher) overall premiums will be. The 2005 hurricane *Katrina* is an example of an unusual event which affected the market cycle and resulted in premium rates for property catastrophe increasing because fewer insurers were prepared to write this type of business!

The Bornhuetter–Ferguson method in brief consists of the following procedure. Given incurred or paid claims data (adjusted for inflation) in triangular form, one uses the basic chain ladder method to determine development factors (or, equivalently, grossing-up factors) for running off the triangle (one does not actually need to run-off the triangle here, it is the development/grossing-up factors which are of primary concern). Next, one turns to the loss ratios, and for each origin year one obtains an initial or prior estimate (on the basis of premiums paid) of the total ultimate loss. For example, if the loss ratio is a constant 0.92 over all years, then the initial estimates of ultimate losses would be 92% of the premiums earned in an origin year. These initial estimates are made independently of the way the claims are developing. In the next step one finds for each year of origin what claims should have been incurred (or paid) at the present time assuming that claims actually do develop to the ultimate according to the calculated development factors. This amount is then subtracted from the initial estimate of ultimate claims to find what is called the *emerging liability*. The emerging liability is then added to the *reported liability* (which is the actual or observed incurred liability) to obtain the Bornhuetter–Ferguson estimate of ultimate liability for a given origin year. If, for example, for a given year of origin we have reported a higher amount

---

[‡]This is also commonly referred to as the *claim ratio*.

than expected (reported liability exceeds predicted current liability), then we adjust our initial estimate of ultimate liability upwards by this amount. These estimates are then used to determine what reserves should be set aside.

The method is best illustrated by an example, but first we give some notation which should assist in understanding the procedure. Assume that our data consists of a delay triangle of cumulative incurred (or paid) claims over k development years, where $C_{i,j}$ represents the cumulative amount of claims by development year j which originate in year i, for $i = 1, \ldots, k+1$ and $j = 0, 1, \ldots, k$. We let r_i, P_i and U_i^I be, respectively, the loss ratio, earned premium and initial estimate of the ultimate liability for claims originating in year i. In many cases, r_i will be assumed to be constant over different origin years, perhaps because underwriting practices aim to charge premiums with this objective in mind (for example, that claims should amount to 92% of premiums). We use $d_j = d_{j|k}$ to denote the development factor from development year j to the ultimate (k in this case) which is determined by using the chain ladder technique on the delay triangle. It is worth noting that since there are different methods for determining these factors (for example, by taking weighted or arithmetic averages of the appropriate link ratios), there are several slightly different ways of proceeding in the Bornhuetter–Ferguson method.

The initial estimate of ultimate liability for claims originating in year i is $U_i^I = (r_i)P_i$. If the development factors are good indicators of how claims should evolve, then we would expect to have incurred approximately U_i^I/d_{k+1-i} in claims at the present time. The difference between the initial estimate of ultimate liability for those in origin year i and this approximation of what we should have incurred is $(1-1/d_{k+1-i})U_i^I$, which is called the emerging liability (claims that we still expect to incur or pay). This emerging liability for origin year i is then added to the observed or reported liability $C_{i,k+1-i}$ to obtain the B–F estimate U_i of ultimate liability for origin year i given by

$$U_i = C_{i,k+1-i} + (1 - 1/d_{k+1-i})\, U_i^I.$$

Table 1.14 shows the cumulative amounts ($000's) of incurred claims over five years in a household insurance portfolio. We assume that the figures are adjusted for inflation, and that earned premiums and loss ratios are also given from which initial estimates of ultimate liabilities are obtained. Note that the loss ratio has increased in origin years 4 and 5 from 86% to 88%. Hence the initial estimate for the ultimate liability arising from origin year 4 is $U_4^I = (0.88)7481 = 6583.3$.

The next step in the B–F method is to determine development ratios for running off the delay triangle, and in this case we use the weighted average or pooled method of link (development) factors as described in the basic chain ladder method. For example, the development factor

$$d_2 = d_{2|4} = d_{2|3}\, d_{3|4} = \frac{4176 + 4608}{3956 + 4527}\, \frac{4271}{4176} = 1.05904.$$

TABLE 1.14

Household insurance data: incurred claims, premiums and loss ratios.

Origin year i	Development year j 0	1	2	3	4	Premium	Loss ratio	U_i^I
1	3264	3762	3956	4176	4271	5025	86%	4321.5
2	3617	4197	4527	4608		5775	86%	4966.5
3	4308	4830	5109			6545	86%	5628.7
4	4987	5501				7481	88%	6583.3
5	5378					7990	88%	7031.2

In a similar way the other *ultimate* development factors are calculated, which are given in Table 1.15.

We now use these development factors to determine what we *should have incurred in claims* at the current time were these factors appropriate. For example, for claims originating in year 3 we reported (in the current year) incurred claims of 5109, yet if the development factor $d_{2|4} = 1.05904$ were appropriate and the ultimate loss is to be 5628.7, then we would have expected to currently report about $5628.7/1.05904 = 5314.91$ in claims. This would mean that we still expect to incur

$$5628.7 - 5314.91 = 5628.7\,(1 - 1/d_{2|4}) = 313.79.$$

This is the emerging liability for the origin year 3. Finally, we take this estimate of emerging liability and add it to the reported liability to obtain the B–F estimate of total liability for this origin year. For example, with origin year 3, we would now add our estimate of emerging liability 313.79 to the currently reported liability of 5109 to obtain the B–F estimate of ultimate liability of 5422.79.

Proceeding in a similar fashion we can calculate emerging and total liabilities for the other origin years, and the results are given in Table 1.15. The current estimated total ultimate liability with respect to these five years is therefore 27,531.76. The currently reported incurred claims amount to 24,867, and therefore we should set aside an additional amount of $27,531.76 - 24,867 = 2,664.76$ in reserves for future payments. If we had used the basic chain ladder method alone (with development factors as determined here), the estimate of additional reserves (beyond reported results) would amount to 2,563.21, which is about 101.55 ($101,550) less than the estimate provided by the B–F method. This is further elaborated on below.

In the B–F method for estimating future liabilities, one is essentially combining projections based on a technique like the chain ladder method and those obtained (somewhat independently) on the basis of loss-ratio information. The B–F estimate for ultimate liability on claims originating in year i takes the form

$$U_i = \text{emerging liability} + \text{reported liability}$$

TABLE 1.15

Household insurance data: emerging and total liabilities.

Origin year	Init. ult. loss	Dev. factor	Emerging factor	Emerging liability	Rep. liab.	B–F est. of liab.
i	U_i^I	$d_{5-i|4}$	$(1 - \frac{1}{d_{5-i|4}})$	$U_i^I (1 - \frac{1}{d_{5-i|4}})$	$C_{i,5-i}$	U_i
1	4321.5	1	0	0	4271	4271
2	4966.5	1.02275	0.02224	110.47	4608	4718.47
3	5628.7	1.05904	0.05575	313.79	5109	5422.79
4	6583.3	1.12553	0.11153	734.25	5501	6235.25
5	7031.2	1.27263	0.21422	1506.25	5378	6884.25

$$= \left(1 - \frac{1}{d_{k+1-i|k}}\right) U_i^I + C_{i,k+1-i}$$

$$= \left(1 - \frac{1}{d_{k+1-i|k}}\right) U_i^I + \left(\frac{1}{d_{k+1-i|k}}\right) d_{k+1-i|k} C_{i,k+1-i},$$

where U_i^I is the initial estimate based on the loss ratio information and $d_{k+1-i|k} C_{i,k+1-i}$ is the estimate based on the basic chain ladder method.

Therefore the B–F estimate is actually a weighted average of these two estimates, and in the parlance of credibility theory (as we will see in Chapter 5) we may say that this represents a credibility liability formula of the type

$$U_i = (1 - \mathbf{Z}) U_i^I + \mathbf{Z} d_{k+1-i|k} C_{i,k+1-i}.$$

Here the credibility factor $\mathbf{Z} = 1/d_{k+1-i|k}$ (which is actually the grossing-up factor $g_{k+1-i|k}$) represents the weight we put on the claims information (or data) to date in the ith row of the delay triangle. The factor $(1 - \mathbf{Z})$ is the weight put on the *prior* information provided by the loss ratio estimates.

If $d_{k+1-i|k}$ is large (the ultimate development factors usually increase with i, that is as time progresses and we project over more years), then less weight is put on the information in the ith row of the delay triangle. For origin year 3 of our household insurance data in Table 1.14, the B–F estimate of ultimate liability can be written in the form

$$U_3 = (1 - \mathbf{Z}) U_3^I + \mathbf{Z} d_{2|4} C_{3,2}$$
$$= (0.05575)(5628.70) + (0.94425)(5410.63) = 5422.79,$$

where $\mathbf{Z} = 0.94425$. In the estimate for the most current year (origin year 5) the so-called credibility factor $\mathbf{Z}$ is smaller and equal to 0.78578. In any case note that all of the $\mathbf{Z}$ values (credibility factors) in our example are quite large, indicating that a considerable amount of weight is being put on the chain ladder method estimates and only a small amount on the loss ratio estimates. This is also evident from Table 1.15 where one notes that the ultimate liability estimates U_i are all considerably less than the corresponding initial estimates U_i^I (being pulled down by the chain ladder estimates).

1.5 An example in pricing products

In general insurance, methods for running off delay triangles are crucial tools in estimating reserves for existing business. However, they also can be quite useful in pricing new business. One may be asked to give a "quick" premium quote for a new class of business on the basis of a limited amount of claims and collateral information (such information might result from having reviewed similar risks in other reserving or pricing reviews). In other situations (for example, employers liability insurance), one might be asked to quote on taking over a discontinued book of business from a broker. In any exercise of this type, the analyst will want to look at all available information and usually in several different ways. He/she will look for outliers and trends, then on the basis of these make decisions with the support of prior experience. Of course, such judgements will be subject to scrutiny and should be justifiable to others, possibly through an audit trail.

In the following example, one is asked to quote a price for a new contract given information on past incurred claims data (over the six years 2001 – 2006) for an employer liability scheme in a large company. Information is also available on the historical payroll in the company, all of which is given in Table 1.16. Claim values are indicated with C, and claim numbers with N. A payroll in the region of $86,746,028 is predicted for next year. In this instance, the figures are given in US $ and have not been adjusted for inflation (which in this case, we assume has been, and continues to be, at the rate of 5% per annum). Up until now another company has had the contract for this business. Given that the contract renews on 1 January 2007, what premium can one quote for this business (that is, to cover all claims which will arise from the year 2007)? One is also asked to consider pricing an *each and every loss deductible* for $5,000. What price can one quote for this option? Finally, what would be the price for a two-year insurance period?

Note that the claims data in Table 1.16 is presented in triangular form, but in a slightly different style than previously. Delay triangles can come in all shapes and sizes, but as long as the axes are correctly identified, they should tell you the same things. Most projections for triangular data are performed on spreadsheets, and normally the first step is to put the data into a familiar customized template. Here we do not have available information on the past premiums paid for this business. Although it might be useful and of interest to know this, it could also be quite misleading in some situations (it would probably be an actual premium as opposed to a pure premium, and therefore

TABLE 1.16
Employer liability data – all claim values (C) in US $.

Y	Payroll		@end 01	@end 02	@end 03	@end 04	@end 05	@end 06
1	68,750,000	C	1,250,735	2,138,375	2,461,406	2,534,169	2,579,006	2,529,192
		N	144	243	265	271	271	271
2	57,165,625	C		1,407,613	1,750,281	1,754,032	1,787,872	1,802,665
		N		75	175	175	177	190
3	61,600,000	C			1,461,649	2,370,228	2,420,278	2,503,030
		N			108	202	209	212
4	68,406,250	C				1,029,650	1,458,871	1,551,980
		N				73	176	181
5	76,037,500	C					1,013,163	1,991,081
		N					118	201
6	84,219,444	C						1,041,075
		N						110

could be subject to unknown expenses, investment alterations and perhaps even political adjustments).

In order to quote a premium for the next year of business, we need to predict future frequency and severity of claims. The past and expected future exposure information on annual payrolls should be useful in giving us a benchmark to predict forthcoming claim numbers. Any other relevant information which is easily accessible should also be considered, and this may vary with the type of business being analyzed. For example, in marine insurance the annual number of ships insured might prove a good measure of exposure, while in airline insurance the number of planes covered and/or the number of passenger miles flown might be useful.

Even given the information in Table 1.16, it should be clear that there is no unique way to proceed in generating a premium estimate. The method illustrated here will use run-off triangles for claim numbers and severity of claims, basically relying on the average cost per claim method. However, this example intends to illustrate how in practice an analyst may (slightly) modify such a well-defined method in making a *final selection* of development factors. This (somewhat subjective) selection often takes into account the sometimes vast *prior knowledge* that an experienced claims analyst may have.

In this pricing exercise, we as of yet have no information on the number of claims for the coming year, and this is where the exposure information (in this case, on payroll) will be used. Inflation will be accounted for in a slightly different manner than that considered before (in the basic chain ladder method), and although it is in some sense less precise, it should still give us reasonable estimates.

One way of proceeding to select development factors for projecting claim numbers is detailed in Table 1.17. In the upper part of the table, reported claim numbers are presented in the usual triangular form. We initially might note that claim numbers arising from origin year 2001 seem to have tailed off, suggesting a five-year delay to the ultimate is reasonable (i.e., claims are reported within five years of origin). One should also note a relatively

TABLE 1.17

Employer liability data – development factors for claim numbers.

Origin year ↓	0	1	2	3	4	5
			Development year			
2001	144	243	265	271	271	271
2002	75	175	175	177	190	
2003	108	202	209	212		
2004	73	176	181			
2005	118	201				
2006	110					

Link ratios	$0 \to 1$	$1 \to 2$	$2 \to 3$	$3 \to 4$	$4 \to 5$	$5 \to Ult$
2001	1.688	1.091	1.023	1.000	1.000	
2002	2.333	1.000	1.011	1.073		
2003	1.870	1.035	1.014			
2004	2.411	1.028				
2005	1.703					

Determination of development/grossing-up factors

Average	2.001	1.038	1.016	1.037	1.000	
Exclude max./min.	1.969	1.032	1.014			
Selection	1.969	1.050	1.025	1.010	1.010	1.000
Cum. selection	2.162	1.098	1.046	1.020	1.010	1.000
% of Ult.	46%	91%	96%	98%	99%	100%

smaller number of claims being reported in development year 0 for those originating in 2002 and 2004, and perhaps as a consequence one might seek further information on this variation. The middle part of the table gives the link ratios between development years for each origin year. One would normally scan this data looking for variability, trends and outliers.

We already know of several methods that may be used to get "representative" factors for use in projecting further development. Here results are given for the (arithmetic) average of the link factors, as well as the more robust choice of a trimmed mean (exclude max./min.). In the calculation of this trimmed mean, the two extremes (minimum and maximum) are omitted before calculating the average. After *eye-balling* § the resulting calculations, the analyst makes an informed *selection* (of the development factors to be used in projections). Here, for example, the analyst was happy to use $d_{0|1} = 1.969$ even though the link factor for this development in the origin year 2004 was 2.411. The choice of $d_{1|2} = 1.050$ may be justified as the link factors for the

---

§A quick visual scan for consistency and spotting outliers and/or trends.

development $(1 \to 2)$ were relatively high in the more recent years.

One would normally expect to observe some smoothness in the selected development factors. This is not the case for those factors determined by arithmetic averages, where one notes in particular that the average 1.016 calculated for development $2 \to 3$ is out of line (with other development years). This may be the main reason why a figure of $d_{2|3} = 1.025$ was selected here by the analyst. The analyst was happy to select $d_{3|4} = 1.010$ in spite of the fact that the corresponding average is 1.037. The development factor $d_{4|5} = 1.010$ was selected perhaps somewhat conservatively to allow for the small possibility of a claim arriving in the fifth year of development (in spite of the fact that this did not occur in the only year 2001 where we had the possibility of observing it). The last development factor $d_{5|Ult}$ is set at 1, indicating satisfaction with the assumption that claims will not be delayed more than five years.

It is perhaps worth emphasizing once again the importance of making well-considered decisions with regard to the ultimate (or tail) factors in the development of claims. One needs to be cautious and perhaps slightly conservative in this regard (how many years does it take for claims to develop from a given year of origin?), but at the same time one wants to be realistic. The cumulative development factors of the form $d_{j|Ult}$ and the corresponding grossing-up factors are given in the last two rows of Table 1.17.

TABLE 1.18
Employer liability data – future count predictions and exposure rates.

Origin year ↓	Counts	Ultimate factor	Est. counts	Future claims	Exposure	Revalued exposure	Rate
2001	271	1.00	271	–	68.8	92	2.94
2002	190	1.01	192	2	57.2	73	2.63
2003	212	1.02	216	4	61.6	75	2.89
2004	181	1.05	189	8	68.4	79	2.39
2005	201	1.10	221	20	76.0	84	2.63
2006	110	2.16	238	128	84.2	88	2.69
					Selected exposure rate →		2.70
2007		Expected future claims ⇒		234	86.7	86.7	

Using the *selected* development factors, the numbers of unreported claims arising from the origin years $2001 - 2006$ are estimated and given in Table 1.18. In particular, note that the expected *total number of future claims* from these origin years is 162. Our main objective, however, is to get a good estimate of the number of claims which will which arise out of the year 2007, and here is where we rely on the exposure (in this case, payroll) information available. First of all, we should adjust this payroll information for inflation and restate values currently. The results of this are given in the column labeled *Revalued exposure* in Table 1.18, where payrolls have been adjusted to mid-2007 prices. We assume that the historical payroll figures given are mid-year values, and so, for example, 68.8 units of exposure in 2001 has become $68.8(1.05)^6 = 92$ in 2007 money. The last column in this table gives the rate of predicted number of claims arising from a given origin year relative to units of exposure. Observe that the highest rate occurs in 2001 where there are 2.94 predicted (for 2001 it is actual) claims per unit of payroll (expressed in 2007 money). After studying the various rates (the arithmetic mean here is 2.70), the analyst here was happy to *select* 2.70 for use in prediction for the year 2007. Given an estimated payroll of 86.7 units in 2007, one would then estimate the number of claims arising from origin year 2007 to be $234 = 2.70(86.7)$.

TABLE 1.19
Employer liability data – development factors for claim severity.

Year ↓	Development year					
	0	1	2	3	4	5
2001	1,250,735	2,138,375	2,461,406	2,534,169	2,579,006	2,529,192
2002	1,407,613	1,750,281	1,754,032	1,787,872	1,802,665	
2003	1,461,649	2,370,228	2,420,278	2,503,030		
2004	1,029,650	1,458,871	1,551,980			
2005	1,013,163	1,991,081				
2006	1,041,075					
Link ratios	0 → 1	1 → 2	2 → 3	3 → 4	4 → 5	5 → *Ult*
2001	1.710	1.151	1.030	1.018	0.981	
2002	1.243	1.002	1.019	1.008		
2003	1.622	1.021	1.034			
2004	1.417	1.064				
2005	1.965					

Determination of development/grossing-up factors

Average	1.591	1.060	1.028	1.013	0.981	
Exclude max./min.	1.583	1.042	1.030			
Selection	1.583	1.060	1.025	1.010	0.990	0.990
Cum. selection	1.702	1.076	1.015	0.990	0.980	0.990
% of Ult.	59%	93%	99%	101%	102%	101%

We now turn to an analysis of claim values, which is presented in Tables 1.19 and 1.20. Our objective is to estimate the average value (severity) of a claim arising from origin year 2007. The procedure is the same as that used for the numbers of claims, where link ratios are calculated and studied prior to making a selection of development factors. Note that the factors selected for developments $4 \rightarrow 5$ and $5 \rightarrow Ult$ are less than 1, unlike the situation for claim counts. There are several reasons why one might expect incurred claim values to decrease slightly near the end of development (this would not usually be the case for paid claims). On some occasions, a few claims that are outstanding for a long period and are expected to be large, might in the end be small (or in fact, nothing) due to consequences of legal action. In other situations, it might happen that case reserves are being constantly overestimated (in this example, we might consider what is being reported in Table 1.16 as case reserves since these figures are for incurred claims). This is a conservative approach to reserving and might seem to be reasonable in order to be on the safe side. However, it can also have dangerous consequences as it might make the business look too costly and result in an excessively high premium quote, a consequence of which might be your company not writing the business despite it being a potentially profitable contract!

We return to our analysis of claim severity. After the selection of development factors between consecutive years, cumulative development factors and grossing-up factors are determined. These are then used to estimate total ultimate claim values, and then claim averages by dividing by projected claim numbers (Table 1.20). For example, we estimate total claims arising in 2004 to be $(1.015)\,1{,}551{,}980 = 1{,}574{,}714$, and that the average severity of such claims is $(1{,}574{,}714)/189 = 8{,}321$. For comparative purposes, we have also calculated a column (Nonprojected $\bar{X}$) of the average claim size based only on claims reported up to the present for each origin year.

The astute reader will note that we have not made any adjustment for inflation yet in our analysis of claim values. We could, of course, make a triangle of incremental incurred claims, adjust for inflation, and construct a table of cumulative predicted claims before dividing by projected claim numbers. Here, however, we have been more approximate in nature and simply adjusted the projected average severity of a given origin year for inflation by assuming its monetary value comes from that year. For example, for origin year 2004, we have a projected claim average of 8,321, which adjusting for inflation to mid-2007 has value $8{,}321(1.05)^3 = 9{,}632$. Similarly, inflation-projected average values are calculated for other origin years and given in the last column of Table 1.20. Again, for comparison purposes, averages over the origin years of these various averages are calculated and a *final selection* of 10,500 is made to be used as an (expected) average severity in 2007. It is worth noting that this is significantly smaller than the arithmetic average of 11,036 for all origin years. This selection could be justified on the grounds that the higher value of 11,036 gives equal weight to the earlier years of the data, and that our method for selecting an expected value should recognize the downward trend in average losses.

On the basis of a predicted 234 claims in 2007 with average severity of

TABLE 1.20

Employer liability data – future claim values.

Origin year ↓	Incurred losses	Dev. factor	Est. losses	No. claims	Proj. $\bar{X}$	Nonproj. $\bar{X}$	Inflation adj. $\bar{X}$
2001	2,529,192	0.990	2,503,900	271	9,239	9,333	12,382
2002	1,802,665	0.980	1,766,792	192	9,207	9,488	11,751
2003	2,503,030	0.990	2,477,752	216	11,457	11,807	13,926
2004	1,551,980	1.015	1,574,714	189	8,321	8,574	9,632
2005	1,991,081	1.076	2,141,462	221	9,704	9,906	10,699
2006	1,041,075	1.702	1,772,182	238	7,453	9,464	7,825
			Averages →		9,230	9,762	11,036
2007			*Selection average for claim size →*				10,500

10,500 we would suggest a pure premium of $2,459,250 = 234(10,500)$ for this business.

For a two-year insurance period, we would have to adjust for inflation to 2008 and make some assumptions about possible changes in payroll. If, for example, we can assume that the payroll will only increase in line with ordinary inflation of 5% from 2007 to 2008, then the predicted number of claims for 2008 would remain at 234 (the unit of exposure would become one million payroll in mid-2008 value), while the average severity would increase by 5%. Hence the segment of the pure premium attributed to the year 2008 in a two-year insurance period would be $(1.05)2,459,250 = 2,582,212$, remembering that this is now in mid-2008 money. The quoted premium (for a one or two-year period) would be modified to take account of various factors including expenses, investment credits, reinsurance arrangements and the competitive nature of the business.

When the premium is to be paid is another important consideration, since the pure premiums above are in mid-2007 (or mid-2008) money. For example, if the premium for a two-year period of insurance is to be paid on 1 January 2007, then the pure premium in mid-2007 money of $2(2,459,250) = 4,918,500$ should be discounted for a six-month period, giving a value of $4,799,965 = 4,918,500/\sqrt{1.05}$. In this case, since the premium would be obtained at such an early stage in development (of claims arising during $2007 - 2008$), the investment credit would presumably have a considerable bearing on the ultimately quoted premium.

Finally, suppose that one is asked to quote a premium for this business (say a one-year contract) where a deductible of $5,000 is in force. The effect of the deductible is that only the excess of any claim over $5,000 is actually paid by the insurer. Strictly speaking, we would be on somewhat shaky grounds to come up with a good quote here, for we do not have information on the sizes of individual claims. Normally, with individual claim information, we would try to pick an appropriate distribution to model losses (say a lognormal or Pareto) and use this as a basis for estimating total claims with various possible deductibles. Chapter 2 on loss distributions describes many useful such distributions. In a situation like the present, let us consider using a lognormal distribution with mean 10,500 and standard deviation of say $2(10,500) = 21,000$ (often one might use a lognormal distribution where the standard deviation is between 75% and 275% of the mean)[¶] to model claim size X. Using

$$E(X) \; = \; e^{\mu+\sigma^2/2} = 10,500 \; \text{ and } \; Var(X) \; = \; e^{\mu+\sigma^2/2}(e^{\sigma^2} - 1),$$

we have that $\log X \sim N(\mu = 8.4544, \sigma = 1.2686)$. Therefore the probability that a claim X exceeds the deductible $5,000$ is

$$P(\log X > \log(5000)) \; = \; 1 - \Phi(0.0495) = 0.4803.$$

Hence with such a deductible in force, we would expect about $234(0.4803) = 112.38$ claims, with average settlement of

$$\int_{5000}^{\infty} x f_X(x)\, dx = 10,500 \left[1 - \Phi\left(\frac{\log 5000 - \mu - \sigma^2}{\sigma}\right)\right] = 9330.36.$$

Hence the pure premium for the business with this deductible would be in the region of $9330.36(112.38) = \$1,048,566$.

1.6   Statistical modeling and the separation technique

The separation technique is a statistical method for running off delay triangles, which directly incorporates a factor for inflation. We will very briefly describe this method, and should you require further details refer to the book by Hossack, Pollard and Zehnwirth [29]. A basic assumption in using this method is that over the various origin years, a constant proportion of claims (in real terms) are paid in the various development years.

The idea in the separation method is to model the incremental claims $P_{i,j}$ originating in origin year i and paid in development year j in terms of three separate factors. More precisely, one assumes that $P_{i,j}$ is of the form

$$P_{i,j} = C_i\, r_j\, \lambda_{i+j},$$

where C_i (which is the quantity of primary interest) represents total claims eventually arising from origin year i, r_j represents the proportional development of total claims in development year j and λ_{i+j} is a factor representing effects in calendar year $i + j$ (such as inflation). In theory, one would like to

---

[¶]But this will vary considerably from class to class of business.

obtain estimates of these factors on the basis of the known data, and then make projections into the future. Of course, in particular one would have to make assumptions about the values of the λ factors for future calendar years, and this is often done by projecting estimated values of the λ's for the current (already observed) calendar years.

To begin with, we do not normally know the values for the total cumulative claims C_i arising from any origin year $i > 0$. Usually, the assumption is then made that these values are proportional to the number of claims N_i eventually arising in each origin year. In turn, given that these are also not known (but perhaps easier to estimate than total claim amounts), one assumes that these are proportional to the number of claims $n_{i,0}$ reported in development year 0. Hence we conclude that $C_i = c\, n_{i,0}$ for some constant c, and therefore in dividing C_i by $n_{i,0}$ we obtain Table 1.21. Assuming that development is complete by development year k, we have that the proportional development factors r_j sum to 1 ($r_0 + r_1 + \cdots + r_k = 1$). Using the observed data on incremental claims, one uses diagonal-type methods to estimate the parameters $c\lambda_{i+j}$ and r_j, and then with suitable assumptions on the development of the further (largely due to inflation) values $\lambda_{k+1}, \ldots, \lambda_{2k}$, one may run-off the triangle of claims to obtain estimates of the C_i for $i = 1, \ldots, k$.

TABLE 1.21

Separation model for cumulative claims.

Origin year ↓	Standardized incurred payment in development year j					
	0	1	2	3	...	k
0	$cr_0\lambda_0$	$cr_1\lambda_1$	$cr_2\lambda_2$	$\cdot$	$\cdot$	$cr_k\lambda_k$
1	$cr_0\lambda_1$	$cr_1\lambda_2$	$cr_2\lambda_3$	$\cdot$	$\cdot$	
2	$cr_0\lambda_2$	$cr_1\lambda_3$	$cr_2\lambda_4$	$\cdot$		
3	$cr_0\lambda_3$	$cr_0\lambda_4$	$cr_2\lambda_5$	$\cdot$		
.		.	.			
.		.				
k	$cr_0\lambda_k$					

1.7 Problems

1. Inflation-adjusted cumulative claims which have been incurred on a general insurance account are given (in \$) in Table 1.22. Annual premiums written in 2006 were \$212,000, and the ultimate loss ratio is being estimated as 86%. Claims are assumed to be fully run-off by the end of development year 3. The actual paid claims to date for the policy year

2006 are only \$31,200. Using the Bornhuetter–Ferguson method, estimate the outstanding claims still to be paid from those policies written in 2006 only.

TABLE 1.22
General insurance cumulative claims.

Policy year	Development year			
	0	1	2	3
2003	47,597	101,093	128,511	138,537
2004	50,230	105,962	132,950	
2005	50,542	107,139		
2006	54,567			

2. Cumulative incurred claim numbers N and annual paid claim amounts C (thousands of dollars) for employer liability in a large car manufacturing plant by year of origin and development up until the end of year 2006 are given in Table 1.23. Use the average cost per claim method (with the average grossing-up technique) to determine what reserves should be set aside for future claims.

TABLE 1.23
Annual paid claims (C) and cumulative incurred numbers (N) for car manufacturing plant.

	Development year					
	1		2		3	
Origin year	C	N	C	N	C	N
2004	2,317	132	1,437	197	582	207
2005	3,287	183	1,792	258		
2006	4,816	261				

3. Inflation-adjusted cumulative incurred claim numbers N and amounts C (\$000's)) for personal liability in a large airline company by year of origin and development up until the end of year 2005 are given in Table 1.24. Use the average cost per claim method to determine what reserves should be set aside for future claims.

4. Malicious damage claims (\$000's) for a collection of policies in successive development years are given in Table 1.25, where in each case it is for the actual amount paid in the given years. It can be assumed that all claims are settled by the end of development year 3. Inflation rates for the 12

TABLE 1.24
Personal liability claims in airline company.

	Development year							
	1		2		3		4	
Origin year	C	N	C	N	C	N	C	N
2002	1,752	104	2,192	120	2,514	126	2,988	130
2003	1,798	110	2,366	114	2,714	116		
2004	1,890	124	2,426	132				
2005	1,948	126						

months up to the middle of a year are given by 2002 (4%), 2003 (2%) and 2004 (4%). Using an inflation-adjusted chain ladder technique, determine the amount of reserves that should be set aside at the end of 2004 (in mid-2004 prices).

TABLE 1.25
Malicious damage claims.

	Development year			
Year of origin	0	1	2	3
2001	2,144	366	234	165
2002	2,231	340	190	
2003	2,335	270		
2004	2,392			

5. Table 1.26 gives (inflation-adjusted) cumulative incurred claim numbers (N) and amounts (C) ($000's) for sporting accidents at a large university by year of origin and development up until the end of year 2005. Use the average cost per claim method to determine what reserves should be set aside for future claims at the end of 2005.

TABLE 1.26
Sporting accident claims.

	Development year							
	1		2		3		4	
Origin year	C	N	C	N	C	N	C	N
2002	876	52	1,096	60	1,257	63	1,494	65
2003	899	55	1,183	57	1,357	58		
2004	945	62	1,213	66				
2005	974	63						

6. The inflation-adjusted claims data in Table 1.27 were available at the end of the year 2006 for a class of business written by a general insurance company. It can be assumed that, for a given accident year, all claims will be reported by the end of development year 2.

TABLE 1.27
Inflation-adjusted claims for general insurer.

Accident year	Reported claims ($ 000's) in development year		
	0	**1**	**2**
2004	500	100	40
2005	590	120	
2006	700		

Accident year	No. claims reported in development year		
	0	**1**	**2**
2004	50	6	2
2005	56	7	
2006	60		

As of December 31, 2006, $1,200,000 had been paid by the company as a result of claims on this block of business. Calculate the outstanding claim reserve at December 31, 2006, using the average cost per claim method. Use the "grossing-up" method to run-off the triangles.

7. In Table 1.28 we have the cumulative payments made from motor insurance claims by accident year and development year. Use the chain ladder method to estimate the reserves necessary at the end of 2006 to pay for outstanding claims for these years. Assume that claims are settled within four years of the accident year and that no discounting is necessary.

TABLE 1.28
Payments in motor insurance portfolio.

Policy year	Development year				
	0	1	2	3	4
2002	1,179	2,115	3,324	3,660	3,780
2003	1,356	2,025	3,773	4,194	
2004	1,493	3,021	4,320		
2005	1,830	3,213			
2006	1,775				

8. Fire insurance claim payments (in $000's) for a portfolio of policies in successive development years are given in Table 1.29, where entries are the actual amounts paid in the given years. It can be assumed all claims are settled by the end of development year 3. Inflation rates for the 12 months up to the middle of a year are given by: 2001 (7%), 2002 (5%) and 2003 (3%). Using an inflation-adjusted chain ladder technique, show that the amount of reserves that should be set aside (mid-2003 prices) at the end of 2003 is 687,000. What would the reserves be if we had used an (average) inflation rate of 5% over these three years?

If the estimated inflation rates for the 12 months up to the middle of 2004, 2005 and 2006 were, respectively, 4%, 8% and 7%, what would have been the predicted amount of payments to be made in 2005 in respect to this claims portfolio?

TABLE 1.29
Incremental fire insurance claims.

Year of origin	Development year			
	0	1	2	3
2000	1,072	158	102	104
2001	1,118	174	104	
2002	1,150	188		
2003	1,196			

9. Incremental claim payments for a household insurance scheme in successive development years are given in Table 1.30. These increments are in each case for the actual amounts paid in the given years. An estimate of reserves for IBNR claims originating in 2002 and which still have yet to be reported after four years is also given by 212 in mid-2006 money.

TABLE 1.30
Household insurance incremental claim payments.

Origin year	Payment in development year					IBNR estimate at June 30, 2006
	0	1	2	3	4	
2002	2,060	520	465	230	95	212
2003	2,100	540	468	217		
2004	2,346	590	485			
2005	2,510	655				
2006	2,750					

Suppose the annual claim payments inflation rates over the 12 months up to the middle of a year are given by

$$
\begin{array}{ll}
2002 & 6.2\% \\
2003 & 5.6\% \\
2004 & 5.2\% \\
2005 & 4.1\% \\
2006 & 2.6\%
\end{array}
$$

Using an inflation-adjusted chain ladder technique, estimate (in mid-2006 prices) the total amount outstanding in respect of these claims.

10. Claim numbers by year of reporting of an insurer with respect to wind damage are given in Table 1.31. Use the basic chain ladder method to estimate the number of IBNR claims on this business.

TABLE 1.31
Counts of wind damage claims by year of reporting.

Policy year	Development year					
	0	1	2	3	4	5
1999	126	118	39	27	15	1
2000	102	101	42	28	13	
2001	133	131	44	17		
2002	151	151	49			
2003	143	142				
2004	152					

11. Table 1.32 gives cumulative incurred claim numbers and amounts in thousands of dollars for personal liability by year of origin and development up to the end of 2003. Inflation over the past three years has been at the rate of 3% per annum. Use the average claim size method to determine what reserves (in 2003 monetary value) should be set aside for future claims. What would your estimate be if you ignored claim numbers and used the basic chain ladder (inflation-adjusted) method with (pooled) weighted development factors?

12. Table 1.33 gives cumulative paid claims in a motor insurance scheme over a five-year period, together with annual premium income and estimated loss-ratios determined by an underwriter. One may assume that the amounts have been adjusted for inflation.

Use the Bornhuetter–Ferguson method to estimate outstanding claims in respect of this scheme. In doing so, use the (pooled) weighted development factors of the basic chain ladder method. By how much does the estimate of outstanding claims determined by the basic chain ladder method exceed that derived by the B–F method? Can you give some

TABLE 1.32

Cumulative incurred amounts (C) and numbers (N) for personal liability claims.

	Development year							
	0		1		2		3	
Origin year	C	N	C	N	C	N	C	N
2000	690	42	856	49	1021	55	1248	57
2001	731	45	907	54	1200	58		
2002	803	53	1091	66				
2003	824	49						

TABLE 1.33

Motor insurance scheme: paid claims, premiums and loss ratios (LR).

Origin year i	Development year j					Prem.	LR	U_i^I
	0	1	2	3	4			
1998	31,766	48,708	62,551	69,003	70,587	76,725	92%	70,587
1999	30,043	45,720	59,883	65,671		77,000	92%	70,840
2000	35,819	54,790	71,209			79,100	90%	71,190
2001	40,108	58,960				86,400	92%	79,488
2002	45,701					98,610	94%	92,693.4

insight into why it is greater? What can one say about the underwriter's insight (based on the estimated loss ratios) into ultimate losses?

13. Determine the reserves that should be set aside at the end of 2006 for future payments in respect of claims arising out of the origin years 2001–2006 for the employer liability data appearing in Table 1.16. Use the average cost per claim method with arithmetic averages for development factors.

14. Table 1.34 gives information from a business which handles baggage claims for an airline company. The claims arise from lost and damaged luggage during transport. These incurred claim amounts do not take account of inflation, which one may assume has been, and will continue to be, at the constant rate of 4% per annum. One is also provided information on the annual number of flights flown by the airline during the period of time 1997 – 2002, which is clearly related to the number of claims. It is predicted that the airline would have 43,373 flights in 2003. One is asked to determine a pure premium for a one-year contract for this business in 2003, expressed in terms of mid-2003 money. In the first instance, you are asked to take account of inflation as in the example on employers liability in Section 1.5. In the second case, take account of inflation by determining yearly incremental incurred claims and adjust them appropriately. In both cases, select as your development factors the averages of the link ratios. Compare the estimates for the pure premiums for the two methods. Are they significantly different?

TABLE 1.34

Airline baggage damage data.

Y	Flights		@end 97	@end 98	@end 99	@end 00	@end 01	@end 02
97	34,375	C	766,084	1,309,770	1,377,629	1,412,197	1,451,284	1,451,284
		N	1940	2430	2620	2710	2712	2712
98	28,583	C		862,173	1,072,059	1,084,357	1,095,084	1,104,324
		N		1303	1750	1785	1795	1841
99	30,800	C			895,270	1,451,782	1,482,438	1,508,623
		N			1734	2120	2190	2258
00	34,203	C				630,668	938,236	950,599
		N				1586	1925	1974
01	38,019	C					643,432	1,030 384
		N					1681	2359
02	42,110	C						640,337
		N						1650

2

Loss Distributions

2.1 Introduction to loss distributions

In this chapter, we study many of the classic distributions used to model losses in insurance and finance. Some of these distributions such as the exponential, gamma and Weibull are likely to be familiar to most readers as they are frequently used in survival analysis and engineering applications. We will, however, also consider distributions such as the Pareto and lognormal which are particularly appropriate for studying losses. In modeling a loss, there is usually considerable concern about the chances and sizes of large claims – in particular, the study of the (right) tail of the distribution. For example, the tails of the gamma (in particular, the exponential) and Weibull distributions vanish at an exponential rate. Is such a decay appropriate when it is important not to underestimate the size and frequency of large losses (for example, claims in insurance or defaulted loans in banking)?

In spite of the fact that one may always work with the empirical distribution function derived from a data set of claims, there is often a natural desire to *fit* a probability distribution with reasonably tractable mathematical properties to such a data set. In any attempt to do so, one would initially perform some exploratory analysis of the data and make use of basic descriptive statistics (such as the mean, median, mode, standard deviation, skewness, kurtosis and various quantiles) and plots. One then might try to fit one of the classic parametric distributions using maximum likelihood (or other) methods to estimate parameters. Various tests (for example, the Kolmogorov–Smirnoff, χ^2 goodness-of-fit, Anderson–Darling or the A.I.C. [Akaike Information Criterion]) may be used to assess the fit of a proposed model. Often one may find that a mixture of various distributions works well. In any case, considerable care and perhaps flexibility should be used in settling on a particular distribution. In Section 2.2 we review basic properties of some of the more commonly used and classic loss distributions, and then in Section 2.3 discuss methods of analyzing fit. In Section 2.4 we discuss various properties of mixture distributions for losses, while in Section 2.5 we consider the impact of reinsurance on losses.

Table 2.1 gives the amounts of 120 theft claims made in a household insurance portfolio. This data set (Theft) is small relative to many which one may encounter in practice; however, it will provide a useful example of how one might search for a loss distribution to model typical claims. The mean and standard deviation of this data are given, respectively, by $\bar{x} = 2020.292$ and $s = 3949.857$. Summary statistics (obtained from the statistical package R) are given by

```
> summary(Theft)
   Min. 1st Qu.  Median   Mean 3rd Qu.    Max.
    3.0   271.0   868.5 2020.0 1733.0 32040.0
```

From Minitab, one finds that the skewness $\gamma_1 = 5.1623$ and the kurtosis $\gamma_2 = 33.0954$. The distribution of this claim data is positively skewed with a reasonably fat right tail. Figure 2.1 gives a histogram of the data set Theft. Note that the three relatively large claims of $(11,453, 22,274, 32,043)$ make it challenging to get a feeling for the spread of the other values. Figure 2.2 is a graph of the histogram of the claims restricted to the range $[0, 8500]$, and gives a better perspective on the shape of the distribution.

TABLE 2.1
120 theft claims.

3	11	27	36	47	49	54	77	78	85
104	121	130	138	139	140	143	153	193	195
205	207	216	224	233	237	254	257	259	265
273	275	278	281	396	405	412	423	436	456
473	475	503	510	534	565	656	656	716	734
743	756	784	786	819	826	841	842	853	860
877	942	942	945	998	1029	1066	1101	1128	1167
1194	1209	1223	1283	1288	1296	1310	1320	1367	1369
1373	1382	1383	1395	1436	1470	1512	1607	1699	1720
1772	1780	1858	1922	2042	2247	2348	2377	2418	2795
2964	3156	3858	3872	4084	4620	4901	5021	5331	5771
6240	6385	7089	7482	8059	8079	8316	11,453	22,274	32,043

2.2 Classical loss distributions

2.2.1 Exponential distribution

The exponential distribution is one of the simplest and most basic distributions used in modeling. If the random variable X is exponentially distributed with parameter λ and density function $f_X(x) = \lambda e^{-\lambda x}$ for $x > 0$, then

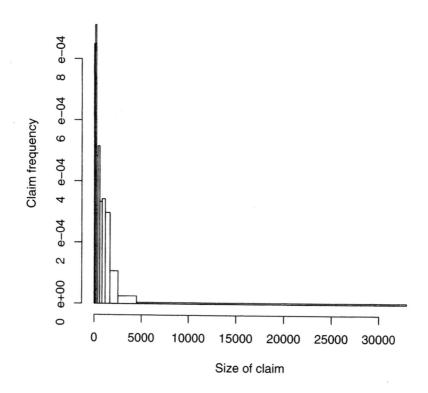

FIGURE 2.1
Histogram of 120 theft claims.

it has survival function $\bar{F}_X(x) = e^{-\lambda x}$, mean $E(X) = 1/\lambda$ and variance $Var(X) = 1/\lambda^2$. The moment generating function of X exists for any $t < \lambda$ and is given by $M_X(x) = \lambda/(\lambda - t)$. Note that for an exponential random variable the mean and standard deviation are the same.

Since the mean $\bar{x} = 2020.292$ and standard deviation $s = 3949.857$ of the 120 theft claims are so different, it is highly unlikely that an exponential distribution will fit the data well. The skewness and kurtosis for any exponential distribution are, respectively, 2 and 6, as compared to the sample estimates of 5.1623 and 33.0954, suggesting that the claims data is both more positively skewed and has a fatter right tail than one would expect from an exponential distribution.

If X has an exponential distribution with $1/\lambda = 2020.292$, then $P(X > 8000) = 0.0191, P(X > 10,000) = 0.0071$, and $P(X > 20,000) = 0.0001$,

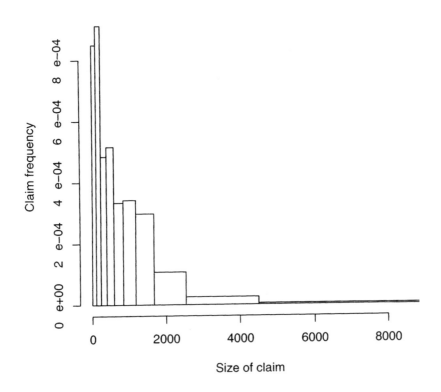

FIGURE 2.2
(Restricted view of) histogram of 120 theft claims.

while the respective observed relative frequencies for the Theft claim data
are $6/120 = 0.05$, $3/120 = 0.025$ and $2/120 = 0.01667$. These observations
suggest that a distribution for the Theft claim data should have a "fatter"
tail than that of an exponential distribution.

An exponential random variable X has the *memoryless* property in that
for any $M, x > 0$, $P(X > M + x \mid X > M) = P(X > x)$. In fact, this memoryless property is shared by no other continuous distribution, and hence
characterizes the family of exponential random variables (similarly, the geometric random variables are the only discrete family with this memoryless
property). The waiting times between events in a homogeneous Poisson process with intensity rate λ are exponential random variables with parameter
λ.

The failure (or hazard) rate function r_X of a random variable X defined at x is the instantaneous rate of failure at time x given survival up to time x. Hence for an exponential random variable with parameter λ this takes the form

$$r_X(x) = \lim_{h \to 0} \frac{F_X(x+h) - F_X(x)}{h} \frac{1}{\bar{F}_X(x)} = \frac{f_X(x)}{\bar{F}_X(x)} = \frac{\lambda e^{-\lambda x}}{e^{-\lambda x}} = \lambda.$$

For an exponential distribution X, the tail probability $\bar{F}_X(x) = P(X > x) = e^{-\lambda x}$ converges to 0 exponentially fast. In many situations, it may be appropriate to try and model a slower vanishing tail distribution. For example, if $P(X > x)$ is of the form $a^\alpha/(b + cx)^\alpha$ for certain positive constants a, b, c and α, then the tail probability of X goes to 0 at a slower (polynomial) rate. For a function of the form $a^\alpha/(b+cx)^\alpha$ to be the survival function of a positive random variable, one must have that $P(X > 0) = (a/b)^\alpha = 1$. This gives rise to the Pareto family of distributions.

2.2.2 Pareto distribution

The random variable X is Pareto with (positive) parameters α and λ if it has density function

$$f_X(x) = \frac{\alpha \lambda^\alpha}{(\lambda + x)^{\alpha+1}}, \text{ or equivalently, survival function } \bar{F}_X(x) = \left(\frac{\lambda}{\lambda + x}\right)^\alpha$$

for $x > 0$. The Pareto distribution is named after Vilfredo Pareto (1848−1923) who used it in modeling welfare economics. Today, it is commonly used to model income distribution in economics or claim-size distribution in insurance. In some circumstances, it may be appropriate to consider a shifted Pareto distribution taking values in an interval of the form $(\beta, +\infty)$.

Like the exponential family of random variables, the Pareto distributions have density and survival functions which are very tractable. Pareto random variables have some nice preservation properties. For example, if $X \sim$ Pareto(α, λ) and $k > 0$, then $kX \sim$ Pareto$(\alpha, k\lambda)$ since

$$P(kX > x) = P(X > x/k) = \left(\frac{\lambda}{\lambda + x/k}\right)^\alpha = \left(\frac{k\lambda}{k\lambda + x}\right)^\alpha.$$

This property is useful in dealing with inflation in claims. Moreover, if $M > 0$, then

$$P(X > M + x \mid X > M) = \left(\frac{\lambda}{\lambda + M + x}\right)^\alpha \bigg/ \left(\frac{\lambda}{\lambda + M}\right)^\alpha = \left(\frac{\lambda + M}{\lambda + M + x}\right)^\alpha,$$

which implies that if $X > M$, then $X - M$ (or the excess of X over M) is Pareto $(\alpha, \lambda + M)$. This property is useful in evaluating the effect of deductibles and/or excess levels for reinsurance in handling losses.

The inverse F_X^{-1} of the distribution function of a Pareto random variable with parameters α and λ has the form

$$F_X^{-1}(u) = \lambda\left[(1-u)^{-1/\alpha} - 1\right] \quad \text{for } 0 < u < 1.$$

For any continuous random variable X, $U = F_X(X)$ is uniformly distributed on $(0, 1)$ (and hence the random variables X and $F_X^{-1}(U)$ have the same probability distribution). Now if $U \sim$ Uniform $(0, 1)$, then likewise $1 - U$ has the same distribution. Therefore

$$X \equiv \lambda[(1-U)^{-1/\alpha} - 1] \sim \lambda(U^{-1/\alpha} - 1)$$

is Pareto with parameters α and λ. This can be usefully employed in simulating values from a Pareto distribution. Using the package R, the following code was used to generate a random sample of size 300 from a Pareto distribution with $\alpha = 3$ and $\lambda = 800$, and then find its sample mean and variance.

```
> sample<-800*((runif(n=300))**(-1/3)-1)
> mean(sample)
[1] 378.1911
> var(sample)
[1] 285857.2
```

When $X \sim$ Pareto(α, λ), one may readily determine the mean (when $\alpha > 1$) and variance (when $\alpha > 2$) by using the expressions $E(X) = \int_0^\infty \bar{F}_X(x)\, dx$ and $E(X^2) = \int_0^\infty 2x\, \bar{F}_X(x)\, dx$. (Of course, one could also use the more traditional expressions $E(X) = \int_0^\infty x\, f_X(x)\, dx$ and $E(X^2) = \int_0^\infty x^2\, f_X(x)\, dx$, but in this case the former expressions are more convenient to use.) Now

$$E(X) = \int_0^\infty \left(\frac{\lambda}{\lambda + x}\right)^\alpha dx$$

$$= -\frac{\lambda^\alpha}{(\alpha - 1)(\lambda + x)^{\alpha-1}} \Big|_0^\infty$$

$$= \frac{\lambda}{\alpha - 1}, \quad \text{and}$$

$$E(X^2) = 2 \int_0^\infty x \left(\frac{\lambda}{\lambda + x}\right)^\alpha dx$$

$$= \frac{2\lambda}{\alpha - 1} \int_0^\infty x\, \frac{(\alpha - 1)\lambda^{\alpha-1}}{(\lambda + x)^\alpha}\, dx$$

$$= \frac{2\lambda^2}{(\alpha - 1)(\alpha - 2)} \quad \text{and therefore}$$

$$Var(X) = \frac{2\lambda^2}{(\alpha - 1)(\alpha - 2)} - \left(\frac{\lambda}{\alpha - 1}\right)^2 = \frac{\alpha\lambda^2}{(\alpha - 1)^2(\alpha - 2)}.$$

Using the method of moments to estimate the parameters α and λ of a Pareto distribution, one could solve the equations

$$\frac{\lambda}{\alpha - 1} = \bar{x} \quad \text{and} \quad \frac{\alpha \lambda^2}{(\alpha - 1)^2(\alpha - 2)} = s^2,$$

yielding

$$\tilde{\alpha} = \frac{2s^2}{s^2 - \bar{x}^2} \quad \text{and} \quad \tilde{\lambda} = (\tilde{\alpha} - 1)\bar{x}.$$

Of course, asymptotically, maximum likelihood estimators are preferred, and for a sample $\mathbf{x}$ of n observations from a Pareto distribution the likelihood function takes the form

$$L(\alpha, \lambda) = \prod_{i=1}^{n} \frac{\alpha \lambda^{\alpha}}{(\lambda + x_i)^{\alpha+1}}.$$

Differentiating the log-likelihood function $l = \log L(\alpha, \lambda)$ with respect to α and λ and then solving for α, one finds that the maximum likelihood estimators must satisfy

$$\frac{\partial}{\partial \alpha} l = 0 = \frac{n}{\alpha} + n \log \lambda - \sum \log(\lambda + x_i)$$

$$\Rightarrow \quad \hat{\alpha} = \frac{n}{\sum \log(1 + x_i/\hat{\lambda})} \quad \text{and} \tag{2.1}$$

$$\frac{\partial}{\partial \lambda} l = 0 = \frac{n\alpha}{\lambda} - (\alpha + 1) \sum \frac{1}{\lambda + x_i}$$

$$\Rightarrow \quad \hat{\alpha} = \frac{\sum 1/(\hat{\lambda} + x_i)}{\sum x_i/(\hat{\lambda}(\hat{\lambda} + x_i))}. \tag{2.2}$$

Hence the maximum likelihood estimator $\hat{\lambda}$ must be a solution of

$$\frac{\sum 1/(\hat{\lambda} + x_i)}{\sum x_i/(\hat{\lambda}(\hat{\lambda} + x_i))} - \frac{n}{\sum \log(1 + x_i/\hat{\lambda})} = 0,$$

which may be solved by numerical methods. $\hat{\alpha}$ may then be found from Equation (2.1) or (2.2).

For the Theft claim data in Table 2.1, the ML (maximum likelihood) estimates for a Pareto distribution are $\hat{\lambda} = 1872.13176$ and $\hat{\alpha} = 1.88047$, while the MM (method of moments) estimators are $\tilde{\lambda} = 3451.911$ and $\tilde{\alpha} = 2.70862$. Figure 2.3 plots the ML fitted Pareto density, as well as the ML fitted exponential density relative to the histogram of the Theft claim data.

If X has an Pareto distribution with $\hat{\lambda} = 1872.13176$ and $\hat{\alpha} = 1.88047$, then the probabilities $P(X > 8000) = 0.0439, P(X > 10,000) = 0.0310$ and $P(X > 20,000) = 0.0098$ are much closer to observed relative frequencies $(0.05, 0.025$ and $0.01667)$ of these events for the Theft claim data than the ML fitted exponential distribution (see Table 2.3).

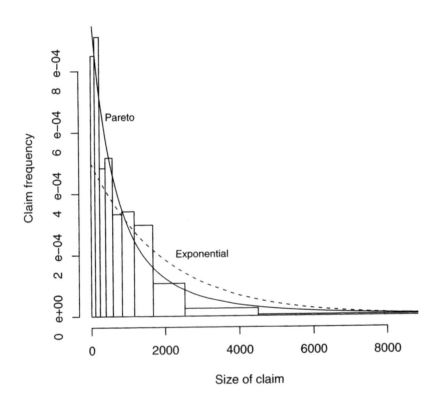

FIGURE 2.3
Maximum likelihood Pareto and exponential densities for Theft data.

2.2.3   Gamma distribution

The gamma family of probability distributions is both versatile and useful. The gamma function is defined for any $\alpha > 0$ by $\Gamma(\alpha) = \int_0^{+\infty} y^{\alpha-1} e^{-y} \, dy$, and has the properties that $\Gamma(n) = (n-1)\Gamma(n-1)$ and $\Gamma(1/2) = \sqrt{\pi}$.

X has a gamma distribution with parameters α and λ ($X \sim \Gamma(\alpha, \lambda)$) if X has density function given by

$$f_X(x) = \frac{\lambda^\alpha}{\Gamma(\alpha)} \, x^{\alpha-1} e^{-\lambda x} \quad \text{for } x > 0.$$

If $X \sim \Gamma(\alpha, \lambda)$, then $M_X(t) = [\lambda/(\lambda - t)]^\alpha$ for $t < \lambda$, $E(X) = \alpha/\lambda$ and $Var(X) = \alpha/\lambda^2$. The parameter α is often called the shape parameter of the gamma distribution, while λ is usually called the scale parameter. In a Poisson process where events are occurring at the rate of λ per unit time, it is well known that the time T_r until the rth event has a gamma distribution with parameters r and λ ($T_r \sim \Gamma(r, \lambda)$). It should be noted that some statistical texts or software use the reciprocal of the rate as the scale parameter of the gamma distribution. For example, in the software R the scale parameter is $1/\text{rate}$. The following R code will generate a plot of a gamma density with shape parameter $\alpha = 5$ and rate $= 0.04$ (or in our terminology scale parameter $25 = 1/0.04$):

```
> x<-seq(0,400,0.01)
> plot(x,dgamma(x,shape=5,scale=25),type="n",
+ ylab="gamma density", main="gamma density + with mean 125 and
variance 3125")
> lines(x,dgamma(x,shape=5,scale=25))
```

When the shape parameter $\alpha = 1$, we obtain the exponential distributions. Moreover, the $\Gamma(r/2, 1/2)$ distribution is precisely the χ^2 distribution with r degrees of freedom, and hence the gamma family includes both the exponential and χ^2 distributions.

Given a set of random observations of X from a gamma distribution, one may obtain the method of moments estimators of α and λ as $\tilde{\alpha} = \bar{x}^2/s^2$ and $\tilde{\lambda} = \bar{x}/s^2$, where $\bar{x}$ and s^2 are, respectively, the mean and variance of the sample. Unfortunately, there are no closed form solutions for the maximum likelihood estimators of α and λ. One method for getting around this is to reparametrize the family. In doing so, one still uses the parameter α, but instead of using λ one uses the mean $\mu = E(X) = \alpha/\lambda$ as the other parameter. This is, of course, just a technique of relabeling the parameters, and one still has the same family of distributions. With this reparametrization, one sets up and solves (resorting to numerical methods) equations to find the maximum likelihood estimates for the parameters α and μ. Then using the invariance property of the method of maximum likelihood, one obtains the maximum likelihood estimates of α and λ. In this instance, having found $\hat{\alpha}$ and $\hat{\mu}$, one obtains $\hat{\lambda} = \hat{\alpha}/\hat{\mu}$.

Example 2.1

Let $X \sim \Gamma(\alpha, \mu = \alpha/\lambda)$. Then

$$f_X(x) = \frac{\alpha^\alpha}{\mu^\alpha} \frac{1}{\Gamma(\alpha)} x^{\alpha-1} e^{-\alpha x/\mu} \quad \text{when } x > 0.$$

Under this new parametrization for the gamma distribution, the likelihood function $L(\alpha, \mu)$ takes the form

$$L(\alpha, \mu) = \prod_{i=1}^n \frac{\alpha^\alpha}{\mu^\alpha} \frac{1}{\Gamma(\alpha)} x_i^{\alpha-1} e^{-\alpha x_i/\mu}.$$

Since

$$\frac{\partial l}{\partial \mu} = \frac{n\alpha}{\mu} \left(\frac{\bar{x}}{\mu} - 1 \right),$$

clearly $\hat{\mu} = \bar{x}$. Using this value for μ in the likelihood, it follows that $\hat{\alpha}$ is the value of α which maximizes

$$l(\alpha, \bar{x}) = \log L(\alpha, \bar{x})$$

$$= n\alpha(\log \alpha - \log \bar{x} - 1) + (\alpha - 1) \sum_{i=1}^n \log x_i - n \log \Gamma(\alpha).$$

Note then that

$$-E \left(\frac{\partial^2}{\partial \mu^2} \log f_X \right) = -E \left(\frac{\partial^2}{\partial \mu^2} \left[\log \frac{\alpha^\alpha}{\mu^\alpha} \frac{1}{\Gamma(\alpha)} x^{\alpha-1} e^{-\alpha x/\mu} \right] \right)$$

$$= E \left(\frac{\partial}{\partial \mu} \left[\frac{\alpha}{\mu} - \frac{\alpha}{\mu^2} x \right] \right) = E \left(\frac{2\alpha}{\mu^3} x - \frac{\alpha}{\mu^2} \right)$$

$$= \frac{2\alpha}{\mu^3} \mu - \frac{\alpha}{\mu^2} = \frac{\alpha}{\mu^2}.$$

Similarly,

$$-E \left(\frac{\partial}{\partial \alpha} \frac{\partial}{\partial \mu} \log f_X \right) = 0,$$

and hence for large n,

$$\begin{pmatrix} \hat{\alpha} \\ \hat{\mu} \end{pmatrix} \dot{\sim} N \left(\begin{bmatrix} \alpha \\ \mu \end{bmatrix}, \left[n \cdot \begin{pmatrix} -E(\partial^2 \log f_X/\partial \alpha^2) & 0 \\ 0 & \alpha/\mu^2 \end{pmatrix} \right]^{-1} \right).$$

In particular, it follows that asymptotically $\hat{\alpha}$ and $\hat{\mu}$ are independent.   □

Using the method of maximum likelihood with the Theft claim data, one obtains the ML estimates for a gamma distribution (using, for example, the procedure *nlm* in R) $\hat{\alpha} = 0.00013$ and $\hat{\lambda} = 1/3244.29450 = 0.00031$, while the MM estimates are $\tilde{\alpha} = 0.26162$ and $\tilde{\lambda} = 0.00013$.

2.2.4 Weibull distribution

A random variable X is a Weibull random variable with parameters $c, \gamma > 0$ ($X \sim W(c, \gamma)$) if it has density function

$$f_X(x) = c\gamma x^{\gamma-1} e^{-cx^\gamma}, \text{ or equivalently, } \bar{F}_X(x) = e^{-cx^\gamma} \text{ for } x > 0.$$

The parameters c and γ are often called the scale and shape parameters for the Weibull random variable, respectively. If the shape parameter $\gamma < 1$, then the *tail* of X is fatter (heavier) than that of any exponential distribution, but not as heavy as that of a Pareto. When $\gamma = 1$, then X is exponential with parameter c. The Weibull distribution is one of the so-called *extreme value distributions* in that it is one of the possible limiting distributions of the mimimum of independent random variables.

The Weibull distribution is named in honor of the Swedish engineer Waloddi Weibull (1887 – 1979). Weibull was an academic, an industrial engineer and a pioneer in the study of fracture, fatigue and reliability. The Weibull distribution was first published in 1939, and has proven to be an invaluable tool in the aerospace, automotive, electric and nuclear power, electronics and biotechnical industries.

A particularly nice property of the Weibull distribution is the functional form of its survival function, which has led to its common use in modeling lifetimes. Another attractive aspect is that the failure or hazard rate function of the Weibull distribution is of polynomial form since

$$r_X(x) = \frac{f_X(x)}{\bar{F}_X(x)} = \frac{c\gamma x^{\gamma-1} e^{-cx^\gamma}}{e^{-cx^\gamma}} = c\gamma x^{\gamma-1}.$$

If $X \sim W(c, \gamma)$ and $Y = X^\gamma$, then

$$P(Y > x) = P(X > x^{1/\gamma}) = e^{-cx} \text{ for any } x > 0,$$

and hence Y is exponential with parameter c. This enables one to easily determine the moments of X since

$$
\begin{aligned}
E(X^k) &= E(Y^{k/\gamma}) \\
&= \int_0^\infty y^{k/\gamma} c e^{-cy} \, dy \\
&= \frac{1}{c^{k/\gamma}} \int_0^\infty w^{k/\gamma} e^{-w} \, dw \quad \text{(using } w = cy) \\
&= \frac{1}{c^{k/\gamma}} \Gamma\left(1 + \frac{k}{\gamma}\right).
\end{aligned}
$$

Example 2.2

The survival time X (in years) for a patient undergoing a specified surgical procedure for bowel cancer is modeled by a Weibull random variable $X \sim$

$W(c = 0.04, \gamma = 2)$. We determine $P(X \leq 5)$, $E(X)$ and $Var(X)$.

$$P(X \leq 5) = \int_0^5 \frac{1}{25}\, 2\, x^{2-1}\, e^{-x^2/25}\, dx = F_X(5) = 1 - e^{-5^2/25} = 0.6321.$$

Moreover, since in this case $c = 0.04$ and $\gamma = 2$, we have

$$E(X) = \Gamma(1 + \frac{1}{2})/(1/25)^{1/2} = 5\frac{1}{2}\Gamma\left(\frac{1}{2}\right) = 2.5\sqrt{\pi} = 4.4311 \quad \text{and}$$

$$Var(X) = \frac{1}{c^{2/\gamma}}\left[\Gamma\left(\frac{2+\gamma}{\gamma}\right) - \Gamma^2\left(\frac{1+\gamma}{\gamma}\right)\right]$$

$$= 25\left[\Gamma(2) - \left(\frac{\sqrt{\pi}}{2}\right)^2\right] = 25\left(1 - \frac{\pi}{4}\right) = 5.365.$$

☐

The gamma function plays an important role in determining the moments of a Weibull random variable, and hence using the method of moments can sometimes be numerically challenging in solving for the parameters c and γ. However, an analogous method, sometimes called the *method of percentiles*, (M%), can be easier to employ. In this (only occasionally used) method, one equates sample quantiles to theoretical ones, and then solves for the unknown parameters. For the Weibull distribution we want to estimate the two parameters c and γ. Let $\tilde{x}_{0.25}$ and $\tilde{x}_{0.75}$ be the first and third sample quartiles of the given data set (hence in particular, 25% of the sample values lie below $x_{0.25}$). Estimates $\ddot{c}$ and $\ddot{\gamma}$ of c and γ, respectively, may be obtained by solving the equations

$$\bar{F}_X(\tilde{x}_{0.25}) = exp(-c\,\tilde{x}_{0.25}^\gamma) = 0.75 \text{ and}$$
$$\bar{F}_X(\tilde{x}_{0.75}) = exp(-c\,\tilde{x}_{0.75}^\gamma) = 0.25.$$

For the 120 Theft claim data (see also summary(Theft) given in Section 2.1),

$$\tilde{x}_{0.25} = 0.25(265) + 0.75(273) = 271 \text{ and}$$
$$\tilde{x}_{0.75} = 0.75(1720) + 0.25(1772) = 1733.$$

On using these sample quantiles one obtains for the Theft data that $\ddot{c} = 0.002494$ and $\ddot{\gamma} = 0.847503$. We will see later (using a chi-square goodness-of-fit test) that the resulting Weibull distribution does not provide a good fit for the Theft claim data.

The maximum likelihood estimates are given by $\hat{c} = 0.00518$ and $\hat{\gamma} = 0.71593$. Figure 2.4 gives a plot of the ML fitted Pareto and Weibull densities for the Theft claim data superimposed on a relative frequency histogram of the data. Both appear to resemble the histogram well, with the Pareto distribution seemingly slightly better (see also Table 2.3).

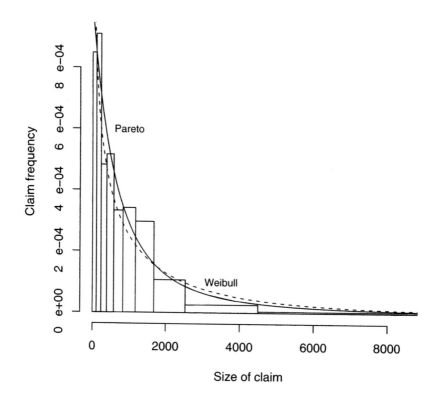

FIGURE 2.4

Maximum likelihood Pareto and Weibull densities for Theft data.

2.2.5 Lognormal distribution

A random variable X has the *lognormal* distribution with parameters μ and σ^2 if $Y = \log X \sim N(\mu, \sigma^2)$. Letting $g(Y) = e^Y = X$, the density function f_X may be determined from that of Y as follows:

$$f_X(x) \;=\; f_Y(\log x)\,|[g^{-1}(x)]'| \;=\; \left[\frac{1}{\sqrt{2\pi}\sigma} e^{-(\log x - \mu)^2/2\sigma^2} \right] \frac{1}{x} \quad \text{for } x > 0.$$

Using the expression for the moment generating function of a normal random variable, one can determine the mean and variance of X as follows:

$$E(X) = E(e^Y) = M_Y(1) = e^{\mu 1 + \sigma^2 1^2/2} = e^{\mu + \sigma^2/2}, \quad \text{and}$$

$$Var(X) = E(X^2) - E^2(X) = E(e^{2Y}) - \left(e^{\mu + \sigma^2/2} \right)^2$$

$$= M_Y(2) - e^{2\mu + \sigma^2} = e^{2\mu + 2\sigma^2} - e^{2\mu + \sigma^2}$$
$$= e^{2\mu + \sigma^2} \left[e^{\sigma^2} - 1 \right] = E^2(X) \left[e^{\sigma^2} - 1 \right].$$

The lognormal distribution is skewed to the right, and is often useful in modeling claim size.

The lognormal density function f_X with parameters μ and σ^2 satisfies the following integral equation, which will be useful in determining excess of loss reinsurance arrangements when claims are lognormal:

$$\int_0^M x\, f_X(x)\, dx = \int_0^M e^{\log x} \frac{1}{\sqrt{2\pi}\sigma} e^{-(\log x - \mu)^2 / 2\sigma^2} \frac{1}{x}\, dx$$

$$= \int_0^M \frac{1}{\sqrt{2\pi}\sigma} e^{-[-2\sigma^2 \log x + (\log x - \mu)^2]/2\sigma^2} \frac{1}{x}\, dx$$

$$= \int_{-\infty}^{\log M} \frac{1}{\sqrt{2\pi}\sigma} e^{-[-2\sigma^2 w + (w - \mu)^2]/2\sigma^2}\, dw \quad (\text{where } w = \log x)$$

$$= \int_{-\infty}^{\log M} \frac{1}{\sqrt{2\pi}\sigma} e^{-[(w - [\mu + \sigma^2])^2 - 2\sigma^2 \mu - \sigma^4]/2\sigma^2}\, dw$$

$$= e^{\mu + \sigma^2/2} \int_{-\infty}^{\log M} \frac{1}{\sqrt{2\pi}\sigma} e^{-(w - [\mu + \sigma^2])^2 / 2\sigma^2}\, dw$$

$$= e^{\mu + \sigma^2/2}\ \Phi\left(\frac{\log M - \mu - \sigma^2}{\sigma} \right), \qquad (2.3)$$

where Φ is the distribution function for the standard normal distribution.

In trying to find a lognormal distribution to model a loss (or claim) distribution, one commonly uses either the method of moments or the method of maximum likelihood to estimate the parameters μ and σ^2. One important observation to make (see Problem 17a) is that when $Y = \log X$ is normal with mean μ and variance σ^2, then given a sample of n observations $\mathbf{x}$, the maximum likelihood estimates of these parameters are $\hat{\mu} = \sum \log(x_i)/n$ and $\hat{\sigma}^2 = \sum (\log x_i - \hat{\mu})^2 / n$.

Revisiting the Theft claim data, let us consider trying to model this data with a lognormal density. The maximum likelihood estimates are given by $\hat{\mu} = 6.62417$ and $\hat{\sigma}^2 = 2.30306$. Figure 2.5 gives a normal quantile plot for the transformed log Theft claim data, while Figure 2.6 gives a plot of the ML estimated lognormal density function overlaying the histogram of the original Theft claim data. Some tail probabilities for the ML fitted lognormal distribution are given in Table 2.3. All of these results give some support for using a lognormal distribution to model the Theft claim data.

Example 2.3

Data (in grouped format) for automobile damage claims in ($000's) during the year 2005 for a fleet of rental cars are given in Table 2.2. We will use the

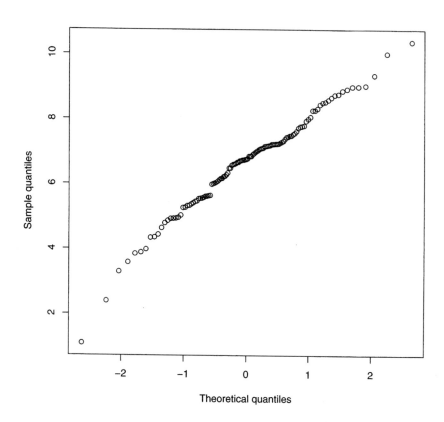

FIGURE 2.5
Normal Q-Q plot of log(Theft) data.

method of moments (MM) to fit a lognormal distribution to the data and use
it to estimate the (future) proportion of such claims which will exceed 20,000
and 15,000, respectively.

As the data is in grouped form, we estimate the mean and variance of a
typical claim X by

$$E(X) \doteq 2\frac{81}{325} + 6\frac{124}{325} + 10\frac{65}{325} + 14\frac{33}{325} + 18\frac{14}{325} + 22\frac{5}{325} + 26\frac{3}{325}$$
$$= 7.563077 \ (000's), \quad \text{and}$$
$$Var(X) \doteq 2^2\frac{81}{325} + 6^2\frac{124}{325} + 10^2\frac{65}{325} + 14^2\frac{33}{325} + 18^2\frac{14}{325} + 22^2\frac{5}{325}$$
$$+ 26^2\frac{3}{325} - (7.563077)^2 = 25.076791 \ (000,000's).$$

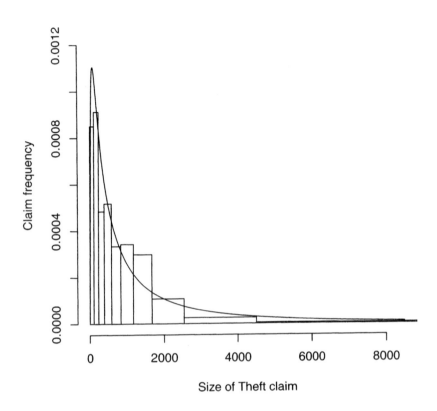

FIGURE 2.6
ML fitted lognormal density and histogram of Theft data.

TABLE 2.2
Grouped data on automobile damage.

Group	Claim interval	Observations
1	[0, 4)	81
2	[4, 8)	124
3	[8, 12)	65
4	[12, 16)	33
5	[16, 20)	14
6	[20, 24)	5
7	[24, 28)	3
8	[28, ∞)	0

Using the method of moments, we solve

$$e^{\mu+\sigma^2/2} = 7563.077 \text{ and } e^{2\mu+\sigma^2}(e^{\sigma^2} - 1) = 25{,}076{,}791$$

to find $\tilde{\mu} = 8.74927$ and $\tilde{\sigma}^2 = 0.36353$.

Therefore we use the model $\log X \sim N(8.74927, 0.36353)$, and estimate

$$P(X > 20{,}000) = 1 - \Phi\left(\frac{\log 20{,}000 - 8.74927}{0.60294}\right) = 0.02779.$$

Similarly, we obtain $P(X > 15{,}000) = 0.07533$. These may be compared with the (approximated) observed frequencies of $8/325 = 0.02462$ and $30.25/325 = 0.09308$, respectively. ☐

2.3 Fitting loss distributions

Fitting a probability distribution to claims data can be both an interesting and a challenging exercise. When trying to fit a distribution to claims data, it is well worth remembering the famous quote of George Box [8] "All models are wrong, some models are useful." In the previous section we have discussed the methods of maximum likelihood (ML), moments (MM) and percentiles (M%) in estimating parameters of some of the more classic loss distributions. But how do we ultimately decide on the particular type of distribution and estimation method to use, and whether or not the resulting distribution provides a *good fit*?

Exploratory Data Analysis (EDA) techniques (histograms, qqplots, boxplots) can often be useful in investigating the suitability of certain families of distributions. In attempting to fit the Theft claim data we have already seen (Figures 2.3 and 2.4) that the ML fitted Pareto and Weibull densities seem to be good approximations to a histogram of the data, while the ML fitted exponential does not. The Q-Q (quantile-quantile) normal plot and the plot of the ML fitted lognormal density in Figures 2.5 and 2.6 give some support to the use of a lognormal distribution for the Theft data. Given the importance of the tails in fitting a loss distribution to data, it can sometimes be useful to compare observed tail probabilities with those determined from various competing fitted distributions. Table 2.3 gives three tail probabilities for the eight distributions we have fitted in the previous section. Although these specific tails have been selected somewhat arbitrarily, they do suggest that the (ML) exponential and (M%) Weibull fitted distributions are doing a poor job of estimating tail behavior, while the fitted Pareto (ML or MM), Weibull (ML) and lognormal (ML) distributions have acceptable behavior.

These techniques for analyzing fit are exploratory, and one would also usually make use of one or more of the traditional classic methods to test fitness such as the Kolmogorov–Smirnoff (K–S), Anderson–Darling (A–D), Shapiro–Wilk (S–W) or chi-square goodness-of-fit tests. The K–S and A–D tests are used to test continuous distributions (the S–W for testing normality), while

TABLE 2.3
Observed frequencies and tail probabilities for distributions fitted to Theft data.

Method	Distribution	$P(X > 8,000)$	$P(X > 10,000)$	$P(X > 20,000)$
ML	exponential	0.0191	0.0071	0.0001
ML	Pareto	0.0439	0.0310	0.0098
MM	Pareto	0.0388	0.0251	0.0056
ML	Weibull	0.0397	0.0230	0.0020
M%	Weibull	0.0063	0.0022	0.0000
ML	gamma	0.0375	0.0190	0.0007
MM	gamma	0.0679	0.0469	0.0088
ML	lognormal	0.0597	0.0442	0.0154
Observed frequency		**0.0500**	**0.0250**	**0.0167**

the chi-square goodness-of-fit test can be used to test both continuous and discrete distributions.

A natural estimator for the theoretical distribution function F underlying any sample **x** is the empirical cumulative distribution function (ecdf) defined by

$$\hat{F}_n(x) = [\#x_i \leq x]/n.$$

The ecdf describes any data set precisely, and when one has a very large amount of data there is certainly justification in using this as a basis for statistical inference. However, there is often considerable aesthetic (and also some practical) appeal in modeling data with a classic loss distribution such as a Pareto, Weibull, gamma or lognormal.

2.3.1 Kolmogorov–Smirnoff test

The Kolmogorov–Smirnoff (K–S) test is useful in testing the null hypothesis H_0 that a sample **x** comes from a probability distribution with cumulative distribution function (cdf) F_0. The (two-sided) K–S test rejects the hypothesis H_0 if the maximum absolute difference d_n between F_0 and the ecdf $\hat{F}_n$ given in Equation (2.4) is large.

$$d_n = sup_{-\infty < x < \infty} | \hat{F}_n(x) - F_0(x) | \qquad (2.4)$$

A strength of the K–S test is that it is nonparametric and the null distribution of d_n is the same for all continuous distribution functions F_0. Hence one set of critical values (which are widely available) are appropriate for using this test. On the other hand, its omnibus nature as a test has its weaknesses, and in particular is often not good at detecting tail discrepancies (the upper tail of a loss distribution is usually of considerable interest).

The K–S test is invariant under transformations – in particular you can test that a data set $\mathbf{x}$ comes from a distribution with *cdf* F_0, or, equivalently, that the transformed sample data $F_0(\mathbf{x})$ comes from a uniform distribution on $[0, 1]$. The Kolmogorov–Smirnoff test that the Theft data comes from an (ML fitted) exponential distribution can be obtained in S^+ or R as follows:

```
> ks.test(1-exp(-Theft/mean(Theft)), "punif")
        One-sample Kolmogorov-Smirnov test
data:   1 - exp(-Theft/mean(Theft)) D =0.2013,p-value=0.0001192
alternative hypothesis: two.sided
%Warning message: cannot compute correct p-values
%with ties in: ks.test(Theft, "pexp", 1/mean(Theft))
```

which gives the same result as testing that the data set Theft is exponential with rate parameter $1/\text{mean}(\text{Theft})$:

```
> ks.test(Theft, "pexp",1/mean(Theft))
        One-sample Kolmogorov-Smirnov test
data:   Theft D = 0.2013, p-value = 0.0001192
alternative
hypothesis:two.sided
```

Note that the K–S test statistic is 0.2013, representing the *distance* between the empirical distribution function for the Theft claim data and the ML fitted exponential distribution. Figure 2.7 shows that this distance occurs at the observation (or claim size) 1395.

The K–S test for the ML Pareto fitted distribution yields:

```
> ks.test(1-(1872.13176/(1872.13176+Theft))**(1.880468),"punif")
        One-sample Kolmogorov-Smirnov test
data:   1 - (1872.13176/(1872.13176 + Theft))^(1.880468)
D = 0.0561, p-value = 0.8443 alternative hypothesis: two.sided
```

The K–S statistic is 0.0561 with a corresponding *p*-value of 0.8443. This suggests a much better fit for the (ML fitted) Pareto distribution, and this is illustrated in Figure 2.8.

The Anderson–Darling (A–D) test is a modification of the Kolmogorov–Smirnoff test which gives more weight to the tails of the distribution. It is therefore also a more sensitive test, but has the disadvantage that it is not a nonparametric test, and critical values for the test statistic must be calculated for each distribution being considered. Many software packages now tabulate critical values for the A–D test statistic when testing the fitness of distributions such as normal, lognormal, Weibull, gamma, etc. The A–D test statistic A_n^2 for a sample $\mathbf{x}$ of size n from the null distribution function F_0 (and corresponding density function f_0) is given by

$$A_n^2 = n \int_{-\infty}^{+\infty} \frac{[F_0(x) - \hat{F}_n(x)]^2}{F_0(x)[1 - F_0(x)]} f_0(x) \, dx.$$

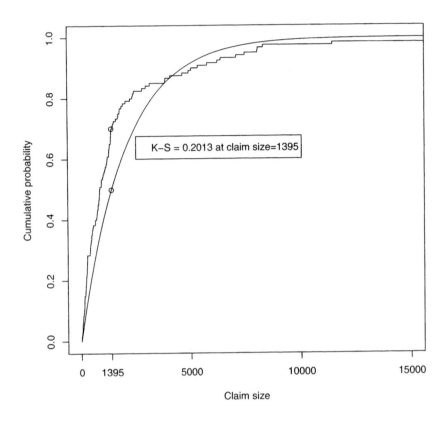

FIGURE 2.7
Kolmogorov–Smirnoff test for ML exponential fit with the Theft data ecdf.

2.3.2   Chi-square goodness-of-fit tests

The chi-square goodness-of-fit test is often used to test the how well a specified probability distribution (either discrete or continuous) fits a given data set. In theory, the test is an asymptotic one where the test of fit for a particular distribution is essentially reduced to a *multinomial* setting. In practice, when testing the fit of a continuous distribution, the data are usually first binned (or grouped) into k intervals of the form $I_i = [c_i, c_{i+1})$, for $i = 1, \ldots, k$, although this clearly involves losing information in the sample! Then, based on the grouped data, the number of expected observations E_i is calculated and compared with the actual observed numbers O_i for each interval. A measure of fit of the hypothesized null distribution is then obtained from the

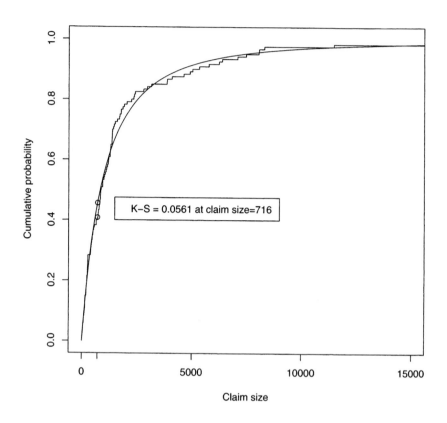

FIGURE 2.8

Kolmogorov–Smirnoff test for ML Pareto fit with the Theft data ecdf.

test statistic

$$\chi^2_{GF} = \sum_1^k (O_i - E_i)^2 / E_i, \qquad (2.5)$$

which compares observed and expected values. Large values of the test statistic χ^2_{GF} lead one to reject the null hypothesis under consideration since they indicate a lack of fit between what was observed and what one might expect. What is meant by a large value in this context is one which is large relative to a χ^2 distribution (introduced by Karl Pearson in 1900) with an appropriate number of degrees of freedom d. If the null hypothesis completely specifies the distribution, then the appropriate number of degrees of freedom is $d = k - 1$. If parameters must be estimated from the data (grouped or not), then the number of degrees of freedom depends on the method of estimation. In practice, one often estimates the r parameters in question for the null distribution from the (original) data, and then subtracts one degree of freedom for each such parameter. One then rejects the null hypothesis if χ^2_{GF} is large relative to the χ^2 distribution with $d = k - 1 - r$ degrees of freedom. Strictly speaking, this approach is valid if, when given the k intervals, one estimates the parameters using maximum likelihood on the grouped data.

For example, if the null hypothesis is that the distribution is exponential with parameter λ, then the maximum likelihood estimate $\hat{\lambda}_G$ using grouped data with $k = 10$ intervals is the value of λ which maximizes the likelihood

$$\prod_{i=1}^{10} \left[e^{-\lambda c_i} - e^{-\lambda c_{i+1}} \right]^{O_i},$$

where there are O_i observations in the ith interval $I_i = [c_i, c_{i+1})$ for $i = 1, \ldots, 10$. Normally, $\hat{\lambda}_G \neq \hat{\lambda} = 1/\bar{x}$. Letting $\hat{\theta}_i$ denote the maximum likelihood estimate of an observation in I_i, one then calculates

$$E_i = n \hat{\theta}_i = n \left[e^{-\hat{\lambda}_G c_i} - e^{-\hat{\lambda}_G c_{i+1}} \right]$$

as the expected number of observations in the ith interval I_i. The chi-square test statistic $\chi^2_{GF} = \sum_1^{10} (O_i - E_i)^2 / E_i$ is calculated and then one finds the probability of a larger (more extreme) result from a $\chi^2_{10-1-1} = \chi^2_8$ distribution.

What often happens in practice is that the probabilities $\hat{\theta}_i$ are calculated using the method of maximum likelihood on the full (as opposed to the grouped, interval or *binned*) data, and then the chi-square statistic is calculated. In reality, when the parameters are estimated in this way, this test is probably conservative (leading to rejection more often than it should). It has been shown (see [13], [16], [31] and [33]), however, that in this situation the appropriate number of degrees of freedom d is bounded by $k - 1$ and $k - 1 - r$ as expressed by

$$\bar{F}_{\chi^2_{k-1-r}}(t) \leq P(\chi^2_{GF} > t) \leq \bar{F}_{\chi^2_{k-1}}(t). \tag{2.6}$$

Hence it is generally advisable to compare the test statistic with both the χ^2_{k-1} and χ^2_{k-1-r} distributions.

In the use of the chi-square test statistic, one normally requires moderately large values of the expected counts E_i, and a frequently used rule of thumb is that each should be at least 5. If this is not the case, then one should consider joining adjacent bins. Moore [42] summarizes other rules of thumb, including the rule where one needs all $E_i \geq 1$ and at least 80% of the $E_i \geq 5$. The chi-square test is also sensitive to the choice (and number) of bins, but most reasonable choices lead to similar conclusions (see [42] for recommendations). The use of equiprobable bins is often suggested as a way of avoiding some of the arbitrariness in choice.

2.3.2.1 Fitting a distribution to the Theft data

Can we find a reasonable fit to our Theft data with one of the classic loss distributions? Previous considerations suggest that the ML fitted Pareto, Weibull or lognormal distributions are still possible candidates! We now proceed to test these (and the others considered) via a chi-square goodness-of-fit test.

As a starting point, the data of 120 theft claims (Table 2.1) was broken into 10 equiprobable intervals as determined by the ML fitted Pareto distribution, and the resulting intervals are given in Table 2.4. That is, using the ML estimates $\hat{\lambda} = 1872.132$ and $\hat{\alpha} = 1.880$ (rounded to 3 decimal places), and solving $(\hat{\lambda}/[\hat{\lambda} + c_i])^{\hat{\alpha}} = (11 - i)/10$ for $i = 1, \ldots, 10$, one obtains the left-hand break points of the intervals. For example, $c_2 = 107.92 = \hat{\lambda}([(11 - 2)/10]^{-1/\hat{\alpha}} - 1)$. Given that the intervals are of equal probability $1/10$ for the ML fitted Pareto distribution, the expected numbers (E) are all equal to 12. Using the same intervals, the expected number of observations for the other proposed distributions are calculated. For instance, using the ML Weibull fitted distribution ($\hat{c} = -0.00518$ and $\hat{\gamma} = 0.71593$), the expected number of observations in the second interval (c_2, c_3) is

$$120 \left[e^{-\hat{c}\, c_2^{\hat{\gamma}}} - e^{-\hat{c}\, c_3^{\hat{\gamma}}} \right] = 10.87614.$$

The values of the χ^2 test statistics and their p-values relative to the $\chi^2_{10-1-2} = \chi^2_7$ (χ^2_8 for the exponential) distribution are given in the last two rows of Table 2.4. These results suggest that the best choice of a model is a Pareto (using ML), but that the lognormal is also a possibility.

TABLE 2.4
Observed and expected values for fitting classic distributions to Theft data.

Distribution → Method →		Pareto ML	Pareto MM	exp ML	gamma ML	gamma MM	Weibull ML	Weibull M%	LN ML
Interval	O	E	E	E	E	E	E	E	E
(0.00, 107.92)	11	12	9.60	6.24	15.88	43.30	16.50	14.82	12.03
(107.92, 235.93)	14	12	10.08	6.98	9.58	9.65	10.88	12.26	14.64
(235.93, 391.11)	9	12	10.60	7.89	8.80	7.24	9.89	11.88	13.29
(391.11, 584.51)	12	12	11.17	9.03	8.78	6.33	9.64	11.93	12.09
(584.51, 834.68)	10	12	11.81	10.47	9.20	6.03	9.81	12.22	11.20
(834.68, 1175.81)	14	12	12.50	12.33	10.05	6.11	10.31	12.65	10.61
(1175.81, 1679.79)	18	12	13.25	14.80	11.43	6.61	11.22	13.13	10.32
(1679.79, 2534.73)	11	12	13.99	18.03	13.64	7.73	12.67	13.40	10.39
(2534.73, 4499.51)	6	12	14.49	21.28	17.12	10.31	14.93	12.37	11.10
(4499.51, +∞)	15	12	12.52	12.94	15.53	16.70	14.16	5.34	14.33
χ^2 Stat →		8.67	10.30	26.78	17.88	67.36	14.42	25.45	10.84
p-value* →		0.28	0.17	***	0.01	***	0.04	***	0.15

*(*** = p-value < 0.001)

2.3.3 Akaike information criteria

Another criterion which is often used in fitting a model is the AIC or Akaike Information Criterion. The AIC of one or several fitted model objects for which a log-likelihood value can be obtained is given by

$$\text{AIC} = -2(\text{log-likelihood}) + s \cdot r,$$

where r represents the number of parameters in the fitted model and $s = 2$ for the usual AIC, or $s = \log n$ (n being the number of observations) for the so-called BIC or SBC (Schwarz's Bayesian criterion). When comparing fitted objects, the smaller the AIC, the better the fit.

2.4 Mixture distributions

There are many situations where a classical parametric distribution may not be appropriate to model claims, but where a *mixture* of several such distributions might do very well! If F_1 and F_2 are two distribution functions and $p = 1 - q \, \epsilon \, (0, 1)$, then the $p : q$ mixture of F_1 and F_2 has the distribution function F defined by

$$F(x) = p \, F_1(x) + q \, F_2(x).$$

If in the above, X, X_1 and X_2 are random variables with respective distribution functions F, F_1 and F_2, then we say that X is a $p : q$ mixture of the random variables X_1 and X_2.

Example 2.4
Let U_1 and U_2 be uniform random variables on the intervals $[0, 1]$ and $[9, 10]$, respectively. We define U to be the $0.5 : 0.5$ mixture of U_1 and U_2, and $V = (U_1 + U_2)/2$. We may imagine that there is a random variable I taking the values 1 and 2 with probability 0.5 each, such that if $I = i$ then $U = U_i$ for $i = 1, 2$. It is important to note that although

$$\begin{aligned}
E(U) = E_I(E(U \mid I)) &= E(U_1)\,(1/2) + E(U_2)\,(1/2) \\
&= 5 \\
&= [E(U_1) + E(U_2)]/2 \\
&= E(V),
\end{aligned}$$

U and V are very different random variables. In fact, the range of U is $[0, 1] \cup [9, 10]$, while that of V is $[4.5, 5.5]$. $\square$

More generally, one may mix any (including an infinite) number of distributions. For example, suppose that for every θ in the set Θ there is a distribution F_θ. If G is a probability distribution on Θ (with corresponding density g), then we can define the mixture distribution F by

$$F(x) \;=\; \int_\Theta F_\theta(x)\,dG(\theta) \;=\; \int_\Theta F_\theta(x)\,g(\theta)\,d\theta.$$

In Example 2.4, there were only two distributions F_1 and F_2, and G put an equal weight on each.

Although in theory one can form mixtures of many types (and numbers) of random variables, in some cases mixtures from classic families yield well-known distributions. The following is an interesting example often used to model the situation when the random variable X represents the number of (annual) claims arising from a randomly selected policyholder. Conditional on knowing the claim rate (say λ) for the individual in question, one might model the number of claims by a Poisson random variable with parameter λ. However, in most cases it is not fair to assume that the claim rate is constant amongst policyholders. One might assume that the possibilities for λ vary over $(0, \infty)$ according to some probability (prior) distribution. The gamma family of distributions is both versatile and mathematically attractive. If one can assume that the variability in the claim rate λ obeys a $\Gamma(\alpha, \beta)$ distribution, then the following shows that the resulting X has a negative binomial distribution.

Example 2.5

$$\begin{aligned}
P(X = x) &= \int_0^\infty P(X = x \mid \lambda)\,dG_\Lambda(\lambda) \\
&= \int_0^\infty \frac{\lambda^x e^{-\lambda}}{x!}\,\frac{\beta^\alpha \lambda^{\alpha-1} e^{-\lambda\beta}}{\Gamma(\alpha)}\,d\lambda \\
&= \frac{\beta^\alpha}{\Gamma(\alpha)\,x!}\,\frac{\Gamma(\alpha + x)}{(\beta + 1)^{\alpha + x}} \\
&= \frac{\Gamma(\alpha + x)}{\Gamma(\alpha)\,x!}\left(\frac{\beta}{\beta + 1}\right)^\alpha \left(\frac{1}{\beta + 1}\right)^x
\end{aligned}$$

Hence X, which is a $\Gamma(\alpha, \beta)$ mixture of Poisson random variables, has in fact a negative binomial distribution with parameters α (which need not be an integer) and $p = \beta/(\beta + 1)$. We denote this by $X \sim NB(\alpha, p)$. If α is an integer, then X may be interpreted as the random variable representing the number of *failures* X until the αth success in a sequence of Bernoulli trials with success probability p.

It is not clear that this interpretation of X (as being negative binomial) is of any practical use in the context of the number of claims a randomly selected individual might make! In Problem 24, the reader is asked to fit both a Poisson (where claim rates are assumed to be constant or homogeneous over policyholders) and a negative binomial distribution to claims data and comment on the relative fits. (Note that there are two commonly used definitions of the negative binomial distribution X with parameters k and p. In our case, X represents the number of *failures* to the kth success while it is also sometimes defined to be the number of *trials* to the kth success (see Subsection 3.2.2.3).   ▯

The following R code generated a sample of $10,000$ "claim numbers" from a portfolio of policyholders where the claim rate parameter λ varies according to a gamma distribution with mean $5/[1/0.04] = 0.2$ and variance $5/[1/0.04]^2 = 0.008$.

```
> x<-c(rep(0,10000))
> for (i in 1:10000){ x[i]<-rpois(n=1,
  lambda=rgamma(n=1,scale=.04,shape=5))}
> table(x)
x
   0    1    2    3    4
8237 1549  192   21    1
> mean(x) [1] 0.2
> var(x) [1] 0.2122212
```

Table 2.5 gives some of the better known mixture distributions. The generalized Pareto distribution X with parameters (k, λ, δ) has density function

$$f_X(x) = \frac{\Gamma(\alpha + k)\, \delta^\alpha}{\Gamma(\alpha)\, \Gamma(k)} \frac{x^{k-1}}{(\delta + x)^{\alpha+k}} \quad \text{for } x > 0.$$

TABLE 2.5
Some common mixture distributions.

θ	$X\mid\theta$ distribution	Mixing distribution	X distribution
λ	Poisson (λ)	$\lambda \sim \Gamma(\alpha, \beta)$	$NB\,(\alpha, p = \beta/(1+\beta))$
p	$B(n, p)$	$p \sim \text{Beta}\,(\alpha, \beta)$	Beta Bin (n, α, β)
λ	Exponential λ	$\lambda \sim \Gamma\,(\alpha, \delta)$	Pareto (α, δ)
λ	$\Gamma(k, \lambda)$	$\lambda \sim \Gamma\,(\alpha, \delta)$	Gen. Pareto (k, α, δ)

2.5   Loss distributions and reinsurance

As policyholders buy insurance to obtain security from risks, so too an insurance company buys reinsurance to limit and control its own exposure to risk. One of the benefits of reinsurance is that it allows the insurer to expand its own capacity to take on risk. In transferring some of its risk to a reinsurer (or in some cases to several reinsurance companies), the insurance company is said to *cede* some of its business to the reinsurer, and hence is sometimes referred to as the *cedant* (although, for the most part, we shall use the term *baseline insurance company*). There are usually various types of reinsurance contracts available to an insurance company which broadly speaking fall into two categories – those (claim based) that are based on the sharing of risk per claim, and those (aggregate based) that are based on an agreement concerning the total or aggregate claims.

In *proportional reinsurance*, the baseline (or ceding) insurance company cedes to the reinsurance company an agreed proportion or percent of *each claim*. When the proportion varies between policies or contracts, this is sometimes referred to as *quota reinsurance*. In proportional or quota reinsurance, the reinsurer is normally involved in all claims, and this may lead to considerable administrative costs for the reinsurer. In some cases, the insurer must be careful not to cede too much of the business to the reinsurer in order to remain solvent. *Surplus reinsurance*, where only a proportion of each of the larger claims are ceded to the reinsurer, is another type of reinsurance which addresses this concern. For example, there may be a retention level M, such that the reinsurer pays only a part of those claims (often subject to limits) which exceed M. Another common type of reinsurance which is individual claim based is *excess of loss* reinsurance. In this type of arrangement, the reinsurance company covers the excess of any individual claim over an agreed amount (called the excess or retention level M), and hence is involved in only a fraction $(\bar{F}_X(M))$ of the claims.

In this section, we investigate the division of a claim resulting from a *claim-by-claim based* reinsurance arrangement, while in Chapter 3 on risk theory we consider the impact of reinsurance on aggregate claims. In a *claim-by-claim based* reinsurance agreement, each individual claim X is split into two components,

$$X \;=\; Y + Z \;=\; h_I(X) + h_R(X),$$

which are, respectively, handled by the insurance $(Y = h_I(X))$ and reinsurance $(Z = h_R(X))$ companies.

2.5.1   Proportional reinsurance

In proportional reinsurance, $h_I(X) = Y = \alpha X$ and $h_R(X) = Z = (1 - \alpha)X$ for some $0 \le \alpha \le 1$. An interesting property shared by the classic loss distributions we have considered (exponential, Pareto, gamma, Weibull and lognormal) is that they are closed under multiplication by a positive scalar factor (and hence are called *scale invariant*). In other words, if the random loss X belongs to one of these families and $k > 0$, then so does $k\,X$. Hence if X belongs to any one of these families, so do both of the proportions $Y = \alpha X$ and $Z = (1 - \alpha)X$ handled by the insurer and reinsurer!

2.5.2   Excess of loss reinsurance

In an excess of loss agreement (or treaty) with a reinsurer, the reinsurer handles the excess of each claim X over an agreed excess level M. We may write $h_I(X) = Y = min(X, M)$ and $h_R(X) = Z = max(0, X - M)$. In other words, $X = Y + Z$, where Y is the amount paid by the (baseline) insurance company and Z is that paid by the reinsurer with

$$Y = \begin{cases} X & \text{if } X \le M \\ M & \text{if } X > M \end{cases} \quad \text{and} \quad Z = \begin{cases} 0 & \text{if } X \le M \\ X - M & \text{if } X > M \end{cases}.$$

In introducing an excess of loss reinsurance agreement with excess level M, the expected payment per claim for the insurer is reduced from $E(X)$ to

$$\begin{aligned}
E(Y) &= \int_0^M x\, f_X(x)\, dx + M\, \bar{F}_X(M) \\
&= E(X) - \int_M^\infty x\, f_X(x)\, dx + M\, \bar{F}_X(M) \\
&= E(X) - \int_M^\infty (x - M)\, f_X(x)\, dx \\
&= E(X) - \int_0^\infty y\, f_X(y + M)\, dy \qquad \text{(letting } y = x - M\text{)}.
\end{aligned}$$

If X is an exponential random variable with parameter λ and M is the excess level, then

$$E(Y) = \frac{1}{\lambda}\left(1 - e^{-\lambda M}\right).$$

Hence by using an excess level of $M = (\log 4)E(X) = (\log 4)/\lambda$, the insurance company can reduce its average claim payment by 25% since

$$E(Y) = \frac{1}{\lambda}\left(1 - e^{-\lambda(\log 4)/\lambda}\right) = (0.75)\frac{1}{\lambda}.$$

When an excess of loss contract has been agreed, the insurer is really only interested in $Y = max(X, M)$ for any loss X, and hence one might view the claims data for the insurer as a censored sample of $n + m$ losses of the form

$$\mathbf{x} = x_1, x_2, M, x_4, M, x_6, x_7, M, \ldots,$$

where m is the number of censored losses (exceeding M) and n is the number of uncensored ($\leq M$) losses. Therefore, in trying to estimate the parameters θ of an appropriate loss distribution, one would maximize the log-likelihood function given by

$$L(\theta) = \prod_1^n f_X(x_i, \theta) \prod_1^m \bar{F}_X(M, \theta).$$

For example, in the exponential case,

$$L(\lambda) = \prod_1^n \lambda e^{-\lambda x_i} \prod_1^m e^{-\lambda M}$$

and hence

$$\frac{\partial}{\partial \lambda} \log L(\lambda) = \frac{\partial}{\partial \lambda} \left[n \log \lambda - \lambda \left(\sum_1^n x_i + mM \right) \right] = 0$$

$$\Rightarrow \hat{\lambda} = \frac{n}{\sum_1^n x_i + mM}.$$

2.5.2.1 The reinsurer's view of excess of loss reinsurance

Let us now consider excess of loss reinsurance from the point of view of the reinsurer. Representing a typical claim X in the form $X = Y + Z$, the part of the claim Z paid by the reinsurer is 0 with probability $F_X(M)$. The reinsurer is, however, more likely to be interested in the positive random variable Z_R, which is the amount of a claim it has to pay in the case (that is, conditional on) $X > M$. One may view Z as a mixture of 0 and Z_R, and hence

$$E(Z) = 0 \cdot F_X(M) + E(Z_R)\bar{F}_X(M) = E(Z_R)P(X > M).$$

Now

$$\bar{F}_{Z_R}(z) = P(X > M + z \mid X > M) = \frac{\bar{F}_X(M + z)}{\bar{F}_X(M)},$$

and on differentiating with respect to z (and multiplying by -1), one obtains

$$f_{Z_R}(z) = \frac{f_X(z + M)}{\bar{F}_X(M)} \quad \text{for } z > 0.$$

In the special case where X is exponential with parameter λ,

$$f_{Z_R}(z) = \frac{\lambda e^{-\lambda(z+M)}}{e^{-\lambda M}} = \lambda e^{-\lambda z},$$

which is not surprising due to the lack of memory property of the exponential distribution.

If X has a Pareto distribution with parameters α and λ, then the density function of Z_R takes the form

$$f_{Z_R}(z) = \frac{\alpha\lambda^\alpha/(\lambda + z + M)^{\alpha+1}}{\lambda^\alpha/(\lambda + M)^\alpha} = \frac{\alpha(\lambda + M)^\alpha}{(\lambda + z + M)^{\alpha+1}}.$$

That is, Z_R is Pareto with parameters α and $\lambda + M$ and mean $(\lambda + M)/(\alpha - 1)$ when $\alpha > 1$. Furthermore, if $X = Y + Z$, then

$$\begin{aligned}
E(Y) = E(X) - E(Z) &= E(X) - \bar{F}_X(M)E(Z_R) \\
&= \frac{\lambda}{\alpha - 1} - \left(\frac{\lambda}{\lambda + M}\right)^\alpha \frac{\lambda + M}{\alpha - 1} \\
&= \frac{\lambda}{\alpha - 1} - \frac{\lambda^\alpha}{\alpha - 1}\left(\frac{1}{\lambda + M}\right)^{\alpha-1},
\end{aligned}$$

from which it is clear that $E(Y)$ increases (respectively, $E(Z)$ decreases) with the excess level M.

2.5.2.2 Dealing with claims inflation

Claim size often increases over time due to inflation, and it is worth investigating how this affects typical payments for the ceding insurer and reinsurer if the same reinsurance treaty holds. For example, suppose that claims increase by a factor of k next year, but that the same excess level M is used in an excess of loss treaty between the insurer and reinsurer. Would one expect the typical payment for the (ceding) insurer to increase by a factor of k, and if not, would it be larger or smaller than k? Consequently, how would the typical payment change for the reinsurer?

On reflection, it is not difficult to see that the typical payment for the insurer should increase by a factor *less* than k (and, therefore, the factor for the reinsurer would be *greater* than k since the total claim size on the average increases by k). One may heuristically argue that typically any (*small*) claim X less than M/k this year will be $kX < M$ next year, and hence the insurer's payment next year on *small* claims will increase by a factor of k. However, for any (*larger*) claim $X > M/k$ this year, the insurer next year will pay $M = min(kX, M) \le kX$. Hence the increase overall in payment by the (ceding) insurer is less than k. This assertion is now more formally established.

Suppose that due to inflation next year, a typical claim $X = Y + Z$ next year will have distribution $X^* = k\,X$, where $k > 1$. If Y is that part of the claim X handled by the (ceding) insurer this year, then next year it will be $Y^* = g(X)$ defined by

$$Y^* = \begin{cases} kX & \text{if } kX \le M \\ M & \text{if } kX > M \end{cases}.$$

Then the amount paid by the insurer next year on a typical claim X^* is

$$E(Y^*) = \int_0^{M/k} kx\, f_X(x)\, dx + \int_{M/k}^{\infty} M f_X(x)\, dx$$

$$= k\left[\int_0^{M/k} x f_X(x)\, dx + \int_{M/k}^{M} (M/k)\, f_X(x)\, dx + \int_M^{\infty} (M/k)\, f_X(x)\, dx\right]$$

$$\leq k\left[\int_0^{M/k} x f_X(x)\, dx + \int_{M/k}^{M} x\, f_X(x)\, dx + \int_M^{\infty} M\, f_X(x)\, dx\right]$$

$$= k\left[\int_0^{M} x f_X(x)\, dx + \int_M^{\infty} M\, f_X(x)\, dx\right]$$

$$= k\, E(Y).$$

This shows that, in general, $E(Y^*) \leq k\, E(Y)$, and one might say that the *actual* or *effective* excess level decreases with inflation for the insurer! The following derivation gives a useful expression for $E(Y^*)$.

$$E(Y^*) = \int_0^{M/k} kx f_X(x)\, dx + \int_{M/k}^{\infty} M f_X(x)\, dx$$

$$= k\int_0^{\infty} x f_X(x)\, dx - k\int_{M/k}^{\infty} x f_X(x)\, dx + M\int_{M/k}^{\infty} f_X(x)\, dx$$

$$= k\, E(X) - k\int_0^{\infty} (y + M/k)\, f_X(y + M/k)\, dy$$

$$\qquad + M\int_0^{\infty} f_X(y + M/k)\, dy \quad \text{(using } y = x - M/k)$$

$$= k\left[E(X) - \int_0^{\infty} y f_X(y + M/k)\, dy\right].$$

In the case where X is exponential with parameter λ,

$$E(Y^*) = k\left[E(X) - \int_0^{\infty} y\, \lambda e^{-\lambda(y + M/k)}\, dy\right] = \frac{k}{\lambda}\left[1 - e^{-\lambda M/k}\right]$$

Example 2.6
A typical claim is modeled by an exponential distribution with mean 100, and an excess of loss reinsurance treaty is in effect with excess level $M = 150$. The expected cost per claim for the insurer under this arrangement is

$$E(Y) = \frac{1}{0.01}(1 - e^{-0.01(150)}) = 77.69.$$

Suppose now that inflation of 6% is expected for next year, and that the excess level remains at 150. Then the expected payment per claim next year for the insurer is

$$E(Y^*) = \frac{k}{\lambda} \left[1 - e^{-\lambda M/k} \right] = \frac{1.06}{0.01} \left[1 - e^{-0.01\,(150)/1.06} \right] = 80.25,$$

which is significantly different from $kE(Y) = 1.06(77.69) = 82.35$. □

2.5.2.3 Policy excess and deductibles

Introducing a *deductible* into a policy is another form of policy modification which a company might use to reduce both the number and amount of claims. One of the most common forms of a deductible is the *fixed amount deductible*. In this case, a deductible D is effected, whereby only claims in excess of D are considered and, therefore, the amount paid by the insurer on a loss of size X is $max(0, X - D)$. In the *proportional deductible*, the insured (or insurant) must pay a proportion α of each claim. For example, in a common form of health insurance in the USA, the insured pays 20% of any claim. Another form of a deductible is the *minimum* or *franchise* deductible. Here the insured is compensated for the entire claim X only if X exceeds a deductible D, otherwise there is no compensation.

The theory behind deductibles is clearly similar to that for claim-based excess of loss reinsurance treaties (where the relationship between the individual policyholder and the insurance company parallels that between the baseline or ceding insurance company and the reinsurer). There are many possible reasons for introducing deductibles. One such reason is to reduce the number of small claims made on the insurer. Since such claims are often administratively relatively expensive, a possible consequence of introducing a deductible is that premiums may be reduced, which in turn makes the product seemingly more attractive to the market.

Suppose that a deductible of size D is in effect, whereby on a loss of X the insurance company pays Y given by

$$Y = \begin{cases} 0 & \text{if } X \leq D \\ X - D & \text{if } X > D. \end{cases}$$

In this situation, the position of the insurer is similar to that of the reinsurer when an excess of loss reinsurance contract is in effect and it follows that

$$E(Y) = \int_D^\infty (x - D)\, f_X(x)\, dx = \int_0^\infty y\, f_X(y + D)\, dy.$$

Note, however, that here $E(Y)$ represents the average amount paid by the insurance company in respect of all losses X, while the average amount paid in respect of claims actually made (that is, with respect to the losses which exceed the deductible D) is given by

$$E(Y \mid X > D) = \int_D^\infty (x - D)\, \frac{f_X(x)}{\bar{F}_X(D)}\, dx.$$

Example 2.7

Claims (losses) in an automobile insurance portfolio had a mean of 800 and a standard deviation of 300 last year. Inflation of 5% is expected for the coming year, and it can be assumed that losses can be modeled by a lognormal distribution. An excess of loss reinsurance level of 1200 will be increased in line with inflation, and a policy excess (deductible) of 500 will be introduced.

If we let X be the lognormal random variable representing a typical loss next year, then $E(X) = (1.05)800 = 840$ and $Var(X) = [(1.05)300]^2 = 99{,}225$. The new policy excess will be $(1.05)1200 = 1260$. Solving the equations

$$e^{\mu+\sigma^2/2} = 840 \qquad \text{and} \qquad e^{2\mu+\sigma^2}e^{\sigma^2-1} = 99{,}225,$$

one finds $\tilde{\sigma} = 0.36273$ and $\tilde{\mu} = 6.66761$.

The proportion of incidents involving the reinsurance company is therefore

$$\begin{aligned} P(X > 1260) &= 1 - \Phi([\log 1260 - 6.66761]/0.36273) \\ &= 1 - \Phi(1.29917) \\ &= 0.09694. \end{aligned}$$

Moreover, the proportion of incidents where no claim will be made (due to the policy excess) is

$$P(X < 500) = \Phi(-1.24886) = 0.10586.$$

If Z is the part of the loss X paid by the reinsurer, then

$$\begin{aligned} E(Z) &= \int_{1260}^{\infty} (x - 1260) f_X(x)\,dx \\ &= 840 - \int_0^{1260} x f_X(x)\,dx - 1260\,P(X > 1260) \\ &= 840 - 840\,\Phi\left(\frac{\log 1260 - 6.66761 - 0.36273^2}{0.36273}\right) - 122.1484 \\ &\qquad \text{(using Equation (2.3))} \\ &= 24.45, \end{aligned}$$

which is the average amount paid by the reinsurer in respect of all incidents. Letting Z_R be the amount paid by the reinsurer if the reinsurer is involved, then

$$E(Z) = E(Z_R) \cdot P(X > 1260) + 0 \cdot P(X \le 1260) = E(Z_R) \cdot (0.09694),$$

from which it follows that $E(Z_R) = 24.45/0.09694 = 252.24$.

If the loss to an insured next year exceeds 500, then she will pay the first 500 while the insurance companies will pay the rest. If U is the part of any loss X borne by the insured (policyholder), then

$$E(U) = \int_0^{500} x f_X(x)\, dx + 500(1 - 0.10586)$$

$$= 840\ \Phi\left(\frac{\log 500 - 6.66761 - 0.36273^2}{0.36273}\right) + 447.0713$$

$$= 492.03.$$

Representing any loss X in the form $X = U + Y + Z$, where Y is the part borne by the insurance company, $E(Y) = 840 - 492.03 - 24.45 = 323.52$. $\square$

2.6   Problems

1. The random variable X represents the storm damage to a premises which has encountered a loss. The insurance company handling such claims will pay only W, the excess of the damage over $40,000, for any such damage (i.e., $W = X - 40,000$ if $X > 40,000$). The payments made by the company in 2005 amounted to: \$14,000, \$21,000, \$6,000, \$32,000 and \$2,000. Assume that the density function for the damage sustained X takes the form

$$f_X(x) = \alpha\, 2^\alpha\, 10^{4\alpha}\, (20{,}000 + x)^{-\alpha - 1} \qquad \text{for } x > 0$$

   where α is an unknown parameter.

   (a) Determine the density function, mean and variance for W, the typical amount paid by the insurance company (in respect of damage in excess of \$40,000 to a premises).

   (b) Using the method of maximum likelihood, find an estimate $\hat{\alpha}$ of α based on the 2005 data. Give an estimate for the standard error of $\hat{\alpha}$.

   (c) Suppose that inflation in 2006 is expected to be 4%. If the excess level remains at \$40,000, what is the average amount the company will pay on a storm damage claim over \$40,000?

2. A claim size random variable X has density function of the form,

$$f_X(x) = \frac{\theta}{400 + x}\left(\frac{400}{400 + x}\right)^\theta, \qquad x > 0$$

   for some unknown θ.

(a) Find the forms of the method of moments estimator $\tilde{\theta}$ and the method of maximum likelihood estimator $\hat{\theta}$ based on a random sample of size n.

(b) A sample of 50 claims from last year gave an average of 200. Use the method of moments to estimate θ. An arrangement with a reinsurance company has been made whereby the excess of any claim over 400 is handled by the reinsurer.

    i. What proportion of claims will be handled by the reinsurer?

    ii. What is the probability distribution and mean value for (positive) claim amounts handled by the reinsurer?

    iii. With this reinsurance arrangement, what is the average amount paid out by the baseline insurance company on claims made?

3. Eire General Insurance has an arrangement with the reinsurance company SingaporeRe, whereby the excess of any claim over M is handled by the reinsurer. Claim size is traditionally modeled by a Pareto distribution with parameters α and $\lambda = 8400$. Show that the maximum likelihood estimator of α based on a sample of $n + m$ claim payments (for Eire General) of the form $(x_1, \ldots, x_n, M, \ldots, M)$ takes the form

$$\hat{\alpha} = n / \left(\sum_1^n \log(1 + x_i/\lambda) + m \log(1 + M/\lambda) \right).$$

If the amounts paid by Erie General based on a sample of size $10 = 7 + 3 = n + m$ were

$$(14.9,\ 775.7,\ 805.2,\ 993.9,\ 1127.5,\ 1602.5,\ 1998.3,\ 2000,\ 2000,\ 2000),$$

what would the maximum likelihood estimate of α be?

4. A sample of 90 hospital claims of X is observed where $\bar{x} = 5010$ and $s^2 = 49{,}100{,}100$. Table 2.6 (of grouped data) was constructed in order to test the goodness-of-fit of: 1) an exponential model for X, and 2) a Pareto model for X (using the method of moments). Complete the table and perform the appropriate χ^2 goodness-of-fit tests. Comment on the adequacy of fit.

5. A claim-size random variable is modeled by a Pareto distribution with parameters $\alpha = 3$ and $\lambda = 1200$. A reinsurance arrangement has been made whereby in future years the excess of any claim over 800 is handled by the reinsurer. If inflation next year is to be 5%, determine the expected amount paid per claim by the insurance company next year.

6. Claims in a portfolio of house contents policies have been modeled by a Pareto distribution with parameters $\alpha = 6$ and $\lambda = 1500$. Inflation

TABLE 2.6

Hospital claims data.

	Interval	O_i (Obs)	E_i (Exp)	E_i (Pareto-MM)
1	0 - 528	14		
2	528 - 1,118	17		
3	1,118 - 1,787	9		
4	1,787 - 2,559	8		
5	2,559 - 3,473	7		
6	3,473 - 4,591	12		
7	4,591 - 6,032	7		
8	6,032 - 8,063	4		
9	8,063 - 11,536	5		
10	11,536 - $+\infty$	7		

for next year is expected to be 5%, but a \$100 deductible is to be introduced for all claims as well. What will be the resulting decrease in average claim payment for next year?

7. A claim-size random variable is being modeled in an insurance company by a Pareto random variable $X \sim \text{Pareto}(\alpha = 4, \lambda = 900)$. A reinsurance arrangement has been made for future years whereby the excess of any claim over 600 is paid by the reinsurer.

 (a) Determine the mean reduction in claim size for the insurance company which is achieved by this arrangement.

 (b) Next year, inflation is expected to be 10%. Assuming the same reinsurance arrangement as for this year, determine the expected amount paid per claim next year by the insurance company.

8. Claims resulting in losses in an automobile portfolio in the current year have a mean of 500 and a standard deviation of $(\sqrt{2})\,500$. Inflation of 10% is expected for the coming year and it can be assumed that a Pareto distribution is appropriate for claim size. A policy excess (or standard deductible), whereby the company pays the excess of any loss over 200, is being considered for the coming year. Using the method of moments, estimate

 (a) the % reduction in claims made next year due to the introduction of the deductible.

 (b) the reduction in average claim payment next year due to the deductible.

 (c) the reduction in average claim payment next year if the deductible was in fact a *franchise* deductible.

9. The following claim data set of 40 values was collected from a portfolio of home insurance policies, where $\bar{x} = 272.675$ and $s = 461.1389$.

10	11	15	22	28	30	32	36	38	48	51
55	56	68	68	85	87	94	103	104	105	106
109	119	121	137	178	181	226	287	310	321	354
393	438	591	1045	1210	1212	2423				

It is decided to fit a Pareto distribution $X \sim$ Pareto (α, λ) to the data using the method of moments. Find these estimates, and use them to perform a χ^2 goodness-of-fit for this distribution by completing Table 2.7.

TABLE 2.7
Interval data on 40 home insurance claims.

Interval	Observed	Expected
0, 42.594	*	8
42.594, 102.270	*	8
102.270, 196.444	*	*
196.444, 322.336	*	*
322.336, $+\infty$	*	*

10. A claim-size random variable X has density function $f_\theta(x) = \theta x e^{-\theta x^2/2}$ for $x > 0$. Determine the method of moments estimator $\tilde{\theta}$ of θ based on a random sample of size n. Show that the maximum likelihood estimator $\hat{\theta}$ of θ based on a sample of size n takes the form $\hat{\theta} = 2n/\sum X_i^2$.

11. A random sample of 120 claims was observed from a portfolio, where $\sum x_i = 9{,}000$ and $\sum x_i^2 = 420{,}000$. It was decided to test the fit of the data to (a) an exponential distribution with density $\theta e^{-\theta x}$, and (b) a Weibull density of the form $f(x) = \theta x e^{-\theta x^2/2}$. In both cases, parameters were estimated using the method of maximum likelihood. Complete Table 2.8 and test the fitness of the resulting distributions using chi-square goodness-of-fit tests. Comment on the adequacy of fit.

12. Household content insurance claims are modeled by a Weibull distribution with parameters $c > 0$ and $\gamma = 2$.

(a) A random sample of 50 such claims yields $\sum_{1}^{50} x_i = 13{,}500$ and $\sum_{i=1}^{50} x_i^2 = 4{,}500{,}000$. Calculate the method of moments estimator $\tilde{c}$ and the method of maximum likelihood estimate $\hat{c}$ of c using this information. Determine an approximate 95% confidence interval for c based on maximum likelihood.

TABLE 2.8
Portfolio of 120 claims.

	Interval	Observed O_i	Expected E_i(Exp)	Expected E_i (Weibull)
1	[0, 7.90]	4	12	2.12
2	[7.90, 16.74]	9	12	7.11
3	[16.74, 26.75]	14	12	12.95
4	[26.75, 38.31]	16	12	18.91
5	[38.31, 51.99]	21	12	23.46
6	[51.99, 68.72]	22	12	24.30
7	[68.72, 90.30]	18	12	19.45
8	[90.30, 103.97]	7	?	?
9	[103.97, 172.69]	8	?	?
10	[172.69, +∞]	1	?	?

(b) If a deductible of 200 is introduced, estimate (using maximum like-lihood) the reduction in the proportion of claims to be made.

13. The 30 claims in Table 2.9 are for vandal damage to cars over a period of six months in a certain community:

TABLE 2.9
Claims for vandal damage to cars.

38	56	77	110	112	138	152	168	188	210
228	241	252	273	283	288	291	299	305	317
321	356	374	422	485	527	529	559	567	656

Use the method of percentiles (based on quartiles) to fit a Weibull distribution of the form $F(x) = 1 - e^{-cx^{\gamma}}$ to the data. Complete Table 2.10 and perform a chi-square goodness-of-fit test for this Weibull distribution.

TABLE 2.10

Interval	Observed	Expected
[0, 145]	*	*
[145, 225]	*	*
[225, 310]	*	*
[310, 420]	*	*
[420, +∞]	*	*

14. If $X \sim W(c, \gamma)$, then determine the form of F_X^{-1}. Use this to write

R code for generating a random sample of 300 observations from a $W(0.04, 2)$ distribution. Run the code and compare your sample mean and variance with the theoretical values.

15. Assume that 3000 claims have occurred in a portfolio of motor policies, where the mean claim size is $800 and the standard deviation is $350. Using both a normal and a lognormal distribution to model claim size, estimate the size of claims, w, such that 150 claims are larger than w and also the expected number of claims in the sample which are less than $125. Comment on the results.

16. An analysis of 3000 household theft claims reveal a mean claim size of 1500 and a standard deviation of 600. Assuming claim size can be modeled by a lognormal distribution, estimate the proportion of claims < 1000, and the claim size M with the property that 1000 of the claims would be expected to exceed M.

17. Suppose that X has a lognormal distribution with parameters μ and σ^2.

   (a) Show that the ML estimators of these parameters based on a random sample of size n take the form:

   $$\hat{\mu} = \frac{\sum_1^n \log x_i}{n} \quad \text{and} \quad \hat{\sigma}^2 = \frac{\sum_1^n [\log x_i - \hat{\mu}]^2}{n}.$$

   (b) A sample of 30 claims from a lognormal distribution gave

   $$\sum_1^{30} \log x_i = 172.5 \text{ and } \sum_1^{30} (\log x_i)^2 = 996.675.$$

   Using the method of maximum likelihood, estimate the mean size of a claim, and the proportion of claims which exceed 400.

   (c) Let $W = kX$ where $k > 0$. Show that W is also lognormal and determine its parameters.

18. On a particular class of policy, claim amounts coming into Surco Ltd. follow an exponential distribution with unknown parameter λ. A reinsurance arrangement has been made by Surco so that a reinsurer will handle the excess of any claim above $10,000. Over the past year, 80 claims have been made and 68 of these claims were for amounts below $10,000; these 68 in aggregate value amounted to $220,000. The other 12 claims exceeded $10,000.

   (a) Let X_i represent the amount of the ith claim from the 68 claims beneath $10,000. Show that the log–likelihood function is

   $$\ell(\lambda) = 68 \log \lambda - \lambda \sum_{i=1}^{68} x_i - 120,000 \, \lambda.$$

Hence find $\hat{\lambda}$ and calculate an approximate 95% confidence interval for λ.

(b) Let Z denote the cost to the reinsurer of any claim X, and hence $X = Y + Z$. Determine an expression for $E(Z)$ in terms of λ. Estimate $E(Z)$ using maximum likelihood.

(c) Next year, claim amounts are expected to increase in size by an inflationary figure of 5%. Suppose that the excess of loss reinsurance level remains at \$10,000. Let Z^* represent the cost to the reinsurer of a typical claim next year. Estimate $E(Z^*)$. Using your answer in (18a) or otherwise, derive a 95% confidence interval for $E(Z^*)$.

19. The typical claim X in an insurance portfolio has density function

$$f_X(x) = 2\,x/10^6 \quad \text{for } 0 \le x \le 1000, \text{ and } 0 \text{ otherwise.}$$

The insurance company handling the claims has made an excess of loss treaty with a reinsurer with excess level $M = 800$. If Y represents the part paid by the ceding company for the claim X, determine $E(Y)$. If claims inflation of 5% is expected for next year and the same reinsurance treaty remains in effect, what will be the expected cost of a claim to the ceding insurer?

20. Suppose that claims resulting from incidents in a certain automobile portfolio had a mean of 400 and a standard deviation of 150 last year. Inflation of 20% is expected for the coming year, and it can be assumed that claims can be modeled by a lognormal distribution. An excess of loss reinsurance level of 800 will be increased in line with inflation, and a policy excess of 300 will be introduced. Estimate

(a) The proportion of incidents where no claim will be made (due to the policy excess).

(b) The proportion of incidents involving the reinsurance company.

(c) The average amount paid by the reinsurer in respect of all incidents.

(d) The average amount paid by the reinsurer in respect of incidents which involve the reinsurer.

(e) The average amount paid by the direct insurer in respect of all incidents.

21. Suppose that claims resulting from incidents in a certain automobile portfolio had a mean of 400 and a standard deviation of 250 last year. Inflation of 10% is expected for the coming year, and it can be assumed that claims can be modeled by a lognormal distribution. An excess of loss reinsurance level of 1000 will be increased in line with inflation, and a policy excess of 200 will be introduced. Estimate

(a) The proportion of incidents where no claim will be made (due to the policy excess).

(b) The proportion of incidents involving the reinsurance company.

(c) The average amount paid by the reinsurer in respect of all incidents, as well as the average amount in respect of incidents with which it is actually involved.

22. Use Kolmogorov–Smirnoff tests to test fitness of the Weibull (ML and M%) and lognormal distributions to the Theft claim data.

23. In a large population of drivers the accident rate Λ of a randomly selected person varies from person to person according to a $\Gamma(2, 10)$ random variable. If $X \mid [\Lambda = \lambda]$ is the number of accidents a person with accident rate λ incurs in a year, then assume $X \mid [\Lambda = \lambda]$ is a Poisson random variable with parameter λ. If X represents the number of accidents a randomly selected person has in a year, what are $E(X)$ and $Var(X)$?

24. Table 2.11 gives the distribution of the number of claims for different policyholders in a general insurance portfolio. Fit both the Poisson and negative binomial distributions to this data, and comment on which model provides a better fit.

TABLE 2.11
Claims in general insurance portfolio.

Number of claims	Frequency
0	65,623
1	12,571
2	1,644
3	148
4	13
5	1
6	0

3

Risk Theory

3.1 Risk models for aggregate claims

In 1930, Harold Cramér (see [18] and [54]) wrote that "The Object of the Theory of Risk is to give a mathematical analysis of the random fluctuations in an insurance business and to discuss the various means of protection against their inconvenient effects." In our modern world, individuals and companies continually encounter situations of risk where decisions must be made in the face of uncertainty. Risk theory can be useful in analyzing possible scenarios as well as options open to the analyst, and therefore assist in the ultimate decision-making process. For example, in contemplating a new insurance product, what is the probability that it will be profitable? What modifications can one make to the price structure of a product in order to enhance its profitability, yet at the same time maintain a reasonable degree of security and competitiveness? In this chapter we investigate various models for the risk consisting of the total or aggregate amount of claims S payable by a company over a fixed period of time. Our models will inform us and allow us to make decisions on, amongst other things: expected profits, premium loadings, reserves necessary to ensure (with high probability) profitability, as well as the impact of reinsurance and deductibles.

Assume that S is the random variable representing the total amount of claims payable by a company in a relatively short fixed period of time from a portfolio or collection of policies. Restricting consideration to shorter periods of time like a few months or a year often allows us to ignore aspects of the changing value of money due to inflation. We shall consider two types of models for S, the *collective* and the *individual* risk models.

In the collective risk model for S, we introduce the random variable N which indicates the number of claims made, and write

$$S = X_1 + \cdots + X_N.$$

In this model, X_i is the random variable representing the amount arising from the ith claim which is made in the time period being considered. Under the collective risk model, S has what is called a *compound distribution*. In some sense, we might say that the model is compounded by the fact that the number of terms in the sum is random and not fixed.

On the other hand, in the individual risk model for S, we let n be the number of policies (in some cases, this may coincide with the number of policyholders) in the portfolio under consideration and write

$$S = Y_1 + \cdots + Y_n,$$

where Y_i is the random variable representing the claim amount arising from the ith policy (or policyholder). We refer to this as the individual risk model for S since there is a term in the sum for each individual policy or policyholder. Since in a short period of time normally only a small proportion of policies give rise to claims, most of the the terms Y_i will be equal to 0. One of the assumptions in the individual risk model is that at most one claim may arise from a policy, while in the collective risk model multiple claims may result from a single policy or policyholder. It is important to understand the difference between the two models for the total claims S, and in particular the difference in meaning for the claim size random variables X_i and Y_i in each case. Both of these models have appealing aspects for modeling, and their appropriateness in any situation will depend on the assumptions one can make. Although this chapter provides a good introduction to risk theory, there are several other books which deal more extensively with the topic ([3], [9], [19], [20] and [55]).

3.2   Collective risk models

In the collective risk model for claims, we model S as a compound distribution of the form $S = X_1 + \cdots + X_N$. We assume that the component terms $X_1, X_2, \ldots,$ are independent identically distributed random variables which are also independent of the random number of terms N in the sum. Often N is assumed to be Poisson, but other distributions such as the binomial or negative binomial can be used. When N is Poisson, S has a *compound Poisson distribution*, and if N is binomial or negative binomial then S has, respectively, a *compound binomial* or *compound negative binomial distribution*.

Compound distributions are used to model many phenomena. For example, we might model the total annual number of traffic fatalities F in a country using a compound distribution where $F = D_1 + \cdots + D_N$, N represents the number of fatal traffic accidents in a year, and D_i is the number of fatalities in the ith fatal traffic accident. The total amount (centimeters) of rainfall $R = C_1 + \cdots + C_M$ in a particular location over a fixed period of time might be modeled by a compound Poisson distribution where C_i is the amount falling in the ith rainfall and M is the number of rainfalls. The daily amount of employee working time W in a factory may be modeled by a compound binomial distribution of the form $W = H_1 + \cdots + H_N$ where H_i is the number of hours worked by the ith arriving employee, n is the total number of employees in total, and $N \sim B(n, q)$ is a binomial random variable denoting the number who actually show up for work on the day.

3.2.1 Basic properties of compound distributions

We initially establish some basic distributional properties for compound distributions. The double expectation theorem (see appendix on *Some Basic Tools in Probability and Statistics*) is useful in obtaining compact formulae for the mean, variance and various generating functions of S. For example:

$$
\begin{aligned}
E(S) &= E_N(E(S \mid N)) \\
&= \sum_{n=0}^{\infty} E(X_1 + \cdots + X_n \mid N = n)P(N = n) \\
&= \sum_{n=0}^{\infty} [nE(X)] \, P(N = n) \\
&= E(X) \sum_{n=0}^{\infty} nP(N = n) \\
&= E(X)E(N).
\end{aligned}
$$

In a similar fashion, we obtain:

$$
\begin{aligned}
Var(S) &= Var_N(E(S \mid N)) + E_N(V(S \mid N)) \\
&= Var_N(E(X) \cdot N) + E_N(N \cdot Var(X)) \\
&= E^2(X)Var(N) + Var(X)E(N),
\end{aligned}
$$

and the moment generating function of S is given by:

$$
\begin{aligned}
M_S(t) &= E(e^{tS}) = E_N[E(e^{tS} \mid N)] \\
&= E_N[E(e^{t[X_1+\cdots+X_n]}) \mid N = n)] \\
&= E_N(M_X^n(t) \mid N = n) \\
&= E_N(M_X^N(t)) \\
&= E_N(e^{N \log M_X(t)}) \\
&= M_N(\log M_X(t)).
\end{aligned}
$$

In the special case when all claims are a constant $X = K$ (and hence $Var(X) = 0$), one clearly has that $E(S) = K\,E(N)$, $Var(S) = K^2\,Var(N)$, and $M_S(t) = M_N(\log e^{tK}) = M_N(tK)$.

3.2.2 Compound Poisson, binomial and negative binomial distributions

One of the main objectives in studying compound distribution models of the form $S = X_1 + \cdots + X_N$ for aggregate claims is that they allow us to incorporate attributes of both the severity of a typical claim (represented by X) and the frequency (represented by N). We consider in some detail compound Poisson, binomial and negative binomial distributions for S, however, the compound Poisson is the most widely used of these. One reason is because it is simpler than the others. It has just one rate parameter λ for the count variable N, while the others have two $((n, q)$ for the binomial and (k, p) for the negative binomial). Formulae for the basic moments of the compound Poisson (mean, variance, skewness, i^{th} central moment) are straightforward as well as easily expressed in terms of λ and the moments of X. Furthermore, it has the important property of being preserved under convolutions. This is very useful in modeling combined risks over different companies or portfolios within a company. For these reasons, we begin our study of compound distributions with the compound Poisson.

3.2.2.1   Compound Poisson distribution

S is compound Poisson when N is Poisson with parameter λ. Since $E(N) = Var(N) = \lambda$ and $M_N(t) = e^{\lambda[e^t - 1]}$, it follows that

$$E(S) = \lambda E(X),$$
$$Var(S) = \lambda E(X^2), \quad \text{and}$$
$$M_S(t) = M_N(\log M_X(t)) = e^{\lambda[M_X(t) - 1]}.$$

These expressions are well worth remembering due to the popularity of the compound Poisson distribution. Using the cumulant moment generating function $C_S(t)$ of a compound Poisson random variable S (which gives central moments of a random variable), one may easily determine the skewness of S. Since $C_S(t) = \log M_S(t) = \lambda[M_X(t) - 1]$, the third central moment of S is

$$E(S - E(S))^3 = C_S'''(0) = \frac{\partial^3}{\partial t^3}\{\lambda[M_X(t) - 1]\}\,|_{t=0}$$
$$= \lambda M_X'''(t)\,|_{t=0} = \lambda m_3,$$

and the ith central moment of S is given by

$$E(S - E(S))^i = \lambda m_i = \lambda E(X^i) \quad \text{for any } i \geq 2.$$

Example 3.1
Total claims in a portfolio of policies are modeled by a compound Poisson distribution with parameter λ where the claim X size is lognormal ($\log X \sim N(\mu, \sigma^2)$). Using $Y = \log X$, the moments of X are easily obtained since $m_i = E(X^i) = E(e^{iY}) = e^{\mu i + (\sigma^2 i^2)/2}$ for $i = 1, \ldots,$. Therefore $E(S) = \lambda m_1 = \lambda e^{\mu + \frac{\sigma^2}{2}}$, $Var(S) = \lambda m_2 = \lambda e^{2\mu + 2\sigma^2}$, and the skewness of S is given by

$$skew(S) = \frac{\lambda e^{3\mu + \frac{9}{2}\sigma^2}}{(\lambda e^{2\mu + 2\sigma^2})^{3/2}} = \frac{1}{\sqrt{\lambda}}\, e^{3\sigma^2/2} \to 0 \text{ as } \lambda \to \infty.$$

One may easily calculate the kurtosis of S to be

$$kurt(S) = \frac{\lambda e^{4\mu + \frac{16}{2}\sigma^2}}{(\lambda e^{2\mu + 2\sigma^2})^2} = \frac{e^{4\sigma^2}}{\lambda}.$$

Hence the kurtosis is very small if the expected number of claims λ is large relative to the variance of $\log X$, while it will be large when the expected number of claims is relatively small. Note that both the skewness and kurtosis of S are independent of the parameter $\mu = E(\log X)$, but not of $E(X)$ itself.
□

3.2.2.2 Compound binomial distribution

The compound binomial distribution $S = X_1 + \cdots + X_N$, where N is binomially distributed with parameters n and q, may be useful when there are n policies, each of which might give rise to a claim in a given period of time with probability q. Note the use of q (instead of the usual p) for the probability of a claim – as the insurance company would certainly not regard a claim as a *success*! The binomial distribution $B(n, q)$ has moment generating function $M_N(t) = (qe^t + p)^n$, therefore

$$M_S(t) = M_N(\log M_X(t)) = (qe^{\log M_X(t)} + p)^n = (qM_X(t) + p)^n.$$

Using $m_i = E(X^i)$ for $i = 1, \ldots$, one may readily establish that

$$E(S) = E(N)E(X) = nqm_1 \tag{3.1}$$
$$Var(S) = E(N)Var(X) + Var(N)E^2(X)$$
$$= nq(m_2 - m_1^2) + nqpm_1^2 = nq(m_2 - qm_1^2) \tag{3.2}$$
$$C_S(t) = \log M_S(t) = n\log(qM_X(t) + p). \tag{3.3}$$

Finding the 3rd derivative of $C_S(t)$ with respect to t and evaluating at 0, one has that $C_S'''(0) = nqm_3 - 3nq^2m_2m_1 + 2nq^3m_1^3$, which enables us to calculate the skewness of S. The skewness of S approaches 0 as the parameter $n \to \infty$ since

$$skew(S) = \frac{nqm_3 - 3nq^2m_2m_1 + 2nq^3m_1^3}{(nqm_2 - nq^2m_1^2)^{3/2}} = \frac{1}{\sqrt{n}} \frac{qm_3 - 3q^2m_2m_1 + 2q^3m_1^3}{(qm_2 - q^2m_1^2)^{3/2}}. \tag{3.4}$$

This is to be expected since for large n the central limit theorem applies (we can view S as a sum of n independent identically distributed random variables each of which is 0 with probability $p = 1 - q$), therefore S is approximately normal and in particular symmetric. The skewness of a compound binomial distribution may be negative. This is true when claims are constant ($X = K$) and $q > 1/2$, since then

$$C_S'''(0) = E[S - E(S)]^3 = K^3 E(N - nq)^3 < 0.$$

Although this is theoretically possible, in most practical applications we encounter q is small and S is positively skewed.

Example 3.2
Consider a collection of 5000 policies each of which has probability $q = 0.002$ of giving rise to a claim in a given year. Assume all policies are for a fixed amount X of $K = 400$. Then $m_i = (400)^i$ for all i, and hence

$$E(S) = nqm_1 = 5000(0.002)(400) = 4000,$$
$$Var(S) = nq(m_2 - qm_1^2)$$
$$= 5000(0.002)[400^2 - (0.002)(400)^2]$$
$$= 1{,}596{,}800 \text{ and}$$
$$skew(S) = [nqm_3 - 3nq^2m_2m_1 + 2nq^3m_1^3]/[nq(m_2 - qm_1)]^{3/2}$$
$$= [5000(0.002)(400)^3 - 3(5000)(0.002)^2(400)^2(400)$$
$$\qquad + 2(5000)(0.002)^3(400)^3]/[1{,}596{,}800]^{3/2}$$
$$= 636{,}165{,}120/[1{,}596{,}800]^{3/2}$$
$$= 0.31527.$$

If there had been only 50 policies in this collection, then the mean, variance and skewness would have been, respectively, $E(S) = 40$, $Var(S) = 15{,}968$, and $skew(S) = 3.1527$. ◻

3.2.2.3 Compound negative binomial distribution

The compound negative binomial distribution $S = X_1 + \cdots + X_N$ where $N \sim NB(k, p)$ may also be effectively used to model aggregate claims on a collective-risk basis. Here N has the negative binomial distribution with parameters k and p, where p is the probability of success in a sequence of Bernoulli trials and N denotes the number of failures until the kth success.

If the parameter $k = 1$, then $N \sim NB(1, p)$ counts the number of *failures* until the 1st success and has the geometric distribution with parameter p. It is important to note that sometimes one defines the geometric random variable with parameter p as the number N^* of *trials* (as opposed to *failures*) until the 1st success. If N^* represents the number of trials and N the number of failures until the 1st success, then of course N^* and N only differ by 1. Although they have different means $(E(N^*) = 1/p$ while $E(N) = 1/p - 1 = q/p)$, they have the same variance q/p^2. In a similar fashion, one could define a negative binomial random variable with parameters (k, p) to be the number of trials until the kth success; however, here we shall continue to use the definition which counts the number of failures until the kth success.

One interpretation of N is that it is the sum (convolution) of k geometric random variables with parameter p. This interpretation gives us some motivation for considering the compound negative binomial distribution as a model for aggregate claims where there are k policies (or policyholders) in a portfolio. We may, for instance, view the number of claims arising from the ith policy (for $i = 1, \ldots, k$) as being represented by the number of failures between successes $i - 1$ and i in the sequence of Bernoulli trials. This allows for the possibility that more than one claim may arise from a policy, which is a restriction on the compound binomial distribution.

There are other reasons for considering the compound negative binomial distribution as a model for aggregate claims. The negative binomial distribution has two parameters while the Poisson has only one, hence it could be considered to be more versatile in modeling claim frequency. One restriction on the use of the Poisson random variable for claim frequency is that the mean and variance are the same. If, for example, we feel the variability in claims is greater than the expected number, then this may be incorporated through use of the negative binomial since if $N \sim NB(k, p)$, then $Var(N) = kq/p^2 > kq/p = E(N)$. Another reason to use the negative binomial for modeling claim frequency is that the negative binomial distribution may be interpreted as a gamma mixture of Poisson random variables.

We now determine basic formulae for the mean, variance and skewness of the compound negative binomial distribution for S. The cumulant generating function for S takes the form $C_S(t) = \log(p/(1 - qM_X(t))^k$, and therefore

$$E(S) = C_S'(t) \mid_{t=0} = \frac{kq\, M_X'(t)}{1 - qM_X(t)} \mid_{t=0} = \frac{kq}{p}\, m_1 \quad \text{and} \tag{3.5}$$

$$Var(S) = C_S''(t) \mid_{t=0} = \frac{kq[M_X''(t)(1 - qM_X(t) + q(M_X'(t))^2]}{[1 - qM_X(t)]^2} \mid_{t=0} \tag{3.6}$$

$$= \frac{kq(pm_2 + qm_1^2)}{p^2}. \tag{3.7}$$

Now

$$C_S'''(t) = kq\left(\, [M_X''(t)(1 - qM_X(t)) + qM_X'^2(t)]'[1 - qM_X(t)]^2 \,+\right.$$
$$\left. 2[1 - qM_X(t)]qM_X'(t)[M_X''(t)[1 - qM_X(t)] + qM_X'^2(t)])/(1 - qM_X(t))^4, \right.$$

and therefore

$$skew(S) = \frac{C_S'''(0)}{[Var(S)]^{3/2}}$$

$$= \frac{kqm_3/p + 3kq^2m_1m_2/p^2 + 2kq^3m_1^3/p^3}{[k(pqm_2 + q^2m_1^2)/p^2]^{3/2}}$$

$$= \frac{p^2qm_3 + 3pq^2m_1m_2 + 2q^3m_1^3}{\sqrt{k}\,(pq\,m_2 + q^2m_1^2)^{3/2}}.$$

Note in particular that the compound negative binomial distribution (unlike the compound binomial) is always positively skewed. Moreover, as $k \to \infty$

(and S can be viewed as the sum of a large number of independent geometric random variables), $skew\,(S) \to 0$.

Example 3.3

Consider a compound negative binomial model for aggregate claims of the form $S_1 = X_1 + \cdots + X_{N_1}$ where $N_1 \sim NB(800, 0.98)$ and the typical claim X is exponential with mean 400. A model of this type might be considered when there are 800 policies and the number of claims arising from any particular policy is geometric with mean $0.02/0.98 = 0.0204$. The first three moments of X are given by $m_1 = 400, m_2 = 2(400)^2$ and $m_3 = 6(400)^3$. Therefore

$$E(S_1) = kqm_1/p = 800(0.02)(400)/(0.98)$$
$$= 6530.612,$$
$$Var(S_1) = \frac{kq(pm_2 + qm_1^2)}{p^2} = \frac{800(0.02)\,[(0.98)2(400)^2 + (0.02)(400)^2]}{(0.98)^2}$$
$$= 5{,}277{,}801 = 2297.346^2,$$
$$skew(S_1) = \frac{(0.98)^2(0.02)6(400)^3 + 3(0.98)(0.02)^2400(2)(400)^2 + 2(0.02)^3(400)^3}{\sqrt{800}\,[0.98(0.02)2(400)^2 + (0.02)^2(400)^2]^{3/2}}$$
$$= (7{,}375{,}872 + 150{,}528 + 1024)/14{,}264{,}868$$
$$= 0.5277.$$

We might also wish to model aggregate claims in this situation using a compound binomial distribution of the form $S_2 = X_1 + \cdots + X_{N_2}$ where $N_2 \sim B(800, 0.02)$ and the typical claim X is exponential with mean 400. Here our interpretation might be that in each of the 800 policies there will be one claim with probability 0.02, and none with probability 0.98. In this case

$$E(S_2) = nqm_1 = 800(0.02)(400)$$
$$= 6400,$$
$$Var(S_2) = nq(m_2 - qm_1^2) = 800(0.02)[2(400)^2 - (0.02)(400)^2]$$
$$= 5{,}068{,}800 = 2251.400^2 \quad \text{and}$$
$$skew(S_2) = \frac{800\,[(0.02)6(400)^3 - 3(0.02)^2(2(400)^2)(400) + 2(0.02)^3(400)^3]}{[800(0.02)\,(2(400)^2 - 0.02(400)^2]^{3/2}}$$
$$= 6{,}021{,}939{,}200/(2251.400)^3$$
$$= 0.5277.$$

The compound negative binomial model has slightly greater mean and more variability, although the two distributions have (to 4 decimal places) the same skewness. Which is the most appropriate distribution to use? This is always one of the challenges in modeling! One usually tries to pick a model that incorporates the important factors of the situation, yet still can be interpreted in a reasonable way.   ☐

3.2.3 Sums of compound Poisson distributions

One of the most useful properties of the compound Poisson distribution is that it is preserved under convolutions. Given that one often wants to bring together claims from different portfolios or companies, this can be useful in studying the distribution of the aggregate claims from different risks.

THEOREM 3.1

Assume that S_i has a compound Poisson distribution with Poisson parameter λ_i and claim (or component) distribution function F_i for $i = 1, \ldots, k$. If the random variables $S_1, \ldots, S_k$ are independent, then the sum or convolution $S = S_1 + \cdots + S_k$ is also compound Poisson with Poisson parameter $\lambda = \lambda_1 + \cdots + \lambda_k$ and claim or component distribution function $F = \sum_{i=1}^{k} (\lambda_i/\lambda) F_i$.

PROOF Let $M_i(t)$ be the moment generating function corresponding to F_i (the component distribution of S_i) for $i = 1, \ldots, k$. Using the expression for the moment generating function of a compound Poisson distribution and the independence of the S_i, one obtains

$$M_S(t) = \prod_{i=1}^{k} M_{S_i}(t) = e^{\sum \lambda_i [M_i(t) - 1]} = e^{\lambda [\sum (\lambda_i/\lambda) M_i(t) - 1]},$$

which is the moment generating function of a compound Poisson distribution with Poisson parameter $\lambda = \sum \lambda_i$ and component distribution function with moment generating function given by $\sum (\lambda_i/\lambda) M_i(t)$. However, by the uniqueness property of moment generating functions, this is the distribution of the mixture of $F_1, \ldots, F_k$ with the respective mixing constants $\lambda_1/\lambda, \ldots, \lambda_k/\lambda$. ∎

Example 3.4
Let $S_1 = U_1 + \cdots + U_{N_1}$, $S_2 = V_1 + \cdots + V_{N_2}$ and $S_3 = W_1 + \cdots + W_{N_3}$ be three independent compound Poisson distributions representing claims in three companies C_1, C_2 and C_3. The Poisson parameters for N_1, N_2 and N_3 are, respectively, $4, 2$ and 6, and the probability distributions for typical claims U, V and W in the three respective companies are given in Table 3.1.

By Theorem 3.1, S is compound Poisson with Poisson parameter given by $\lambda = 4 + 2 + 6 = 12$, and the typical component X is a mixture distribution of U, V and W with respective mixing weights given by $(1/3, 1/6, 1/2)$. For example

$$P(X = 200) = \frac{1}{3}(0.5) + \frac{1}{6}(0) + \frac{1}{2}(0.2) = \frac{16}{60},$$

and the rest of the distribution is given in Table 3.1. Hence $E(S) = \lambda E(X) = 12(336.67) = 4040$, and $Var(S) = \lambda E(X^2) = 12(126,000) = 1,512,000$.

TABLE 3.1

Probability distributions for U, V, W and X.

x	$P(U = x)$	$P(V = x)$	$P(W = x)$	$P(X = x)$
200	0.5	0	0.2	16/60
300	0.3	0.3	0.3	18/60
400	0.2	0.4	0.3	17/60
500	0	0.3	0.1	6/60
600	0	0	0.1	3/60

Letting $N = N_1 + N_2 + N_3$, one finds, for example, that

$$P(S \leq 400) = P(N = 0) + P(X \leq 400)P(N = 1) + P^2(X = 200)P(N = 2)$$
$$= e^{-12}\{1 + (51/60)(12) + (16/60)^2(12^2/2)\}$$
$$= 0.0001.$$

▯

Example 3.5

Claims in a company are grouped into two portfolios and modeled by compound Poisson distributions. Those in portfolio 1 are modeled by a compound Poisson distribution with rate parameter $\lambda_1 = 3$/month and where claims are exponentially distributed with mean 500. The rate parameter for those in portfolio 2 is $\lambda_2 = 7$/month, and claims are exponentially distributed with mean 300. By Theorem 3.1, total *annual* claims S in the two portfolios are modeled by a compound Poisson distribution with rate parameter $\lambda = 12(3 + 7) = 120$ claims per *year* and component or claim distribution X which is a 30% : 70% mixture of exponential distributions with means 500 and 300, respectively. In particular

$$E(S) = \lambda E(X) = 120[(0.3)500 + (0.7)300]$$
$$= 432{,}000$$
$$Var(S) = \lambda E(X^2) = 120[(0.3)2(500)^2 + (0.7)2(300)^2]$$
$$= 33{,}120{,}000 \text{ and}$$
$$skew\,(S) = 120[(0.3)6(500)^3 + (0.7)6(300)^3]/(33{,}120{,}000)^{3/2}$$
$$= 0.2130.$$

The moment generating function of S is given by $M_S(t) = e^{120[M_X(t)-1]}$ where

$$M_X(t) = 0.3\,\frac{1}{1 - 500t} + 0.7\,\frac{1}{1 - 300t}.$$

▯

It is worth noting that in *some* cases convolutions of compound binomial (negative binomial) distributions are also compound binomial (negative binomial). For example, suppose that $S_1, \ldots, S_r$ are independent and compound binomial (negative binomial) distributed with common claim size random variable X, and where for some common value of q (common value of p) the random number of claims in S_i is $N_i \sim B(n_i, q)$ (respectively, $N_i \sim NB(k_i, p)$). Then S has compound binomial (negative binomial) distribution with typical claim distribution X where the number of claims $N \sim B(n_1 + \cdots + n_r, q)$ ($N \sim NB(k_1 + \cdots + k_r, p)$).

3.2.4 Exact expressions for the distribution of S

If the number of claims N in the collective risk model $S = X_1 + \cdots + X_N$ has either a Poisson, binomial or negative binomial distribution, and the claim random variable X takes positive integer values, then we may establish an exact recursive expression for $P(S = r)$ in terms of the probabilities $P(S = j)$ for $j = 0, 1, \ldots, r - 1$ and the distribution of X. This expression is often referred to as Panjer's recursion formula (see [20], [47] and [48]), and it can be of considerable practical use because it may be easily implemented with basic computer programming.

Let us assume N is a random variable with the recursive property that for some constants α and β,

$$P(N = n) = (\alpha + \beta/n) \; P(N = n - 1) \tag{3.8}$$

holds for $n = 1, \ldots, max(N)$. The Poisson, binomial and negative binomial distributions satisfy this property (and, in fact, are the only nonnegative random variables which do). When $N \sim \text{Poisson}(\lambda)$, using $\alpha = 0$ and $\beta = \lambda$, one has that

$$P(N = n) = \frac{\lambda^n e^{-\lambda}}{n!} = (0 + \lambda/n) \; P(N = n - 1),$$

while when $N \sim B(m, q = 1 - p)$ (and using $\alpha = -q/(1 - q)$ and $\beta = (m + 1)q/(1 - q)$) it follows that

$$P(N = n) = \binom{m}{n} q^n (1 - q)^{m-n}$$

$$= \frac{m - n + 1}{n} \frac{q}{1 - q} \binom{m}{n - 1} q^{n-1} (1 - q)^{m-(n-1)}$$

$$= \left[\frac{-q}{1 - q} + \frac{(m + 1)q/(1 - q)}{n} \right] P(N = n - 1) \tag{3.9}$$

for $n = 1, \ldots, m$.

We use S_n to denote the probability distribution of $X_1 + \cdots + X_n$ (in particular then $S_n = S_{[N=n]}$), f_X to denote the density function of the claim random variable X which takes only positive integer values, and f_{S_n} and f_S to be the density functions of S_n and S, respectively. Then the following recursive formula for f_S results:

THEOREM 3.2
For the collective risk model $S = X_1 + \cdots + X_N$ *where N has the recursive property (3.8) and X takes positive integer values, one has that $f_S(0) = f_N(0)$ and*

$$f_S(r) = \sum_{j=1}^{r} (\alpha + \frac{\beta j}{r}) f_X(j) f_S(r-j) \quad for \quad r = 1, 2, \ldots . \tag{3.10}$$

PROOF   The key is to consider two different but equivalent expressions for the conditional probability $E(X_1 \mid S_{n+1} = r)$, and to use the fact that $\sum_1^r f_X(j) f_{S_n}(r-j) = f_{S_{n+1}}(r)$ for any $n \geq 0$ and $r \geq 1$.

The terms in S are independent and identically distributed random variables. Hence given that the sum of $n+1$ of them is equal to r, the conditional expected value of each of them must be the same, or in other words that $E(X_1 \mid S_{n+1} = r) = r/(n+1)$. On the other hand, this can also be expressed using the standard definition of the conditional expectation of X_1, given that $S_{n+1} = r$ (that is, by summing over the values of X_1 multiplied by the appropriate conditional probabilities). Therefore by setting the two expressions equal to one another, one obtains

$$\frac{\sum_{j=1}^{r} j f_X(j) f_{S_n}(r-j)}{f_{S_{n+1}}(r)} = E(X_1 \mid S_{n+1} = r) = r/(n+1). \tag{3.11}$$

Therefore for any integer $r = 1, 2, \ldots$, it follows that

$$\sum_{j=1}^{r} (\alpha + \frac{\beta j}{r}) f_X(j) f_S(r-j) = \sum_{j=1}^{r} (\alpha + \frac{\beta j}{r}) f_X(j) \left[\sum_{n=0}^{\infty} f_{S_n}(r-j) \, P(N=n) \right]$$

$$= \sum_{n=0}^{\infty} \alpha \, P(N=n) \sum_{j=1}^{r} f_X(j) f_{S_n}(r-j)$$

$$+ \sum_{n=0}^{\infty} \beta P(N=n) \sum_{j=1}^{r} \frac{j}{r} f_X(j) f_{S_n}(r-j)$$

$$= \sum_{n=0}^{\infty} \alpha P(N=n) f_{S_{n+1}}(r) + \beta P(N=n) \frac{f_{S_{n+1}}(r)}{n+1}$$

$$\text{(using (3.11))}$$

$$= \sum_{n=0}^{\infty} [\alpha + \beta/(n+1)] \, P(N=n) \, f_{S_{n+1}}(r)$$

$$= \sum_{n=0}^{\infty} P(N=n+1) \, f_{S_{n+1}}(r) \quad \text{(using (3.8))}$$

$$= f_S(r).$$

■

Example 3.6

The total amount of claims S for a general insurance portfolio over a fixed period of time is being modeled by a compound Poisson distribution where

$$S = X_1 + \cdots + X_N,$$

X is uniformly distributed on $\{100, 200, 300, 400, 500, 600, 700, 800, 900\}$ and N has Poisson parameter λ. Since S is compound Poisson, it follows that

$$E(S) = \lambda E(X) = \lambda (100) (1 + \cdots + 9)/9 \;\; = 500\lambda, \quad \text{and}$$

$$Var(S) = \lambda E(X^2) = \lambda (100^2) \left[\sum_{i=1}^{9} i^2/9 \right] = (2{,}850{,}000)\,(\lambda/9)$$

$$= 316{,}666.7\,\lambda.$$

We determine the cumulant moment generating function of S in order to calculate its skewness. Now

$$C_S(t) = \log M_S(t) = \lambda[M_X(t) - 1] = \lambda \left[\sum_{i=1}^{9} e^{100i\,t}/9 - 1 \right].$$

Taking subsequent derivatives of $C_S(t)$, we find

$$C_S'(t) = (\lambda/9)(100) \sum_{i=1}^{9} i\, e^{100i\,t},$$

$$C_S''(t) = (\lambda/9)(100)^2 \sum_{i=1}^{9} i^2\, e^{100i\,t} \quad \text{and}$$

$$C_S'''(t) = (\lambda/9)(100)^3 \sum_{i=1}^{9} i^3\, e^{100i\,t}.$$

Therefore $E(S - E(S))^3 = (\lambda/9)(100)^3 \left[\sum_{i=1}^{9} i^3 \right] = (\lambda/9)(20.25)\,10^8$ and hence

$$skew(S) = (\lambda/9)(20.25)\,10^8/[Var(S)]^{3/2} = 1.2626/\sqrt{\lambda}.$$

Note that the skewness of S converges to 0 as $\lambda \to \infty$.

Consider the specific case where $\lambda = 3$. Then $E(S) = 1500$, $Var(S) = 974.68$ and $skew\,(S) = 0.7290$, indicating that S is positively skewed. Is it appropriate to assume S is approximately normal, i.e., is $S \sim N(1500, 974.68^2)$? One way to answer this is to calculate the exact distribution for S and then to compare it to that of a normal distribution.

Working in units of 100, we let $S^\star = S/100$ and $X^\star = X/100$. Since the probability distribution of $X^\star$ is uniform on the set $\{1, 2, \ldots, 9\}$, the recursion formula (3.10) reduces to

$$P(S^\star = r) = \sum_{j=1}^{min(r,9)} \frac{\lambda j}{r} \, f_{X^\star}(j) \, P(S^\star = r - j)$$

$$= \frac{3}{9r} \sum_{j=1}^{min(r,9)} j \, P(S^\star = r - j)$$

for $r \geq 1$. Given that $\lambda = 3$, we have

$P(S^\star = 0) = P(S = 0) = e^{-3} = 0.049787$ and hence
$P(S^\star = 1) = (1/3)\, 1\, P(S^\star = 0) = 0.016596,$
$P(S^\star = 2) = (1/6)\, [1\, P(S^\star = 1) + 2\, P(S^\star = 0)] = 0.019362,$
$P(S^\star = 3) = (1/9)\, [1\, P(S^\star = 2) + 2\, P(S^\star = 1) + 3\, P(S^\star = 0)] = 0.022435,$
$P(S^\star = 4) = (1/12)\, [1\, P(S^\star = 3) + 2\, P(S^\star = 2) + 3\, P(S^\star = 1) + 4\, P(S^\star = 0)]$
$\qquad = 0.025841.$

Such calculations are easily implemented (in, for example, Excel or R), and one may readily establish Table 3.2 for the probability distribution of $S^\star$ (where we have rounded off probabilities to 4 decimals). Figure 3.1 gives a histogram of its distribution. Note the modest positive skewness for $S^\star$ and the spiked nature of the left tail due to the effect of the probability that $N = 0$.

Of course, Table 3.2 also gives us a distribution table for S. For example, the probability that S falls within 1 standard deviation of its mean 1500 is

$$P(525 \leq S \leq 2475) = P(5.25 \leq S^\star \leq 24.75) = 0.6280.$$

Similarly, the probability that S is within 2 standard deviation units of 1500 is 0.9613. These are very close to the corresponding values of 0.6826 and 0.9544 for the normal distribution $N(1500, 974.68^2)$.

☐

Example 3.7
Assume $S = X_1 + \cdots + X_N$ has a compound binomial distribution where $N \sim B(50, 0.04)$ and the typical claim random variable X (in units of 10,000) has distribution as given in Table 3.3.

Working in units of 10,000, one may verify (using (3.1), (3.2) and (3.4)), that $E(S) = 6.2$, $Var(S) = 37.8312 = (6.1507)^2$, and $skew\,(S) = 1.3633$. Letting $\alpha = -0.04167$ and $\beta = 2.125$, we have according to Equations (3.10) and (3.9) that

$f_S(0) = (1 - q)^{50}$ and

$$f_S(r) = \sum_{j=1}^{r} (-0.04167 + \frac{2.125j}{r}) \, f_X(j) \, f_S(r - j) \qquad \text{for} \quad r = 1, 2, \ldots, 50,$$

TABLE 3.2

Exact (compound Poisson) probability for $S^\star$.

r	0	1	2	3	4
$P(S^\star = r)$	0.0498	0.0166	0.0194	0.0224	0.0258
r	5	6	7	8	9
$P(S^\star = r)$	0.0296	0.0338	0.0383	0.0434	0.0489
r	10	11	12	13	14
$P(S^\star = r)$	0.0383	0.0394	0.0402	0.0406	0.0405
r	15	16	17	18	19
$P(S^\star = r)$	0.0400	0.0388	0.0371	0.0345	0.0311
r	20	21	22	23	24
$P(S^\star = r)$	0.0295	0.0277	0.0258	0.0238	0.0218
r	25	26	27	28	≥ 29
$P(S^\star = r)$	0.0197	0.0177	0.0158	0.0141	0.1095

the results of which are given in Table 3.4.

From Figure 3.2 we can see that the distribution of S is bimodal. The normal density function with mean $E(S) = 6.2$ and variance 6.15^2 is also plotted, and it is clear that the normal approximation to aggregate claims S is not particularly good. The normal approximation for the probability that claims are greater than or equal to 100,000 (that is, $S \geq 10$) is 0.2958, while the actual value is 0.2877. These tail probabilities are good even though the normal approximation to S is not.

TABLE 3.3

Distribution of claim size X in Example 3.7.

Claim amount C	$j = C/10,000$	$f_X(j) = \text{Prob}[X = j]$
10,000	1	0.40
20,000	2	0.35
50,000	5	0.10
100,000	10	0.15

TABLE 3.4

Exact (compound binomial) distribution for S (in $0,000's$) of Example 3.7.

r	0	1	2	3	4	5
$f_S(r)$	0.1299	0.1082	0.1389	0.0891	0.0671	0.0626
r & 6	7	8	9	≥ 10		
$f_S(r)$	0.0422	0.0373	0.0220	0.0150	0.2877	

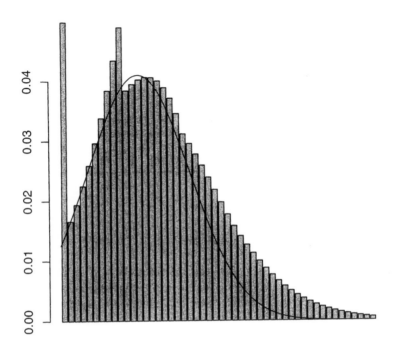

FIGURE 3.1
Probability distribution for $S^\star$ of Example 3.6.

[]

3.2.5 Approximations for the distribution of S

The algorithm of the previous section is very useful for calculating the exact distribution of S when the claim size distribution is discrete and known, and N has either the Poisson, binomial or negative binomial distribution. In some cases, use of this approach may involve a considerable number of calculations, particularly when it is used as a simulation tool for investigating various models. A quick approximation to the distribution of S can prove very useful, and in many situations a normal approximation to the distribution of S may be used. As we have already seen this is usually justified in the case of the compound Poisson (binomial or negative binomial) when λ (nq or

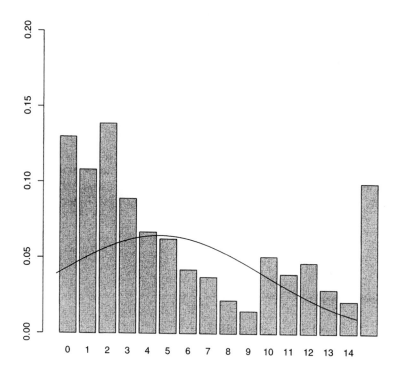

FIGURE 3.2

Normal approximation to compound binomial distribution of Example 3.7.

kp) is reasonably large. On the other hand, the normal distribution can in theory take negative values and is symmetric, while aggregate claims S are always nonnegative and often positively skewed. One alternative to the normal approximation which does not have these deficiencies and is sometimes utilized is the *shifted* or *translated gamma* distribution. We denote by $\Gamma(\alpha, \delta)$ the gamma distribution with mean α/δ, variance α/δ^2, and skewness $2/\sqrt{\alpha}$. The gamma family of distributions is very versatile. A random variable has the shifted gamma distribution with parameters (α, δ, τ) if it is distributed as $\tau + \Gamma(\alpha, \delta)$.

If we have approximate knowledge of the mean μ, variance σ^2 and skewness κ of S, then we may consider approximating the distribution of S with a shifted gamma distribution $\tau + \Gamma(\alpha, \delta)$ where α, δ and τ are chosen to satisfy $\mu = \tau + \alpha/\delta$, $\sigma^2 = \alpha/\delta^2$ and $\kappa = 2/\sqrt{\alpha}$. Solving these three equations, one obtains

$$\alpha = 4/\kappa^2, \quad \delta = 2/(\sigma\kappa) \quad \text{and} \quad \tau = \mu - 2\sigma/\kappa. \tag{3.12}$$

Example 3.8

Consider a compound Poisson risk model $S = X_1 + \cdots + X_N$ where N has Poisson parameter $\lambda = 4$ and the typical claim X has density function $f_X(x) = x/5000$ for $0 \le x \le 100$. Then

$$E(S) = \lambda m_1 = 4(100^3/15{,}000) = 266.6667$$
$$Var(S) = \lambda m_2 = 4(5000) = 20{,}000 = (141.4214)^2, \quad \text{and}$$
$$skew(S) = \lambda m_3/[Var(S)]^{3/2} = 4(400{,}000)/(141.4214)^3 = 0.5657.$$

Using Equations (3.12), we find $\alpha = 12.5$, $\delta = 0.025$ and $\tau = -233.3333$ as estimates for the parameters of a shifted gamma distribution to approximate S. The shifted gamma density function with these parameters as well as the $N(266.6667, (141.4214)^2)$ density function are plotted in in Figure 3.3.

$\Box$

3.3 Individual risk models for S

In the individual risk model for total claims, we assume that there are n individual risks. The claim amount arising from the jth risk is denoted by Y_j for $j = 1, \ldots, n$, and we use $S = \sum_1^n Y_j$ to denote the aggregate or total amount of claims in a given fixed period of time (say a year). In most applications the majority of the Y_j will be equal to 0, since only a small proportion of the risks will give rise to claims. The so-called *individual* risks may be those individuals insured by a company, or the individual policies in a company. Normally, we assume the Y_j are independent random variables, although they are not necessarily identically distributed. It is important to remember that Y_j refers to the claim amount (which may be 0) of the jth individual, and not to the jth claim which is made during the period of time being considered (in the collective risk model where $S = X_1 + \cdots + X_N$, X_j referred to the jth *claim* made in time). We let I_j be the indicator random variable which is 1 if the jth risk gives rise to a nonzero claim (which happens with probability q_j), and otherwise is 0.

A basic assumption in the individual risk model is that an individual makes at most one claim in the (often relatively short) time period being considered. If in fact the jth risk gives rise to a claim, then the size of the claim will be denoted by X_j and hence we write $Y_j = X_j \cdot I_j$. We let $\mu_j = E(X_j)$ and $\sigma_j^2 = Var(X_j)$ for $j = 1, \ldots, n$. Therefore we may express total claims S by

$$S = Y_1 + Y_2 + \cdots + Y_n = X_1 \cdot I_1 + X_2 \cdot I_2 + \cdots + X_n \cdot I_n.$$

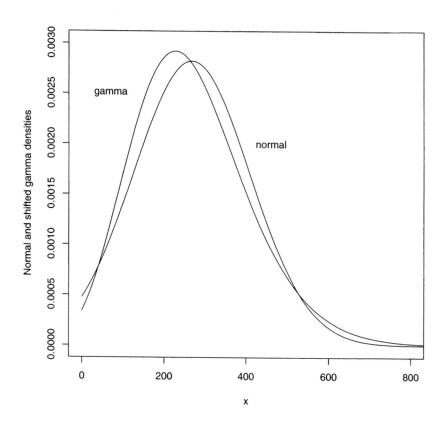

FIGURE 3.3
Normal and shifted gamma approximations for the compound Poisson
distribution of Example 3.8.

3.3.1 Basic properties of the individual risk model

The following results give basic formulae for the mean and variance of S:

THEOREM 3.3
In the individual risk model for total claims where $S = \sum_1^n Y_j$,

$$E(S) = \sum_1^n q_j \mu_j \quad \text{and} \tag{3.13}$$

$$Var(S) = \sum_1^n \{q_j \sigma_j^2 + q_j(1 - q_j)\mu_j^2\}. \tag{3.14}$$

PROOF   By the double expectation theorem, $E(Y_j) = E_{I_j}(E(Y_j \mid I_j))$ where $E(Y_j \mid I_j)$ is the random variable taking the value μ_i when $I_j = 1$ and 0 otherwise. Table 3.5 gives the probability distribution of $E(Y_j \mid I_j)$. Therefore $E(Y_j) = q_j \cdot \mu_j + p_j \cdot 0 = q_j \mu_j$, from which Equation (3.13) follows. We may also determine $Var(Y_j)$ by conditioning on I_j for any j, since

$$Var(Y_j) = E(V(Y_j \mid I_j)) + Var(E(Y_j \mid I_j)).$$

$V(Y_j \mid I_j)$ is the random variable which is determined once we know I_j, and whose distribution is also given in Table 3.5. Thus $Var(Y_j) = \sigma_j^2 q_j + \mu_j^2 q_j p_j$, and using the independence of the Y_j, Equation (3.14) follows.   ∎

TABLE 3.5
Distributions for $E(Y_j \mid I_j)$ and $V(Y_j \mid I_j)$.

i	$P(I_j = i)$	$E(Y_j \mid I_j = i)$	$Var(Y_j \mid I_j = i)$
1	q_j	$\mu_j = E(X_j)$	$\sigma_j^2 = Var(X_j)$
0	$p_j = 1 - q_j$	0	0

Example 3.9
The employees in a hospital are offered (one-year) term life insurance on the basis of summary data on their annual salaries. The employees may be divided into the three categories of nurses, doctors and administrators. We assume that salaries within a given category are normally distributed, with details given in Table 3.6. If a premium of 850,000 is collected to handle this group scheme for the coming year, what is the probability that the premium will cover claims?

TABLE 3.6
Salary information on hospital employees of Example 3.9.

Category	Number	Mortality $= q_j$	Mean salary	Salary sd
Nurse	400	0.02	25,000	2,000
Doctor	60	0.06	75,000	20,000
Administrator	80	0.04	30,000	5,000

Let $S = \sum_{j=1}^{540} Y_j$ be the random variable representing total annual claims from this group, where Y_j is the claim made (if any) by the jth individual. Hence $S = S_N + S_D + S_A$ where $S_N = \sum_1^{400} Y_j$ is the total claims from the nurses, and similarly $S_D = \sum_{401}^{460} Y_j$ and $S_A = \sum_{461}^{540} Y_j$ are, respectively, those for doctors and administrators. In using the individual risk model for S, we are assuming that deaths (and hence resulting claims) are independent events. Hence we have that

$$E(S) = E(S_N) + E(S_D) + E(S_A)$$
$$= 400(0.02)(25{,}000) + 60(0.06)(75{,}000) + 80(0.04)(30{,}000)$$
$$= 566{,}000 \text{ and}$$

$$Var(S) = Var(S_N) + Var(S_D) + Var(S_A) \ = \ \sum_{1}^{540}\{q_j\sigma_j^2 + q_j(1-q_j)\mu_j^2\}$$

$$= 400\,[\,0.02(2000)^2 + 0.02(0.98)(25{,}000)^2\,]$$
$$+60\,[\,0.06(20{,}000)^2 + 0.06(0.94)(75{,}000)^2\,]$$
$$+80\,[\,0.04(5000)^2 + 0.04(0.96)(30{,}000)^2\,]$$
$$= (168{,}082.7)^2.$$

S_N, S_D and S_A are by the central limit theorem approximately normal, and since they are independent $S \overset{.}{\sim} N(566{,}000, (168{,}082.7)^2)$. Therefore the probability that the group premium will cover claims is

$$P[S < 850{,}000] \doteq P[N(0,1) < (850{,}000 - 566{,}000)/168{,}082.7]$$
$$= P[N(0,1) < 1.6896] = 0.9545.$$

$\square$

3.3.2 Compound binomial distributions and individual risk models

The individual risk model is closely related to (and in some sense equivalent to a generalization of) the compound binomial model. Let us consider a *homogeneous* version of the individual risk model. In particular, assume there exists a $q > 0$ and a claim random variable X such that for all $j = 1, \ldots, n$ we have $q_j = q$ and $X_j \sim X$. Furthermore, let I be the indicator random variable where $P(I = 1) = q$, and $Y = X \cdot I$. We use $S^I = Y_1 + \cdots + Y_n$ to denote total claims under the individual risk model. This is to be compared with the compound binomial model $S^C = X_1 + \cdots + X_N$, where $N \sim B(n, q)$. Note that

$$E(S^I) = (nq)\,\mu = E(N)\,E(X) = E(S^C)$$

and

$$Var(S^I) = n[q\sigma^2 + q(1-q)\mu^2]$$
$$= nq[m_2 - m_1^2 + (1-q)m_1^2] \ = \ nq[m_2 - qm_1^2]$$
$$= Var(S^C)$$

where we interchangeably use $\mu = E(X) = m_1$ and $\sigma^2 = Var(X) = m_2 - m_1^2$ to link the formulae used for the mean and variance of a compound binomial distribution and an individual risk model distribution. In fact, in this case S^I and S^C not only have the same probability distribution, but take the same value in any given realization. There is, however, a subtle difference in their representation. In the individual risk model for S^I, Y_j refers to the amount of claim (which may be 0) made by the jth individual (in a given list of individuals $j = 1, \ldots, n$), while in the compound binomial collective risk model for S^C, X_j refers to the amount of the jth claim which is made in order of time ($j = 1, \ldots, N$).

Another way of looking at an individual risk model is to consider it as the sum of independent compound binomial distributions (where individuals with equal claim probabilities q_j and claim distributions X_j have been combined).

3.3.3 Compound Poisson approximations for individual risk models

In the individual risk model where $S = Y_1 + \cdots + Y_n = X_1 \cdot I_1 + \cdots + X_n \cdot I_n$, one would expect to observe about $\lambda = \sum_{j=1}^{n} q_j$ claims in total, and the claims themselves would be a selection from the claim types of the individuals. This might suggest comparing an individual risk model with a compound Poisson (collective risk) model with Poisson parameter $\lambda = \sum_{j=1}^{n} q_j$ where the typical claim random variable W is a mixture of the X_j.

Corresponding to each Y_j in the individual risk model, we define $\tilde{Y}_j$ to be the random variable having the compound Poisson distribution with Poisson parameter $\lambda_j = q_j$ and where the component distribution is X_j. One advantage this approach has over the individual risk model is that it incorporates the possibility that an *individual* can make more than one claim. It follows from Theorem 3.1 that $\tilde{S} = \sum_{j=1}^{n} \tilde{Y}_j$ also has a compound Poisson distribution with Poisson parameter $\lambda = \sum_{j=1}^{n} \lambda_j$ and where the component distribution W is the $\{\lambda_j/\lambda, \quad j = 1, \ldots, n\}$ mixture of the $\{X_j, \ j = 1, \ldots, n\}$.

We may view the collective risk model $\tilde{S}$ as a compound Poisson approximation to the individual risk model S. Of course, they represent different approaches to modeling the same thing (aggregate claims), and one may naturally ask how do they differ in their mathematical properties? In fact as the following shows, they have the same mean but the variance of $\tilde{S}$ is slightly greater than that of S.

$$E(\tilde{S}) = \lambda \, E(W) \;=\; \lambda \sum_{j=1}^{n} \frac{\lambda_j}{\lambda} \, E(X_j) \;=\; \sum_{j=1}^{n} \lambda_j \, E(X_j)$$

$$= \sum_{j=1}^{n} q_j \, \mu_j \;=\; E(S),$$

while

$$Var(\tilde{S}) = \lambda E(W^2) = \lambda \sum_{j=1}^{n} \frac{\lambda_j}{\lambda}(\sigma_j^2 + \mu_j^2)$$

$$= \sum_{j=1}^{n} q_j(\sigma_j^2 + \mu_j^2) \geq \sum_{j=1}^{n} q_j\sigma_j^2 + q_j(1 - q_j)\mu_j^2 = Var(S).$$

The added variability in the compound Poisson approximation $\tilde{S}$ to the individual risk model S is essentially due to the possibility of allowing more than one claim per individual. In most cases, however, this difference $(Var(\tilde{S}) - Var(S) = \sum_{j=1}^{n} q_j^2\mu_j^2)$ is very small.

3.4 Premiums and reserves for aggregate claims

Having analyzed a random risk S, an insurance company will want to decide how much it should charge to handle (take responsibility for) the risk, and whether or not it should set aside reserves in case of extreme or unlikely events occurring. These problems have to be considered in the light of a very competitive market for insurance.

3.4.1 Determining premiums for aggregate claims

Given a risk S, we refer to its expected value $E(S)$ as the *pure* or *office premium* for the risk. Clearly, an insurance company must charge more than the pure premium to cover expenses, allow for variability in the number and amount of claims, and make a profit. When an allowance is made for security or safety (due to the variability of S) in determining a premium for a risk, one speaks of the *net premium*, and when one also takes into account administrative costs, one obtains the *gross premium*. In most of our modeling, we will make the (somewhat naive) assumption that administrative costs are nil, and therefore concentrate on the pure and net premiums. In fact, administrative costs are clearly important in practice, and changes in policy details (like the introduction of a deductible) often influence both claim and administrative costs.

In a simple model for determining premiums, assume that we use a loading (safety or security) factor θ, whereby the net premium charged is of the form $(1 + \theta)E(S)$. A large value of θ will give more security and profits, but also could result in a decrease in the amount of business done (policies in force) because of the competitive nature of the insurance business. This principle for premium calculation is sometimes referred to as the *expected value principle*, and we shall generally use this method in our modeling. In spite of its common usage, this principle for premium calculation takes no account of the variability in a risk, and in particular two risks with the same expected value but widely differing variabilities would be assigned the same net premium using this method. After all, it is the variability in a risk which often motivates an individual to buy insurance in the first place! Two methods that do take into account the variability of the risk S are the *standard deviation principle* (where premium calculation is based on $E(S) + \theta\sqrt{Var(S)}$), and the *variance principle* (where it is based on $E(S) + \theta Var(S)$). For an interesting discussion of the principles of premium calculation, one may refer to [56].

Example 3.10

Fifteen hundred structures are insured against fire by a company. The amounts insured ($000's), as well as the chances of a claim, vary as indicated in Table 3.7. We let q_k be the chance of a claim for a structure in category k, and assume the chance of more than one claim on any individual structure is negligible.

TABLE 3.7
Fire insurance on 1500 structures.

Category k	Amount insured (000's)	q_k	No. Structures
1	20	0.04	500
2	30	0.04	300
3	50	0.02	500
4	100	0.02	200

Assume fires occur independently of one another, and that for a structure insured for $\$A$ the amount of a claim X (conditional on there being a claim) is uniformly distributed on $[0, A]$ (we write $X \sim U[0, A]$). Let N be the number of claims made in a year and S the amount (in units of $1000). Using an individual risk model for S, we determine the mean and variance of N and S. If we wish to use a security loading of 2θ for structures in categories 1 and 2, and θ otherwise, we find the value of θ which gives us a 99% probability that premiums exceed claims. We also find what the corresponding value of θ would be if the number of structures in categories 1 and 2 were doubled.

We may write $N = \sum I_j$ as the sum of 1500 independent Bernoulli random variables, and hence

$$E(N) = 500(0.04) + 300(0.04) + 500(0.02) + 200(0.02) = 46 \text{ and}$$
$$Var(N) = 500(0.04)(0.96) + 300(0.04)(0.96)$$
$$+ 500(0.02)(0.98) + 200(0.02)(0.98)$$
$$= 44.44.$$

As X is uniformly distributed on the interval $[0, A]$, $E(X) = A/2$ and $Var(X) = A^2/12$. We work in units of 1000, and write $S = S_1 + S_2 + S_3 + S_4$ where S_i represents claims from structures of type i. Now

$$E(S_1) = 500(0.04)(10) = 200 \quad \text{and}$$
$$Var(S_1) = 500 \left[0.04(20^2)/12 + 0.04(0.96)10^2\right] = 2586.667.$$

Similar calculations for $i = 2, 3$ and 4 yield:

$$E(S) = E(S_1) + E(S_2) + E(S_3) + E(S_4)$$
$$= 200 + 180 + 250 + 200 = 830 \quad \text{and}$$
$$Var(S) = Var(S_1) + Var(S_2) + Var(S_3) + Var(S_4)$$
$$= 2586.667 + 3492 + 8208.333 + 13,133.333$$
$$= 27,420.333 = (165.5909)^2.$$

Premium income P_I will amount to

$$P_I = (1 + 2\theta)[E(S_1) + E(S_2)] + (1 + \theta)[E(S_3) + E(S_4)]$$
$$= (1 + 2\theta)\,380 + (1 + \theta)\,450 = 1210\,\theta + E(S),$$

and we want θ such that

$$0.99 = P(S < 1210 \cdot \theta + E(S))$$
$$\doteq P(\,N(0,1) < 1210 \cdot \theta/165.5909\,).$$

Therefore $\theta = z_{0.99}\,(165.5909/1210) = 0.3184$.

Suppose now that the numbers of structures in categories 1 and 2 were to be doubled. Let S^* represent the claims which result, and use θ^* to denote the new security factor. Then clearly, $E(S^*) = 2[E(S_1) + E(S_2)] + E(S_3) + E(S_4) = 1210$, new premiums P_I^* are

$$P_I^* = (1 + 2\theta^*)(2)(E(S_1) + E(S_2)) + (1 + \theta^*)(E(S_3) + E(S_4))$$
$$= 4\theta^*(380) + \theta^*(450) + E(S^*)$$
$$= 1970\,\theta^* + E(S^*) \quad \text{and}$$
$$Var(S^*) = Var(S_1) + Var(S_2) + Var(S)$$
$$= 2586.667 + 3492 + 27,420.333$$
$$= 33,499 = (183.0273)^2.$$

Therefore proceeding as with S and using a normal approximation, it follows that $\theta^* = z_{0.99}\,(183.0273/1970) = 0.2161$. $\quad\square$

Generally speaking, the security factor will decrease when the volume of business increases if the relative frequency and severity (or type) of claims remains the same. In this example, we considered doubling the amount of business in some but not all of the categories of business. In general, if business across the board increases by a factor of k, then (assuming other aspects remain the same, including the degree of confidence required for premiums to cover claims) the necessary security factor decreases by a factor of $1/\sqrt{k}$ (see Problem 9).

Example 3.11

Insurance for accidents is provided to all employees in a large factory. The employees have been categorized into three types by virtue of their work. It can be assumed that claims of individuals are independent. The claim incidence rate is given for each type in Table 3.8, together with the number of employees of each type and the corresponding claim size distribution B_k ($k = 1, 2, 3$).

TABLE 3.8
Accident insurance for employees in a large factory.

Class type k	Number	Claim probability q_k	B_k
1	2000	0.01	Pareto ($\beta = 3, \delta = 800$)
2	2000	0.02	exponential ($\mu = 500$)
3	1000	0.01	$U[320, 680]$

Using an individual risk model, we determine the security factor θ which should be used in setting premium levels in order to ensure that the probability claims exceed premiums is 0.02. We also determine what it would be if one were to approximate this individual risk model with a compound Poisson model, and comment on the relationship between the two security factors.

From basic properties of the Pareto, exponential and uniform distributions, one may determine Table 3.9.

TABLE 3.9
Summary statistics for accident claims by class type.

Class type k	Number	Mean μ_k	Variance σ_k^2	$m_2 = E(X_k^2)$
1	2000	400	480,000	640,000
2	2000	500	250,000	500,000
3	1000	500	10,800	260,800

We use S^I to model aggregate claims with an individual risk model and S^C to be the corresponding compound Poisson approximation. S^C has compound Poisson parameter $\lambda = 2000(0.01) + 2000(0.02) + 1000(0.01) = 70$, with the *typical* claim W being the mixture

$$W = \begin{cases} \text{Pareto } (3,800) & \text{with probability } 2/7 \\ \text{exponential } (\mu = 500) & \text{with probability } 4/7 \\ U[320, 680] & \text{with probability } 1/7. \end{cases}$$

Therefore $E(S^I) = E(S^C)$ where

$$E(S^I) = 2000(400)(0.01) + 2000(500)(0.02) + 1000(500)(0.01) = 33{,}000.$$

For the individual risk model we have

$$\begin{aligned} Var(S^I) = {} & 2000[(400)^2(0.01)(0.99) + (0.01)480{,}000] \\ & + 2000[(500)^2(0.02)(0.98) + (0.02)250{,}000] \\ & + 1000[(500)^2(0.01)(0.99) + (0.01)10{,}800] \\ = {} & 12{,}768{,}000 + 19{,}800{,}000 + 2{,}583{,}000 \\ = {} & 35{,}151{,}000 = (5928.828)^2, \end{aligned}$$

while for the compound Poisson approximation

$$\begin{aligned} Var(S^C) = {} & 70\, E(W^2) \\ = {} & 70\,[\tfrac{2}{7}\,(640{,}000) + \tfrac{4}{7}\,(500{,}000) + \tfrac{1}{7}\,(260{,}800)] \\ = {} & 35{,}408{,}000 = (5950.462)^2. \end{aligned}$$

The standard deviation of S^C is only marginally bigger than that of S^I. If one wanted to put a security loading on premiums in order to be 98% sure that premiums exceed claims, then using the individual risk model to determine this loading one would obtain $\theta^I = (2.0537)(5928.828)/33{,}000 = 0.3690$, while for the compound Poisson approximation it would be the marginally larger $\theta^C = (2.0537)(5950.462)/33{,}000 = 0.3703$.   ▯

3.4.2   Setting aside reserves for aggregate claims

Normally, a certain amount of reserves U must be set aside to cover situations when large numbers and/or aggregate amounts of claims occur. There should be enough reserves to ensure with high probability that premiums plus reserves should exceed claims. On the other hand, putting too much into reserves can be both costly and wasteful.

In a large portfolio of policies, the central limit theorem can be very useful in helping us to determine the appropriate amount of reserves for a given situation. In the collective risk model where $S = X_1 + \cdots + X_N$, we may often well approximate the distribution of S by a normal distribution if N is *reasonably large*. This is basically guaranteed by a generalization of the central limit theorem. For example, in a compound Poisson distribution, if

λ is large then we expect a large number of terms in the sum for S and we have already seen that the skewness of S is inversely proportional to $\sqrt{\lambda}$. For the compound binomial distribution, we have seen that if n is large where $N \sim B(n, q)$, then we may also interpret S as the sum of a large number of independent and identically distributed random variables and hence use the central limit theorem directly.

Suppose that total annual claims are being modeled by a compound Poisson distribution with Poisson parameter λ and random claim size X. If we want to determine the amount of reserves U which should be held in order to be $100(1 - \epsilon)\%$ sure that premiums plus reserves cover claims, then (using a normal approximation for S)

$$
\begin{aligned}
1 - \epsilon &= P(\ S < U + (1 + \theta)E(S)\ ) \\
&= P(\ [S - E(S)]/\sqrt{Var(S)}\ <\ [U + \theta E(S)]/\sqrt{Var(S)}\ ) \\
&\doteq P(\ N(0,1)\ <\ [U + \theta \lambda E(X)]/\sqrt{Var(S)}\ ).
\end{aligned}
\tag{3.15}
$$

Hence we want $z_{1-\epsilon} = [U + \theta \lambda E(X)]/\sqrt{\lambda E(X^2)}$, or

$$
U = z_{1-\epsilon}\sqrt{\lambda E(X^2)} - \theta \lambda E(X).
\tag{3.16}
$$

Equation (3.16) establishes an important relationship between necessary reserves U, degree of confidence $1 - \epsilon$, claim rate λ, security or loading factor θ, and type of claim X.

Example 3.12

Suppose that claims in company A can be modeled by a compound Poisson distribution where the typical claim is $\Gamma(2, 0.02)$ distributed and about 200 claims are expected annually. If a loading factor of $\theta = 0.02$ is to be used on premiums, then in order to be 98% sure that premiums plus reserves U_A exceed annual claims, one should set aside reserves of

$$
\begin{aligned}
U_A &= 2.0537\sqrt{200 E(X^2)} - 0.02(200)E(X) \\
&= 2.0537\sqrt{200\ [(2/0.02)^2 + 2/(0.02)^2]} - 4(100) \\
&= 3157.11.
\end{aligned}
$$

When ϵ, θ and the claim distribution are fixed, one can see from Equation (3.16) that U is a quadratic function of $\sqrt{\lambda}$. Initially, as λ increases, so do the necessary reserves U. However, there exists a unique positive solution λ_0 to $0 = z_{1-\epsilon}\sqrt{\lambda E(X^2)} - \theta \lambda E(X)$, and for any $\lambda > \lambda_0$ no reserves are actually needed to be $100(1 - \epsilon)\%$ sure premiums exceed claims. This demonstrates one of the advantages that big companies (holding large numbers of policies and hence with a corresponding large λ) have over smaller ones. Figure 3.4 gives a plot of U_A as a function of the expected number of annual claims λ for

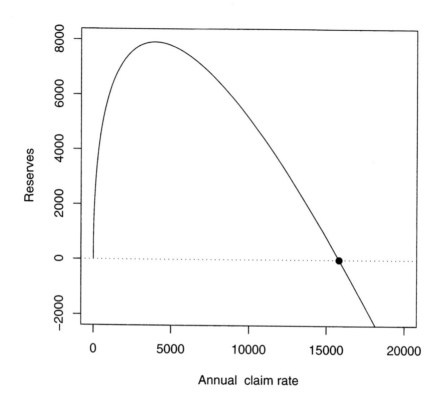

FIGURE 3.4

Necessary reserves for company A in Example 3.12.

company A. Note that if in fact λ_A were as large as 15,816, then no reserves would be needed to meet claims (with 98% confidence).

Assume now that company A is considering merging with company B. Company B has a portfolio of policies where annual aggregate claims are modeled by a compound Poisson distribution with rate parameter $\lambda_B = 100$/year, the security loading is $\theta_B = 0.03$, and claim size is modeled by $X_B \sim \Gamma(2, 0.01)$. On its own, company B would need reserves of

$$U_B = 2.0537\sqrt{100\left[(2/0.01)^2 + 2/(0.01)^2\right]} - 3(200)$$
$$= 4430.52$$

to be 98% confident that reserves plus premiums meet claims for a year. Another advantage that big business has (in this case the bigger business resulting from merging) is that both risks and resources (reserves) can be pooled.

If companies A and B merge, then we can model total claims $S = S_A + S_B$ from policyholders in both companies with a compound Poisson distribution where the expected number of claims in a year is $\lambda = \lambda_A + \lambda_B = 300$ and a typical claim X is a $(2/3, 1/3)$ mixture of X_A and X_B. Another interpretation of X is that $2/3$ of the time the claim will come from a person formerly holding a policy with company A and otherwise with company B. Proceeding as in the derivation of (3.16), one may establish that the amount of reserves U_{A+B} necessary to be $100(1 - \epsilon)\%$ sure that all claims are met by premiums plus reserves is given by

$$U_{A+B} = z_{1-\epsilon}\sqrt{\lambda_A E(X_A^2) + \lambda_B E(X_B^2)} - [\theta_A \lambda_A E(X_A) + \theta_B \lambda_B E(X_B)]. \quad (3.17)$$

Therefore if companies A and B merge, the amount of reserves necessary to be 98% sure that together with premiums all claims will be met is

$$U_{A+B} = 2.0537\sqrt{200(15,000) + 100(60,000)} - [0.02(200)100 + 0.03(100)200]$$
$$= 5161.1.$$

Note then that when the companies merge, the necessary reserves are considerably less than the sum of the reserves $3157.11 + 4430.52 = 7587.63$ needed separately. In one of the problems you are asked to show that if both λ_A and λ_B were 10 times larger, then $U_{A+B} = 9483.11 < U_B = 9907.89$. ☐

Example 3.13
A company has n personal health policies where the probability of a claim is assumed to be q in each case. Let X represent a typical claim and $S = X_1 + \cdots + X_N$ be the total amount of claims in one year as modeled by a compound binomial distribution. If θ is the loading factor for premiums and U is the necessary initial reserves to be 99% sure that all claims will be paid in the coming year, then

$$U \doteq 2.3263\sqrt{nq(m_2 - qm_1^2)} - \theta n q m_1.$$

Assume now that three companies A, B and C are to merge, and that each have annual claim structures as indicated below for portfolios of personal health policies and total claims S_A, S_B and S_C, respectively. What combined reserves are necessary to be 99% sure all claims will be met?

	Company A	Company B	Company C
Claim size X	$\Gamma(2, 0.02)$	$\Gamma(2, 0.01)$	$\Gamma(1, 0.01)$
n number of policies	500	1000	2000
q probability of claim	0.02	0.04	0.01
θ security loading	0.20	0.10	0.10

Using a compound binomial distribution to model total claims S we have that $E(S) = nqm_1$ and $Var(S) = nq(m_2 - m_1^2) + nqpm_1^2$ from Equations (3.1) and (3.2). By appealing to the central limit theorem (reasonable when nq is relatively large) and using an argument similar to that used to establish (3.15), one obtains that $U \doteq 2.3263\sqrt{nq(m_2 - qm_1^2)} - \theta nqm_1$. Similar to the compound Poisson situation (where U initially increases and then decreases as a function of λ), it is clear that U initially increases and then decreases as a function of n.

We model total claims for the three companies as the sum $S = S_A + S_B + S_C$ of compound binomial distributions. Approximating S by a normal distribution, it follows that the necessary reserves $U = U_{A+B+C}$ satisfy

$$U = 2.3263\sqrt{Var(S_A) + Var(S_B) + Var(S_C)}$$
$$-[\,\theta_A E(S_A) + \theta_B E(S_B) + \theta_C E(S_C)\,]$$
$$= 232.63(100)$$
$$\sqrt{5(.02)[150 - .02(1)] + 10(.04)[600 - .04(2)] + 20(.01)[200 - .01(1)]}$$
$$-(0.2(500)(0.02)100 + 0.1(1000)(0.04)200 + 0.1(2000)(0.01)100)$$
$$= 2.3263(1717.452) - 1200$$
$$= 2795.31.$$

◻

3.5 Reinsurance for aggregate claims

In Chapter 2 on loss distributions, we introduced various types of *claim-by-claim* reinsurance, while here we will discuss types of reinsurance for the aggregate amount of claims S. Many such arrangements are based on individual claims. In a claim-by-claim based reinsurance agreement on S, each individual claim X is split into two components

$$X = Y + Z = h_I(X) + h_R(X),$$

which are, respectively, handled by the insurance $(Y = h_I(X))$ and reinsurance $(Z = h_R(X))$ companies. There are, of course, many possibilities of nonnegative functions h_I and h_R with the property that $X = h_I(X) + h_R(X)$. In proportional reinsurance, $h_I(X) = \alpha X$ and $h_R(X) = (1 - \alpha)X$ for some $0 \leq \alpha \leq 1$, while for excess of loss reinsurance we have $h_I(X) = min(X, M)$ and $h_R(X) = max(0, X - M)$ for some claim excess level M.

If S denotes total claims for a portfolio of policies in a collective risk model, then in claim-by-claim based reinsurance we express the amounts paid by the insurer and reinsurer as S_I and S_R, respectively, where

$$S = \sum_1^N X_i = \sum_1^N Y_i + \sum_1^N Z_i$$

$$= \sum_1^N h_I(X_i) + \sum_1^N h_R(X_i)$$

$$= S_I + S_R.$$

Note that in the collective risk model where $S = X_1 + \cdots + X_N$, each of the claims X_i is positive. However, for some forms of reinsurance, it is clearly possible that one of Y_i or Z_i in the decomposition $X_i = Y_i + Z_i$ is actually 0. For example, in excess of loss reinsurance with retention or excess level M, if $X_i < M$ then the reinsurer will not be involved with the claim and hence $Z_i = 0$.

In *stop-loss reinsurance* for S, the reinsurance company handles the *excess of the aggregate claims S over an agreed amount M* with the baseline company taking responsibility for the remainder (the total amount up to this agreed cut-off or stop-loss value). The term stop-loss refers to the fact that the loss of the insurance company in this case is *stopped* or limited to M. In stop-loss reinsurance, we write $S = min(S, M) + max(0, S - M) = S_I + S_R$ for some stop-loss level M.

Reinsurance companies will charge for sharing in the risk of an insurance company, and this could affect both the level and type of agreement that an insurance company may make with a reinsurer. We will continue to use θ to represent the security loading used by a baseline insurance company in determining premiums for its policy holders, and will use ξ to represent the corresponding loading which the reinsurer uses to cover its risk S_R. Hence net premium income over a fixed time period for the insurance company takes the form $(1 + \theta)E(S) - (1 + \xi)E(S_R)$. Normally, the reinsurer will use a heavier loading ($\xi > \theta$), and hence the cedant must balance the advantage of sharing the risk with the reinsurer vis-a-vis the cost involved. The net premiums will be used to pay claims, and hence in particular should normally exceed the expected amount of claims payable. We will use $P_\$$ in this setting to represent the profit or net premiums minus claims, and hence expected profit for the insurer takes the form

$$E(P_\$) = (1 + \theta)E(S) - (1 + \xi)E(S_R) - E(S_I)$$
$$= \theta E(S_I) - (\xi - \theta)E(S_R). \qquad (3.18)$$

This is clearly nonnegative if and only if

$$E(S_I)/E(S_R) \geq (\xi - \theta)/\theta = \xi/\theta - 1. \qquad (3.19)$$

In many situations, one may need to hold sufficient reserves U to cope with situations where claims exceed net premiums.

In this section, we shall investigate in some detail aspects of proportional, excess of loss and stop-loss reinsurance agreements. We will gain some insight into how these arrangements affect reserve requirements, the profitability and the security of an insurance company when using collective risk models for aggregate claims. In general, we will see that if the objective of an insurance company is to maximize expected profits, then reinsurance would rarely be used because it is relatively expensive. On the other hand, security and solvency are of crucial importance, and here the role of the reinsurer is vital. We will address this issue further in the chapter on ruin theory where we consider the probability of ruin as a criterion in evaluating types and levels of reinsurance.

3.5.1 Proportional reinsurance

In proportional reinsurance, an agreed proportion α $(0 \leq \alpha \leq 1)$ of each claim is retained by the baseline insurance company, and the remaining proportion $(1 - \alpha)$ is ceded to the reinsurer. Hence total aggregate claims S can be represented as

$$S = S_I + S_R = \alpha S + (1 - \alpha) S.$$

S_I and S_R are perfectly correlated since $cov(S_I, S_R) = \alpha(1 - \alpha) cov(S, S) = \alpha(1 - \alpha) Var(S)$, from which it follows that $corr(S_I, S_R) = 1$. Note that

$$Var(S_I) + Var(S_R) = [\alpha^2 + (1 - \alpha)^2] \, Var(S) < Var(S),$$

and hence the sum of the variances of the shared risks has been reduced by proportionally sharing S. This may be considered advantageous to both parties! Since skewness and kurtosis are scale-invariant descriptive measures of a random variable, they remain the same for both the insurer and reinsurer.

If S is compound Poisson with Poisson parameter λ and typical claim X, then clearly S_I (S_R) is compound Poisson with Poisson parameter λ and typical claim αX ($(1 - \alpha)X$). Similar statements can be made for the situation when S is compound binomial or compound negative binomial.

Example 3.14

Suppose we are modeling collective risks $S = X_1 + \cdots + X_N$ with a compound binomial distribution where $N \sim B(800, q = 0.025)$ and the typical claim X has gamma distribution $\Gamma(2, 0.04)$. We have agreed on a proportional reinsurance agreement where the baseline insurance company retains 60% of any claim. Now $m_1 = E(X) = 2/0.04 = 50$, and similarly one obtains $m_2 = E(X^2) = Var(X) + E^2(X) = 3750$ and $m_3 = E(X^3) = 375{,}000$. Using Equations (3.1), (3.2) and (3.4) we obtain

$$E(S_I) = \alpha n q m_1 = (0.60)(800)(0.025)(2/0.04)$$
$$= 600$$

$$Var(S_I) = \alpha^2 \, nq(m_2 - qm_1^2)$$
$$= (0.60)^2 \, (800)(0.025)(3750 - (0.025)50^2)$$
$$= 26{,}550 \quad \text{and}$$
$$skew(S_I) = skew(S)$$
$$= \frac{nqm_3 - 3nq^2 m_2 m_1 + 2nq^3 m_1^3}{(Var(S_I))^{3/2}}$$
$$= \frac{20(375{,}000) - 3(20)(0.025)(3750)50 + 2(20)(0.025)^2(50^3)}{(26{,}550)^{3/2}}$$
$$= 1.6694.$$

▯

If the baseline insurance company retains a proportion α of each claim, then its net premium income will be of the form $(1+\theta)E(S) - (1+\xi)(1-\alpha)E(S)$. Net premium income will therefore be positive if and only if $\alpha > (\xi-\theta)/(1+\xi)$. On the other hand, the expected profit (3.18) is of the form

$$E(P_\$) = (1+\theta)E(S) - (1+\xi)(1-\alpha)E(S) - \alpha E(S) \;=\; [\theta - \xi(1-\alpha)]E(S),$$

which from (3.19) is nonnegative if and only if $\alpha/(1-\alpha) \geq (\xi - \theta)/\theta$ or equivalently $\alpha \geq 1 - \theta/\xi$. Since $1 - \theta/\xi \geq (\xi - \theta)/(1 + \xi)$, the insurance company should retain at least $100(1 - \theta/\xi)\%$ of the business (that is, ensure $\alpha \geq 1 - \theta/\xi$). In the unlikely but theoretically possible situation where $\theta \geq \xi$, there is no such restriction on α, and in this case reinsurance is so cheap that the insurance company might consider passing on all of the business (i.e., use $\alpha = 0$).

Note that the expected profit is an increasing function of the retention proportion α, hence if the objective is to maximize expected profits the insurer would select $\alpha = 1$ and not use the option of reinsurance. However, as we know in (insurance) business one usually desires to achieve a balance between security and maximizing expected profits. We will consider this issue more extensively in Chapter 4 on ruin theory.

Example 3.15

Total claims in company A for a period of one year can be modeled by a compound Poisson distribution. Individual claim sizes are exponential in nature with mean 100, and a security loading of $\theta = 0.1$ is used in determining premiums. One would expect 18 claims to be made during the year. We assume that $1000 is available for claims reserving.

1. Claim-by-claim proportional reinsurance is available at a cost of $1.2 = (1 + \xi)$ per unit of coverage. If company A wants to be 99% sure of meeting all claims for which it is responsible at the end of the year, at

what level should proportional reinsurance be taken in order to maximize expected profit? What would the result be if an inflation rate of 5% is forecast for the coming year?

2. Company A believes that it can achieve the desired security (of being 99% sure of meeting all claims) without reinsurance if it increases the volume of business appropriately. By what factor will business have to be increased to achieve this?

If S_I represents the aggregate claims for company A under a proportional reinsurance treaty where it retains a proportion α of any claim, then $S_I = \alpha S$. We let U_α be the reserves necessary for company A to be $100(1-\epsilon)\%$ confident that claims are met by net premiums plus reserves. Using a normal approximation to S, U_α must satisfy

$$(1-\epsilon) = P(\alpha S < U_\alpha + [1 + \theta - (1+\xi)(1-\alpha)] E(S))$$
$$\doteq P\left(N(0,1) < [U_\alpha + [\theta - \xi(1-\alpha)] \lambda E(X)]/\sqrt{\lambda E(\alpha X)^2} \right),$$

implying that

$$U_\alpha \doteq z_{1-\epsilon}\sqrt{\lambda E(\alpha X)^2} - \lambda E(X)[\theta - \xi(1-\alpha)]. \qquad (3.20)$$

In our situation,

$$U_\alpha = z_{0.99}\, \alpha\, \sqrt{(18)2(100)^2} - 18\,(100)\,[0.1 - 0.2(1-\alpha)]$$
$$= 1035.809\alpha + 180,$$

and therefore $U_\alpha \le 1000 \Leftrightarrow \alpha \le 0.7917$. We know that expected profit is an increasing function of α, and that in this instance expected profit is nonnegative if and only if $\alpha \ge (1 - \theta/\xi) = 0.5$. The desired security is met only if $\alpha \le 0.7917$, and hence the optimal choice here is $\alpha = 0.7917$.

If an inflation rate of 5% is expected for claim size, then from Equation (3.20) it is clear that the necessary reserves which are denoted by $U_\alpha^{1.05}$ must satisfy $U_\alpha^{1.05} = (1.05)U_\alpha$, hence the optimal value of α would be $0.7917/1.05 = 0.7540$. Without reinsurance, the relationship of reserves to volume of business (as indicated by λ) is given in this instance (see Equation (3.16)) by $1000 \ge z_{0.99}\sqrt{\lambda\, 2(100)^2} - \lambda(0.1)(100)$. This is a quadratic inequality in $\sqrt{\lambda}$. Solving we find that this holds only if $\lambda \le 3.3886^2 = 11.4824$ (which involves a reduction in business), or $\lambda \ge 29.5110^2 = 870.8967$, representing approximately a very large 48-fold increase in business in order to be confident of meeting claims without reinsurance! $\Box$

3.5.2 Excess of loss reinsurance

In claim-by-claim excess of loss reinsurance with excess level M, any claim X is broken into that part paid by the insurer $Y = \min(X, M)$ and that paid by the reinsurer $Z = \max(X - M, 0)$. The reinsurer will only become involved in a claim with probability $\bar{F}_X(M) = P(X > M)$, hence many of the terms in the representation $S_R = Z_1 + \cdots + Z_N$ will usually be zero. For example, if the claims in a (claim-by-claim) excess of loss reinsurance arrangement with excess level $M = 500$ were $\{390, 765, 1200, 320, 505\}$, then

$$S_I = 390 + 500 + 500 + 320 + 500 = 2210 \quad \text{and}$$
$$S_R = 0 + 265 + 700 + 0 + 5 = 970.$$

We may therefore interpret $S_R = Z_1 + \cdots + Z_N$ as a compound risk model for the reinsurer with a random number of terms N, as long as we accept that with probability $F_X(M)$ any of the terms in this representation will be 0. Realistically, however, the reinsurer is only interested in claims in which it will actually become involved (those exceeding M), and consequently, a more appropriate representation for S_R is of the form $S_R = W_1 + \cdots + W_{N_R}$ where N_R is the number of claims which exceed M, and the random variable W represents the excess over M of a claim X. In other words, $W \sim (X - M)|_{[X>M]}$. Another way of viewing the relationship between Z and W is to note that Z is a mixture of W and 0, where Z is equal to 0 with probability $F_X(M)$ and otherwise (with probability $\bar{F}_X(M)$) equal to the random variable W.

Essentially, we have two equivalent representations for the aggregate amount S_R paid by the reinsurer. Take, for example, the situation where S is compound Poisson with parameter λ and claim distribution X. Then S_I is compound Poisson with parameter λ and claim distribution $Y = min(X, M)$. On the other hand, S_R is also compound Poisson with two different representations. In the first instance, we may express S_R as a compound Poisson distribution of the form $S_R = Z_1 + \cdots + Z_N$ with Poisson parameter λ and claim distribution $Z = max(0, X - M)$. Equivalently, it may be represented as a compound Poisson distribution of the form $S_R = W_1 + \cdots + W_{N_R}$ with Poisson parameter $\lambda \bar{F}_X(M)$ and claim distribution W. Hence in particular,

$$E(S_R) = \lambda E(Z) = \lambda \bar{F}_X(M) E(W) \quad \text{and}$$
$$Var(S_R) = \lambda E(Z^2) = \lambda \bar{F}_X(M) E(W^2).$$

For some special random variables, X and $W \sim (X - M)|_{[X>M]}$ have the same distributional form. If X is exponentially distributed with mean m_1, then (because of the memoryless property of the exponential), so is $W = (X - M)|_{[X>M]}$. Another example is the Pareto distribution, since if $X \sim$ Pareto (β, δ), then $W \sim (X - M)|_{[X>M]} \sim$ Pareto $(\beta, \delta + M)$. For the uniform distribution $(X \sim U[a, b])$ on the interval $[a, b]$, it is easy to see that $W \sim (X - M)|_{[X>M]} \sim U[0, b - M]$.

Example 3.16
Annual aggregate claims in a company are modeled by a compound Poisson distribution where the typical claim is uniform on the interval $[0, 1200]$ and

about 60 claims are expected. An excess of loss reinsurance arrangement is being considered whereby the reinsurer handles the excess of any claim over $M = 800$.

Here $\bar{F}_X(M) = P(X > 800) = 1/3$, and $W \sim (X - M)\,|_{[X>800]} \sim U[0, 400]$. Hence S_R, the aggregate claims for the reinsurer, is compound Poisson with parameter $60/3 = 20$ and typical claim uniform on $[0, 400]$. In general, if U is uniformly distributed on $[0, b]$, then $E(U^i) = b^i/(i + 1)$. Therefore

$$E(S_R) = 20(400)/2 = 4000,$$
$$Var(S_R) = 20(400)^2/3 = 1{,}066{,}667 \quad \text{and}$$
$$skew(S_R) = (20(400)^3/4)/(Var(S_R))^{3/2} = 0.2905.$$

$\Box$

Excess of loss reinsurance with excess level M does not affect the frequency or rate of claims for the ceding company or insurer, but it does reduce the amount paid on larger claims. A convenient tool in analyzing excess of loss reinsurance is the *limited expected value function* (or LEV) $L_X(M)$ of the random variable X (see [32]). For a nonnegative claim distribution X, this function is the expected value of $Y = h_I(X) = min(X, M)$ and is given by

$$L_X(M) = E(Y) = \int_0^M x\, dF_X(x) + M\bar{F}_X(M), \qquad (3.21)$$

where the integral in Equation (3.21) should be interpreted as a sum when X is discrete. For the exponentially distributed random variable X with mean μ, it is easy to see that the limited expected value function takes the form $L_X(M) = \mu\,[1 - e^{-M/\mu}]$. This can be derived directly from Equation (3.21) using integration by parts. However, it can also be seen by noting that

$$\begin{aligned} L_X(M) &= E(X - Z) \\ &= \mu - \bar{F}_X(M)\, E(W) \\ &= \mu - e^{-M/\mu}\, \mu \\ &= \mu\,[1 - e^{-M/\mu}]. \end{aligned}$$

When X has continuous density function f_X, then by differentiating with respect to M one has that $L'_X(M) = \bar{F}_X(M)$ and $L''_X(M) = -f_X(M)$, from which it follows that $L'_X(M)$ is an increasing concave function of M with the property that $\lim_{M \to \infty} L_X(M) = \mu_X$.

How is the limited expected value function affected by a transformation which replaces X by aX for some positive scalar a? This might arise, for example, when, because of an inflationary factor of k, the typical claim changes from X to kX from one year to the next. Naively, one might initially think that $L_{aX}(M) = aL_X(M)$ for positive a, but this is rarely the case. In fact, what is true is that when $a > 0$ and b are constants,

$$L_{aX+b}(M) = a\, L_X\left(\frac{M-b}{a}\right) + b. \tag{3.22}$$

We sketch a proof for the case where X has density function $f_X(x)$. In this case, $P(aX + b \leq x) = F_{aX+b}(x) = F_X([x-b]/a)$, and hence $f_{aX+b}(x) = (1/a)f_X([x-b]/a)$. Therefore using the substitution $u = (x-b)/a$,

$$
\begin{aligned}
L_{aX+b}(M) &= \int_0^M x f_{aX+b}(x)\,dx + M\,\bar{F}_{aX+b}(M) \\
&= \int_0^M \frac{x}{a}\, f_X\left(\frac{x-b}{a}\right)\,dx + M\,\bar{F}_X\left(\frac{M-b}{a}\right) \\
&= a\left[\int_0^{(M-b)/a} u\, f_X(u)\,du + \frac{M-b}{a}\,\bar{F}_X\left(\frac{M-b}{a}\right)\right] + b \\
&= a\, L_X\left(\frac{M-b}{a}\right) + b.
\end{aligned}
$$

Therefore if X is exponential with mean 1200 and next year an inflation rate of 6% is expected, we would consider the claim random variable $U = (1.06)X$ where

$$L_{(1.06)X}(M) = 1.06\, L_X(M/1.06) = 1.06\,(1200)\left[1 - e^{-M/(1.06\cdot 1200)}\right].$$

When the claim random variable X has a lognormal distribution with parameters μ and σ^2, then one has the following expression for the limited expected value function:

$$L_X(M) = e^{\mu + \frac{\sigma^2}{2}}\, \Phi\left(\frac{\log M - \mu - \sigma^2}{\sigma}\right) + M\left[1 - \Phi\left(\frac{\log M - \mu}{\sigma}\right)\right]. \tag{3.23}$$

Figure 3.5 gives a graph of $L_X(M)$ (or equivalently $E(Y) = E[\min(X, M)]$) as a function of M for a lognormal random variable X with mean 900 and standard deviation 300. Note, for example, that $L_X(500) = 497$, while $L_X(1000) = 821$. In Problem 18, you are asked to plot the limited expected value function when X has been increased by an inflationary factor of 7%.

If the reinsurer is involved in a claim (that is, $X > M$), then the average amount paid by the reinsurer is $E(W)$. It is worth noting that as a function of M, this is what is referred to in the language of survival theory as the *mean residual life function*. That is, if X represents a lifetime, then $E(W) = E(X - M \mid X > M)$ is the expected amount of remaining life given survival to age M.

We know from (3.18) and (3.19) that in order for expected net profits to be positive, the insurer must retain a minimal amount of the business. In excess

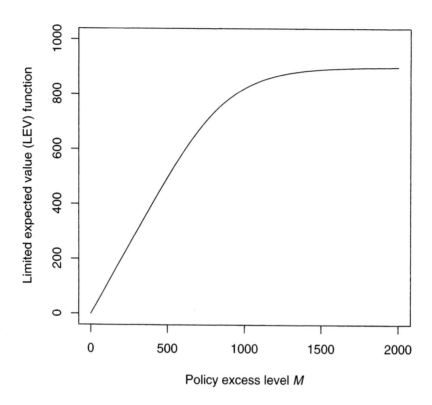

FIGURE 3.5
LEV function for lognormal X where $E(X) = 900$ and $Var(X) = 300^2$.

of loss reinsurance, this is equivalent to the excess level M satisfying

$$E(S_I)/E(S_R) = E(Y)/E(Z)$$
$$= \frac{\int_0^M x\, dF_X(x) + M\bar{F}_X(M)}{\int_M^\infty (x - M)\, dF_X(x)}$$
$$= \frac{E(X) - \int_M^\infty (x - M)\, dF_X(x)}{\int_M^\infty (x - M)\, dF_X(x)}$$
$$= \frac{E(X) - E(Z)}{E(Z)}$$
$$= \frac{E(X) - \bar{F}_X(M)E(W)}{\bar{F}_X(M)E(W)} \qquad (3.24)$$

$$\geq (\xi - \theta)/\theta \;=\; \xi/\theta - 1. \tag{3.25}$$

Note that in general, the minimum excess level M^* which should be considered is independent of the expected claim incidence $E(N)$. Of course, in the unlikely event that $\xi \leq \theta$, any value of M might be considered.

Example 3.17

Consider a compound Poisson model $S = X_1 + \cdots + X_N$ for aggregate claims where the typical claim $X \sim$ Pareto (β, δ). The loading factor is θ, and excess of loss reinsurance is available with a loading factor of ξ. We will determine the minimum value of the excess level M which guarantees that expected (net) profits are nonnegative as a function of the relevant parameters.

Since $X \sim$ Pareto (β, δ), we know that $E(X) = \delta/(\beta - 1)$. Moreover, for any excess level M, $\bar{F}_X(M) = \delta^\beta/(\delta + M)^\beta$ and the random variable $W \sim$ Pareto $(\beta, \delta + M)$. From Equations (3.24) and (3.25) we know that M must satisfy

$$
\begin{aligned}
E(Y)/E(Z) &= \left[\frac{\delta}{\beta - 1} - \left(\frac{\delta}{\delta + M} \right)^\beta \left(\frac{\delta + M}{\beta - 1} \right) \right] \Big/ \left[\left(\frac{\delta}{\delta + M} \right)^\beta \left(\frac{\delta + M}{\beta - 1} \right) \right] \\
&= \left(1 + \frac{M}{\delta} \right)^{\beta - 1} - 1 \\
&\geq \xi/\theta - 1,
\end{aligned}
$$

or, equivalently, that

$$
M \geq \delta \left[\left(\frac{\xi}{\theta} \right)^{1/(\beta - 1)} - 1 \right] \equiv M^*.
$$

Note that for a given θ, M^* is an increasing function of ξ. Table 3.10 gives values for the minimum excess levels M^* which should be considered when $X \sim$ Pareto $(\beta = 3, \delta = 1200)$ for various values of θ and ξ. In Problem 20, you are asked to show that if X is exponentially distributed, then $M^* = E(X) \log \xi/\theta$. $\square$

3.5.3   Stop-loss reinsurance

Much of the theory for stop-loss reinsurance parallels that of excess of loss reinsurance on a claim-by-claim basis, but where now we are dealing with just the *total claim amount* S. In practice, a reinsurer may put an upper limit on the amount it will cover. For great risks, the insurance or ceding company may use several reinsurers or a reinsurer may look for other reinsurers to share the risk.

In making a stop-loss treaty with a reinsurer, the insurer is putting a maximum M on its risk whereby its expected aggregate claim payment will be

TABLE 3.10

Minimum excess level M^* as a function of loadings θ and ξ for insurer when $X \sim$ Pareto (β, δ).

$\theta\backslash\xi$	0.1	0.2	0.3	0.4	0.5
0.1	0	497	878	1200	1483
0.2	0	0	270	497	697
0.3	0	0	0	185	349
0.4	0	0	0	0	142

$L_S(M) = E(S_I) = E(min[S, M])$. The price for this treaty is the *stop-loss premium*, which we assume takes the form

$$(1 + \xi)E(S_R) \equiv (1 + \xi)\,E(max[S - M, 0])$$

where ξ is the reinsurer's loading factor. If θ is the insurer's loading on policyholders, then we have seen that a potential stop-loss level M should only be considered (in order that expected profits are nonnegative) if $E(S_I)/E(S_R) = L_S(M)/[E(S) - L_S(M)] \geq \xi/\theta - 1$, or equivalently, that

$$L_S(M) \geq \left(1 - \frac{\theta}{\xi}\right) E(S). \tag{3.26}$$

In many cases, where there are large volumes of business, the distribution of S can be well approximated by a normal distribution (for example, when using a compound Poisson distribution with large λ). Hence it is useful to consider the limited expected value function of the normal distribution. Up to now we have only considered the limited expected value function for nonnegative (claim) random variables, but clearly the concept can be extended to random variables in general. The limited expected value function for the standard normal distribution takes the particularly nice form (which can be easily checked by differentiation):

$$L_{N(0,1)}(M) = \int_{-\infty}^{M} x\,\phi(x)\,dx + M\,[1 - \Phi(M)]$$
$$= -\phi(M) + M\,[1 - \Phi(M)],$$

where ϕ and Φ are, respectively, the density and distribution functions of the standard normal distribution. $L_{N(0,1)}(M)$ is an increasing concave function of M which approaches 0 as $M \to \infty$. From Equation (3.22), it follows that if $S \dot\sim N(\mu, \sigma^2)$, then

$$L_S(M) \doteq \sigma\,L_{N(0,1)}\left(\frac{M - \mu}{\sigma}\right) + \mu$$
$$= -\sigma\,\phi\left(\frac{M - \mu}{\sigma}\right) + \sigma\,\frac{M - \mu}{\sigma}\left[1 - \Phi\left(\frac{M - \mu}{\sigma}\right)\right] + \mu.$$

Of course, one must take considerable care in using this tool in determining a stop-loss treaty, for S is often skewed and we are assuming the right-hand tail of its distribution is similar to that of a normal.

Example 3.18
Suppose that the distribution of aggregate claims S can be well approximated by a normal distribution with mean $\mu = 50,000$ and $\sigma = 10,000$. A loading factor of $\theta = 0.1$ is used by the insurance company on policyholders, and stop-loss reinsurance with stop-loss level M is being considered at a cost of $(1 + \xi)$ per unit of coverage. We determine the minimum stop-loss level M^* which the insurance company should consider in order that expected profits are nonnegative when $\xi = 0.2, 0.3$ and 0.4, respectively.

From the discussion above (3.26), M^* must satisfy

$$L_{N(0,1)}\left(\frac{M-\mu}{\sigma}\right) = -\phi\left(\frac{M-\mu}{\sigma}\right) + \frac{M-\mu}{\sigma}\left[1 - \Phi\left(\frac{M-\mu}{\sigma}\right)\right]$$

$$= -\frac{\mu}{\sigma}\frac{\theta}{\xi},$$

or equivalently,

$$M^* = \mu + \sigma\, L_{N(0,1)}^{-1}\left(-\frac{\mu}{\sigma}\frac{\theta}{\xi}\right).$$

In this situation, it is clear that M^* is an increasing function of ξ when the other parameters are fixed, since the more expensive (relatively speaking) reinsurance is, the more business the insurer should retain. The limited expected value function $L_{N(0,1)}(M)$ for the standard normal distribution is plotted in Figure 3.6. When $\xi = 0.2$, we find that

$$M_{0.2}^* = 50,000 + 10,000\, L_{N(0,1)}^{-1}\left(-5 \cdot \frac{0.1}{0.2}\right)$$

$$= 50,000 + 10,000\,(-2.49798) = 25,020.17,$$

and similarly, $M_{0.3}^* = 33,541.78$ and $M_{0.4}^* = 38,069.01.$ □

Example 3.19
We return to Example 3.7 and consider the limited expected value function for a risk $S = X_1 + \cdots + X_N$ which is modeled by a compound binomial distribution where $N \sim B(50, 0.04)$. Here S (see Table 3.4 where the exact distribution was calculated in units of 10,000) has mean 6.2 and standard deviation 6.15. Being discrete, the limited expected value function of S is of the form

$$L_S(M) = \sum_0^M x\, P(S = x) + M\, P(S > x).$$

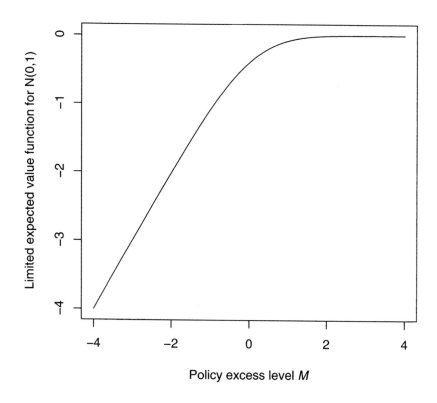

FIGURE 3.6

Limited expected value function for the standard normal distribution.

Using a stop-loss reinsurance level of M, the expected profit (in units of 10,000) when the loadings for the insurer and reinsurer are, respectively, θ and ξ is given by $E(P_\$^M) = \theta L_S(M) - (1 + \xi)\,[6.2 - L_S(M)]$. Figure 3.7 gives a plot of $L_S(M)$. Note that the limiting value of the limited expected value function is clearly $6.2 = E(S)$.

Figure 3.8 gives a plot of the expected profit function $E(P_\$^M)$ when $\theta = 0.3$ and $\xi = 0.4$ or $\xi = 0.8$. It is clear that expected profits are greater for the case when the stop-loss reinsurance is cheaper ($\xi = 0.4$). When the stop-loss level of $M = 15$ is used, expected profits are, respectively, 1.08 and 0.90 in the cases where $\xi = 0.4$ and $\xi = 0.8$. Note that in both cases the limiting value (as $M \to \infty$) of expected profits is $0.3\,E(S) = 1.86$, corresponding to the situation where there is no reinsurance.

☐

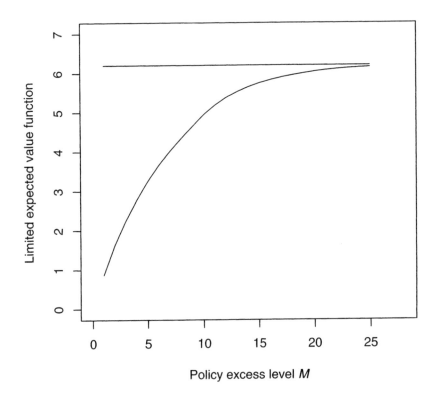

FIGURE 3.7
LEV function for the compound binomial distribution of Example 3.19.

3.6 Problems

1. $S = X_1 + \cdots + X_N$ has a compound Poisson distribution where λ is the
   Poisson parameter for N and X is the typical claim random variable. If
   X is gamma distributed with parameters 2 and β (mean $= 2/\beta$), derive
   expressions for the variance and skewness of S in terms of λ and β.

2. Any claim made from a portfolio of term life policies is for a constant
   amount C. It is decided to model aggregate claims S with a compound
   distribution of the form $S = X_1 + \cdots + X_N$ where $E(N) = 100$ but N
   may be either Poisson, binomial or negative binomial. Determine the

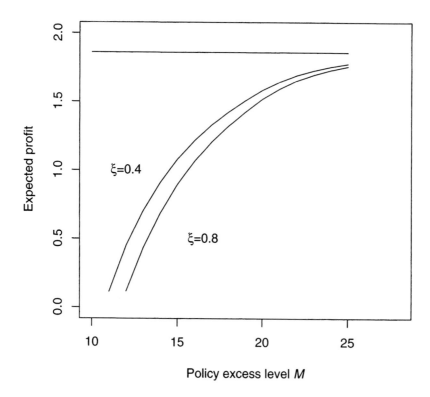

FIGURE 3.8
Expected stop-loss profit for different reinsurance loadings in Example 3.19.

mean, variance and skewness of S for these three models.

3. Assume total annual claims S are modeled by a compound binomial distribution where $N \sim B(200, 0.001)$ and all claims are a constant 500. Determine the mean, variance and skewness of S, and find the probability that S exceeds 600 exactly.

4. Monthly aggregate claims S are modeled by a compound Poisson distribution where N has Poisson parameter 12 and a claim X takes the values $\{1, 5, 10\}$ with respective probabilities $(0.2, 0.3, 0.5)$. Determine the probabilities $P(S = r)$ for $r \leq 200$ and hence find the median, mode, 95th and 99th percentiles of S. What are the mean and variance of S?

5. Total aggregate claims $S = X_1 + \cdots + X_N$ are modeled by a compound

binomial distribution where $N \sim B(4, q = 1/2)$ and $P(X = j) \propto j$ for $j = 1, 2, 3, 4, 5$. Determine the mean, variance, skewness S and find the exact probability distribution for S using Panjer's recursive formula.

6. Consider the collective risk model $S = X_1 + \cdots + X_N$ where $N \sim B(3, q = 2/3)$ (q being the probability of a claim) and the claim size X is uniformly distributed on $\{1, 2, 3\}$. Determine $E(S)$, $Var(S)$ and find $f_S(r)$ for $r = 0, 1, 2, \ldots$.

7. Show that the negative binomial distribution $N \sim NB(k, p)$ satisfies the following recursive property for $n = 1, \ldots, max(N)$.

$$P(N = n) = (\alpha + \beta/n) \, P(N = n - 1)$$

8. S_1 and S_2 are random variables representing total claim amounts in two portfolios, and both can be well modeled by compound Poisson distributions. $S_1 = Y_1 + \cdots + Y_{N_1}$ where N_1 is Poisson with parameter $\lambda_1 = 2$ and the claim size random variable Y has a distribution given by

$$Y = \begin{cases} 200 \text{ prob} = 0.5 \\ 300 \text{ prob} = 0.3 \\ 400 \text{ prob} = 0.2. \end{cases}$$

Similarly, $S_2 = Z_1 + \cdots + Z_{N_2}$ where N_2 is Poisson with parameter $\lambda_2 = 3$ and the claim size random variable Y has a distribution given by

$$Z = \begin{cases} 300 \text{ prob} = 0.1 \\ 400 \text{ prob} = 0.3 \\ 500 \text{ prob} = 0.6. \end{cases}$$

If S_1 and S_2 are independent, what is the probability distribution of $S = S_1 + S_2$? What are its mean, variance and moment generating function? What is $P[S \leq 400]$? Use the recursive formula to find $P[S = r]$ for $r \leq 2000$.

9. In the individual risk model, show that if business increases by a factor of k, then the security loading necessary to ensure premiums meet claims (assuming other things remained constant like claim distribution, claim frequency and degree of confidence) is reduced by the factor $1/\sqrt{k}$.

10. Insurance for accidents is provided to all employees in a large company. Employees are classified into three types for purposes of this insurance. It may be assumed that the claims made in a given year are independent random variables where a maximum of one claim is made annually per person. The claim incidence rate for each class is given below, together with the relevant claim size distribution B_k ($k = 1, 2, 3$) appropriate for each type. B_1 is uniform on $[70, 130]$, B_2 is exponential with mean 150

and B_3 is $\Gamma(2, 0.02)$ (hence has mean 100). The following summarizes characteristics of the situation.

Class type k	Number in class	Claim probability	B_k
1	500	0.02	uniform [70,130]
2	500	0.01	exp. (mean 150)
3	250	0.02	gamma [2,0.02]

It is desired that the probability that total claims exceed the premium income be set at 0.01. If the security loadings for the three class types are to be θ, 2θ and 3θ, respectively, determine θ and the premium for each of the three classes using an individual risk model. What would the appropriate value of θ be if the numbers in each category were doubled?

11. If S is compound Poisson with $\lambda = 1$ and typical claim size X which is exponential with mean 4, what values of $\tau, \alpha,$ and δ would you use to approximate it with a shifted gamma distribution of the form $\tau + \Gamma(\alpha, \delta)$? Determine the exact probability that S exceeds 6 and 8, and compare these with the probabilities found using the shifted gamma and normal approximations to S.

12. A portfolio of 400 insurance policies for house contents (one year) is summarized in Table 3.11 where claim size has been appropriately coded. Note, for example, that there are 280 policies, each of which will give rise to a claim with probability 0.03, and in particular for 160 of these when a claim is made it is equally likely to be anywhere in the interval $[0, 24]$. We are interested in modeling the annual aggregate claims for this portfolio, using both an individual risk model (where aggregate claims are denoted by S^I) and a compound Poisson (collective risk) model (where aggregate claims are denoted by S^C).

(a) Find the mean and variances of the random variables S^I and S^C and comment on their difference. Determine what security loading θ^I (respectively, θ^C) is necessary to be 95% sure premiums exceed claims when using the individual (compound Poisson collective) risk model.

(b) If the numbers of policies in each of the four categories in Table 3.11 were doubled, what security loadings (θ^I and θ^C, respectively) would be necessary to be 95% sure premiums exceed claims?

(c) A reinsurance arrangement has been made whereby the excess of any claim over 36 is handled by the reinsurer. Using the numbers of policies in Table 3.11, find the 95th percentiles of the total amount of claims handled by the reinsurer under the two models.

TABLE 3.11
Policies by incidence and claim distribution.

Incidence	Claim size distribution	
	Uniform [0,24]	Uniform [24,48]
0.03	160	120
0.06	80	40

13. Suppose that in Example 3.12 the volume of business in each company was in fact 10 times larger (that is, $\lambda_A = 2000$ and $\lambda_B = 1000$). Determine the reserves necessary for the separate companies to be 98% sure claims are met by premiums and reserves, and also determine what reserves would be necessary if the companies merged. Comment on the results.

14. If aggregate claims in Example 3.13 were modeled by a compound Poisson distribution instead of a compound binomial, what reserves would be necessary to be 99% sure of meeting claims? Comment on the difference between the two amounts.

15. Total claims S made in respect of a portfolio of fire insurance policies can be modeled by a compound Poisson distribution where the Poisson parameter is λ and the typical claim is X. Let us assume that the claim random variable X is a 40/60 mixture of claims of type I and II, respectively. Claims of type I are Pareto (3, 600), while those of type II are Pareto (4, 900). Calculate $P(X > 400)$, $E(X)$ and $Var(X)$. If the security loading of $\theta = 0.15$ is used for determining premiums and $\lambda = 500$, what reserves are necessary in order to be 99.9% sure of meeting claims? What would be the effect of doubling the security loading?

    Let Y be a Pareto random variable with the same mean and variance as X. What is $P(Y > 400)$? What would be the reserves necessary to be 99.9% sure all claims will be met (from premium income plus reserves) if we had used Y instead of X in our model?

16. Consider a compound Poisson risk model for aggregate claims $S = X_1 + \cdots + X_N$ where N is Poisson with parameter $\lambda = 200$ and the typical claim is exponential with mean 5000. A proportional reinsurance agreement is made whereby the insurer retains 60% of each claim and hence has total risk $S_I = (0.60)S$. Find the mean, variance and skewness of S_I, and compare $Var(S_I) + Var(S_R)$ with $Var(S)$.

17. The aggregate annual claims S is modeled by a compound Poisson distribution where $\lambda = 100$ and the typical claim X is lognormal with $E(X) = 10^4$ and $Var(X) = 3 \cdot 10^8$. Proportional reinsurance is available at a cost of 1.3 per unit of coverage, and the baseline security loading is

$\theta = 0.2$. Determine the maximum value of α which should be considered in order to be 98% confident that reserves of 200,000 plus net premiums meet claims for the baseline insurance company. For this value of α, what is the probability that the net premiums of the reinsurer will meet its claims?

18. In Equation (3.23) we considered an expression for the LEV function of a claim random variable X which was lognormal with mean 900 and standard deviation 300. Its limited expected value function was plotted in Figure 3.5. Suppose now claims have been increased by an inflationary factor of 7%. Plot its limited expected value function and find the value of this function at 500 and 1000.

19. Employees in a factory have subscribed to a group term life insurance arrangement with details for the coming year in Table 3.12. It is possible to arrange (claim-based) excess of loss reinsurance on this group whereby the reinsurer pays the excess of any claim over M, for some agreed value in the interval [100,000, 150,000]. Reinsurance is available at a cost of $(1 + \xi) = 1.4$ per unit of coverage. If S_I^M represents the amount of claims payable by the insurer, P_R^M is the reinsurance premium and M is the excess level, find the value of M which minimizes the probability $P(S_I^M + P_R^M > 2,500,000)$.

TABLE 3.12
Term life insurance for factory employees.

Amount insured	Number of employees	Claim probability
25,000	2000	0.0030
50,000	2500	0.0025
100,000	1500	0.0040
150,000	1000	0.0050

20. Aggregate claims are being modeled by a compound distribution of the form $S = X_1 + \cdots + X_N$ where $X \sim$ exponential. If excess of loss reinsurance with excess level M is available from a reinsurer (at a cost of $(1+\xi)$ per unit cover where θ is the loading factor used by the insurer on policyholders), show that the minimum excess level which should be considered is given by $M^* = E(X) \log \xi / \theta$. Construct a table similar to Table 3.10 for M^* when X has mean 600.

21. Assume that aggregate claims are modeled by a compound Poisson process and that the excess of any claim over M is handled by a reinsurer who uses a security loading ξ (while the insurance company uses a loading of θ on policy holders). The typical claim X has a Pareto distribution

with parameters (β, δ), that is

$$f_X(x) = \frac{\beta \delta^\beta}{(\delta + x)^{\beta+1}}.$$

Assume that the annual expected number of claims in this process is $\lambda = 300, \beta = 3, \delta = 1200, \theta = 0.2$ and $\xi = 0.3$. Determine the minimum excess level M^* which may be considered by the insurance company if it is desired that expected net profit is nonnegative, and complete the following table for a relationship between possible values of M and expected annual net profit.

Retention limit M	Expected annual profit
300	⋆
800	⋆
⋆	28,406.25

22. A portfolio of 200 one-year fire insurance policies is summarized below:

	Claim size distribution	
	Uniform [0,48]	Uniform [48,96]
Claim incidence 0.02	80	60
0.04	40	20

One can see for example that there are 140 policies where the chance of a claim being made is 0.02, and for 80 of these if a claim is made it is equally likely to be anywhere in the interval $[0, 48]$.

(a) Let S denote total claims from this portfolio during the year. Using a compound Poisson distribution to model S, determine the security factor θ which is necessary to be 95% sure premiums exceed claims.

(b) If the number of policies were to triple in each of the four categories, what would be the necessary security loading?

(c) An agreement is made with a reinsurer to handle the excess of any claim over 72. If the reinsurer uses a security factor of $\xi = 0.7$ on premiums, what should it charge the insurance company for this arrangement?

23. In Example 3.11 total claims S arising from accidents of employees in a large factory were modeled by an individual risk model with mean $E(S) = 33,000$ and variance $Var(S) = 35{,}151{,}000$. Approximating this distribution by a normal distribution with the same mean and variance, plot the limited expected value function $L_S(M)$. The insurance company is presently using a security loading of $\theta = 0.37$, and is considering

stop-loss reinsurance with stop-loss level M. Determine the minimal values of the stop-loss level M which should be considered to ensure expected profits are nonnegative when the stop-loss premium loading ξ of the reinsurer is both 0.5 and 0.7.

24. Plot the limited expected value function for the compound Poisson distribution studied in Example 3.6. Consider a stop-loss reinsurance treaty at stop-loss level M where the security loadings for the insurer and reinsurer are, respectively, $\theta = 0.2$ and ξ. Plot the expected profit function $E(P_\$^M)$ as a function of M for the insurer when the security loading ξ for the reinsurer is both 0.3 and 0.5.

25. A motor insurance company sells two types of policies. Claims of the first type arise as a Poisson process with parameter λ_1, and those of the second from an independent Poisson process with parameter λ_2. The (annual) aggregate claim amounts on the respective policy types are denoted S_1 and S_2, and we let $S = S_1 + S_2$.

The insurance company sells 800 policies in total, 200 of type 1 and 600 of type 2. Claims arise on each policy of type 1 at a rate of one claim per 10 years and on those of type 2 at a rate of one claim per 20 years. The distributions of the two types of claims are given by:

$$\text{Type 1} = \begin{cases} \$1000 & \text{prob} = 0.4 \\ \$1500 & \text{prob} = 0.2 \\ \$2500 & \text{prob} = 0.4 \end{cases}$$

$$\text{Type 2} = \begin{cases} \$1500 & \text{prob} = 0.2 \\ \$2000 & \text{prob} = 0.8 \end{cases}$$

(a) Compute $E(S)$, $Var(S)$, $skew(S)$, and the moment generating function of S.

(b) Given that the insurance company uses a security loading of $\theta = 0.15$ and the normal distribution as an approximation to the distribution of S, find the initial reserve required to be 99% sure that premiums plus reserves will cover claims.

(c) The insurance company decides to buy reinsurance cover with aggregate retention $\$50,000$, so that the insurance company pays no more than this amount in claims each year. In the year following the inception of this reinsurance, the numbers of policies in each of the two groups remain the same but, because of changes in the motor insurance contracts, the probability of a claim of type 2 falls to zero. Using the normal distribution as an approximation to the distribution of S, calculate the probability of a claim being made on the reinsurance treaty.

26. Show that if X has the Pareto distribution with parameters β and δ, then its limited expected value function takes the form

$$L_X(M) = \frac{\delta}{\beta - 1}\left[1 - \left(\frac{\delta}{\delta + M}\right)^{\beta - 1}\right].$$

4

Ruin Theory

4.1 The probability of ruin in a surplus process

If the expression $U(t)$ represents the net value of a portfolio of risks or policies at time t, then one would certainly be interested in studying the possible behavior of $U(t)$ over time. In a technical sense, we might say that ruin occurs if at some point t in the future the net value of the portfolio becomes negative. The probability of this event is called the *probability of ruin*, and it is often used as a measure of security for a portfolio. $U(t)$ will take into account relatively predictable figures such as initial reserves U and premium income up to time t, but it also must take account of claim payments that are more random in nature as well as being harder to predict.

We study stochastic models of the so-called surplus process $\{U(t)\}_t$, which represents the surplus or net value of a portfolio of policies throughout time. Although in most cases it is not possible to give an explicit expression for the probability of ruin of a surplus process, an inequality of Lundberg [40] provides a useful upper bound. A technical term known as the *adjustment coefficient* provides an alternative and useful surrogate measure of security for a surplus process. In many situations, simulation can be a useful tool in estimating the probability of ruin. In this chapter, we investigate how the probability of ruin in a surplus process (in both finite and infinite time) is affected by factors such as the premium rate c, the initial reserves U, a typical claim X, the claim arrival rate λ, and various levels and types of reinsurance.

4.2 Surplus and aggregate claims processes

We study the collective risk model over time, taking into account initial reserves U, incoming premiums, and the aggregate claims that are made on a portfolio or collection of policies. In our basic model, we will assume that premium payments are coming into the company at a constant rate of c per unit time. For any given point in time t, we let $S(t)$ be the aggregated claims up to time t. Hence if U is the amount of initial reserves, then the surplus (or balance) $U(t)$ at time t is given by

$$U(t) = U + c \cdot t - S(t).$$

We call $\{U(t)\}_t$ the surplus process, and $\{S(t)\}_t$ is the aggregate claims process where

$$S(t) = \sum_{i=1}^{N(t)} X_i,$$

$N(t)$ is the number of claims made in the interval $(0, t]$, and the $X_i, i = 1, \ldots,$ are independent identically distributed claim random variables.

There are several characteristics of the surplus process $\{U(t)\}_t$ that are naturally of interest to us:

- $T = min\{t : t > 0, U(t) < 0\}$. T is a random variable (which may be infinite), and is called the *time of ruin*.

- $\psi(U)$ is the probability that the time of ruin T is in fact finite when the initial reserves are U.

- $\psi(U, t)$ is the probability of ruin at some point in the time interval $(0, t]$, given initial reserves of U.

In practice, for a given surplus process, the probability of ruin $\psi(U, t)$ in the specified time interval $(0, t]$ can be a useful indicator of the security of the process and it can often be approximated through simulation. However, $\psi(U)$ is often more tractable in a mathematical sense. Clearly, $\psi(U, t)$ is increasing in t, and $\lim_{t \to +\infty} \psi(U, t) = \psi(U)$. When the counting process $\{N(t)\}_t$ for the number of claims is Poisson, then it may be shown that

$$\psi(U) = \frac{e^{-RU}}{E(e^{-RU(T)} \mid T < +\infty)}, \tag{4.1}$$

where the adjustment (or Lundberg's) coefficient R is the unique positive solution to $\lambda M_X(r) - \lambda - cr = 0$. The expression for $\psi(U)$ given by (4.1) is unfortunately not easy to determine (see [21]), and in any case is of limited practical use.

Figure 4.1 gives a possible realization of a surplus process $U(t)$ where the initial reserves are $U = U(0) = 4$, $c = 1.1$, $N(8) = 5$, the times of the claims in the interval $[0, 8]$ are given by

$$\mathbf{T} = (T_1, T_2, T_3, T_4, T_5) = (1, 1.5, 4, 5, 5.6),$$

and the corresponding claim sizes are

$$\mathbf{X} = (X_1, X_2, X_3, X_4, X_5) = (3.1, 1.05, 2.4, 2.1, 3.06).$$

Ruin occurs at time $T = T_5$!

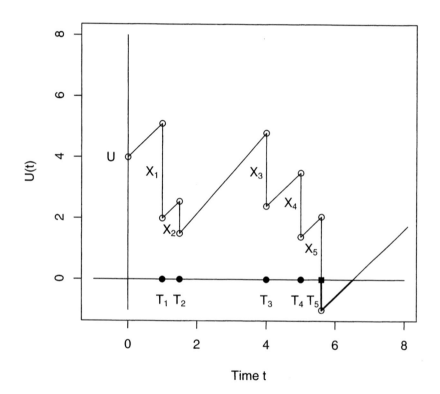

FIGURE 4.1
Realization of a continuous surplus process $U(t)$.

4.2.1 Probability of ruin in discrete time

Often it is only possible to check the status of a surplus process at discrete periods of time. For example, we might want to check a surplus process every 10 minutes, but in other situations we might be interested in observing it only every hour, day, month or even year. Suppose that we are interested in observing a surplus process at times that are multiples of some $h > 0$. Then we define the probability of ruin $\psi_h(U)$ by

$$\psi_h(U) = P\left[\, U(j) < 0 \text{ for some } j = h, 2h, \ldots \,\right],$$

and similarly, $\psi_h(U, t)$ is defined to be the probability of ruin for some j where $j \leq t$. It should be clear that the more often we observe a surplus process, the more often we are likely to observe ruin. In other words, for any integer $k > 1$,

$$\psi_{kh}(U) \leq \psi_h(U) \leq \psi(U).$$

In Figure 4.2, we can see a realization of a surplus process where ruin occurs between (months) 4 and 5. This would be noted if the state of the process were observed every month, but not so if it were observed only every two months (thus indicating why $\psi_{2h}(U) \leq \psi_h(U)$).

4.2.2 Poisson surplus processes

One of the most basic surplus processes is the Poisson surplus process which occurs when the counting process $\{N(t)\}_t$ for claims is a Poisson process. In this case, the aggregate claims process $\{S(t) = \sum_1^{N(t)} X_i\}_t$ is called a *compound (aggregate claims)* Poisson process. Note that when we study processes like $\{N(t)\}_t$, $\{S(t)\}_t$, or $\{U(t)\}_t$, we are inherently interested in a range or possibly all points in time t, hence it is an infinite number of random variables with which we are concerned. If we focus attention on a particular point in time, say t_0, then $N(t_0)$ has a Poisson distribution and $S(t_0)$ has a compound Poisson distribution.

Example 4.1

Let c be the rate of premium income per year in a compound Poisson surplus process where $c = \lambda(1 + \theta)E(X)$, $\lambda = 20, \theta = 0.2, U = 2000$ and the typical claim X is exponential with mean 500. The random variable $U(3)$ represents the surplus at the end of three years, and takes the form

$$U(3) = U + c \cdot 3 - S(3) = 2000 + 20(1.2)\,500 \cdot 3 - \sum_1^{N(3)} X_i.$$

$S(3)$ has a compound Poisson distribution where

$$E[U(3)] = 2000 + 36{,}000 - 3(20)E(X) = 8000 \text{ and}$$

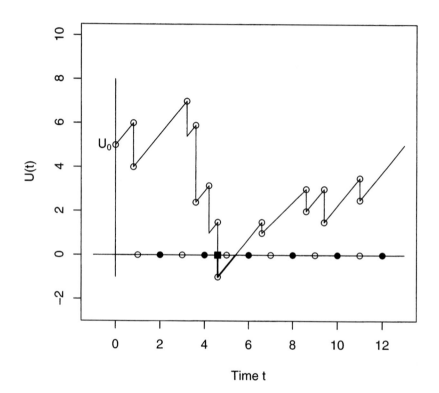

FIGURE 4.2
Realization of a discrete surplus process $U(t)$.

$$Var[U(3)] = Var[S(3)] \;=\; 3\lambda E(X^2) \;=\; 3(20)2(500^2)$$
$$= 30{,}000{,}000 = (5477.226)^2.$$

Since $\{S(t)\}_t$ is a compound Poisson process, it follows that for any $t > 0$, $Var[U(t+3) - U(t)] = (5477.226)^2$ while $E[U(t+3) - U(t)] = 6000$. ⬜

4.3   Probability of ruin and the adjustment coefficient

Most random variables X possess a moment generating function $M_X(r) = E(e^{rX})$ defined in a neighborhood of 0, although the Cauchy distribution (or t-distribution with 1 degree of freedom) with density function $f(x) = 1/[\pi(1+x^2)]$ is a classic example of a fat-tailed distribution that does not! In what follows we shall assume that X has a moment generating function and that there exists a γ_X (which may be positive or $+\infty$) such that

$$\gamma_X = sup\{r : M_X(r) < +\infty\} \quad \text{and} \quad \lim_{r \to \gamma_X^-} M_X(r) = +\infty. \qquad (4.2)$$

Here γ_X is the *sup* (or *supremum*) of all values of r for which $M_X(r)$ exists, and $r \to \gamma_X^-$ represents convergence to γ_X from the left.

  If, for example, X has a gamma distribution with moment generating function $M_X(r) = (\beta/(\beta - r))^\delta$, then $\gamma_X = \beta$. If $X \sim N(\mu, \sigma^2)$, then $M_X(r) = e^{\mu r + \sigma^2 r^2/2}$, and hence $\gamma_X = +\infty$. The following technical lemma is useful in establishing the existence of the so-called *adjustment coefficient* for surplus processes with claim random variable X.

LEMMA 4.1
Let $X \geq 0$ be a claim random variable where $\gamma_X > 0$. Then for any numbers $\lambda, c > 0$,

$$\lim_{r \to \gamma_X^-} [\lambda M_X(r) - cr] = +\infty.$$

PROOF   If $\gamma_X < +\infty$, then the Lemma is clearly true by Equation (4.2). If $\gamma_X = +\infty$, then one may find $a > 0$ such that $P(X \geq a) = b > 0$. Hence $M_X(r) = E(e^{rX}) \geq e^{ra}b$, and therefore

$$\lim_{r \to +\infty} [\lambda M_X(r) - cr] \;\geq\; \lim_{r \to +\infty} [\lambda e^{ra} b - cr] = +\infty.$$

∎

4.3.1   The adjustment equation

The rate of premium income per unit time c can be modeled in many ways, but since it is reasonable that premium income should exceed expected claim payments (or the pure premium) per unit time, we normally assume in a Poisson surplus process that $c > \lambda E(X)$. A simple model is where $c = (1 + \theta)\lambda E(X)$, where θ is interpreted as a security or loading factor on premiums. Given a (Poisson) surplus process with parameters c, λ and a claim distribution X, we define the *adjustment function* to be $A(r) = \lambda M_X(r) - \lambda - cr$ and the adjustment equation by

$$A(r) = \lambda M_X(r) - \lambda - cr = 0. \tag{4.3}$$

Note that the function $A(r)$ has the following properties:

- $A(0) = \lambda M_X(0) - \lambda - c \cdot 0 = 0$ ($r = 0$ is always a root of $A(r) = 0$).
- $A'(r) = \lambda M_X'(r) - c$, and in particular $A'(0) = \lambda E(X) - c < 0$.
- $A''(r) = \lambda M_X''(r) = \lambda \int_0^\infty x^2 e^{rx} f_X(x)dx > 0$, and hence A is convex.
- $\lim_{r \to \gamma_X^-}[\lambda M_X(r) - \lambda - cr] = +\infty$.

Therefore it follows that $A(r)$ as a function of r on $[0, \gamma_X)$ is convex, initially 0 and decreasing, and then increasing to $+\infty$. In particular, it will have a unique positive root R, which is defined to be the *adjustment coefficient* for the surplus process.

Note that in the simple model where the premium income is a multiple of the claim rate λ, the adjustment coefficient R is independent of λ. Some insight into why this is the case for Poisson processes will be given later. In most situations, one would use numerical methods to find the adjustment coefficient. Figure 4.3 is a graph of the adjustment function $A(r)$ for the Poisson surplus process where $X \sim \Gamma(2, 0.01)$, $\lambda = 30$, and $c = (1 + 0.2)\lambda E(X) = 7200$. Therefore by solving

$$A(r) = 30 \left(\frac{0.01}{0.01 - r} \right)^2 - 30 - 7200\, r = 0,$$

one may determine that the adjustment coefficient $R = 0.001134$.

If the claim size distribution X in a (Poisson) surplus process is exponential, then one may solve explicitly for the adjustment coefficient R. If X is exponentially distributed with parameter β (that is, $E(X) = 1/\beta$) then

$$A(r) = \lambda \frac{\beta}{\beta - r} - \lambda - cr = 0 \tag{4.4}$$

has roots $r = 0$ and $r = \beta - \lambda/c$, and hence when $c = \lambda(1 + \theta)/\beta$, the adjustment coefficient takes the form

$$R = \frac{\beta\theta}{1 + \theta} = \beta\theta/(1 + \theta). \tag{4.5}$$

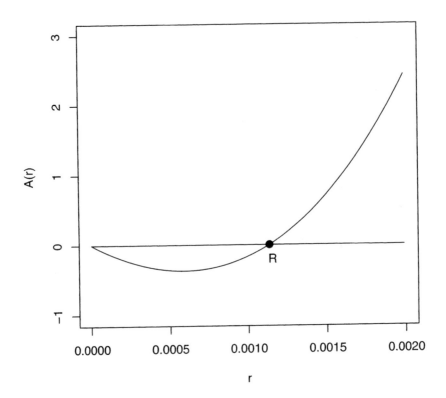

FIGURE 4.3
Plot of the adjustment function $A(r) = 30[1/(1 - 100r)^2 - 1 - 240r]$.

Note that by definition of the adjustment coefficient R,

$$\lambda + cR = \lambda M_X(R)$$
$$= \lambda \int e^{Rx} f_X(x)\, dx$$
$$\geq \lambda \int_0^{+\infty} \left[1 + Rx + \frac{R^2 x^2}{2} \right] f_X(x)\, dx$$
$$= \lambda \left[1 + R E(X) + \frac{R^2 E(X^2)}{2} \right]$$
$$\Rightarrow R \leq 2\, \frac{c - \lambda E(X)}{\lambda E(X^2)}$$

$$= \frac{2\theta E(X)}{E(X^2)}.$$

This provides a useful upper bound (independent of λ) for the adjustment coefficient R. In fact, $2\theta E(X)/E(X^2)$ often serves as a good approximation to R when R is small.

Example 4.2

The typical claim in a Poisson surplus process is modeled by a lognormal random variable X where $\log X \sim N(\mu = 8.749266, \sigma^2 = 0.363535)$. If a premium loading of $\theta = 0.15$ is used, then an upper bound for the adjustment coefficient is given by

$$R \leq R_0 = 2(0.15) \frac{e^{\mu + \frac{\sigma^2}{2}}}{e^{2\mu + 2\sigma^2}} = 0.000028.$$

▯

4.3.1.1 The Newton–Raphson method

The Newton–Raphson method is a basic technique in numerical analysis for finding roots of an equation, and it can often be useful in finding the adjustment coefficient for a surplus process. Let us suppose that we are trying to solve $A(r) = 0$, where A is a differentiable function and we have a reasonable first approximation R_0 to a zero R of the function A. The basic idea behind the Newton–Raphson method is that the tangent line at $(R_0, A(R_0))$ should be a good local approximation to $A(r)$ near R_0, and hence we can probably get an even better approximation to R by finding the point R_1 where this tangent line crosses the r axis. In other words, solving

$$\frac{A(R_0) - 0}{R_0 - R_1} = A'(R_0)$$

for R_1, we obtain $R_1 = R_0 - A(R_0)/A'(R_0)$, which is the second approximation to R.

Proceeding in this way, we may obtain a sequence of approximations $R_k, k = 1, \ldots,$ given by $R_k = R_{k-1} - A(R_{k-1})/A'(R_{k-1})$, which often converges quickly to R.

Example 4.3

Consider a Poisson surplus process where $\lambda = 50$, $\theta = 0.20$ and claims are constant with value 25. Then the adjustment equation takes the form

$$A(r) = 50 M_X(r) - 50 - 50(1 + \theta)E(X)\, r = 50 \left[e^{25r} - 1 - 30r \right] = 0.$$

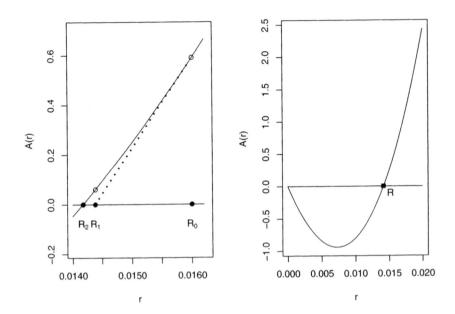

FIGURE 4.4
Newton–Raphson method for finding R in Example 4.3.

A first approximation (and upper bound) to the adjustment coefficient R is given by $R_0 = 2\theta E(X)/E(X^2) = 2(0.2)25/25^2 = 0.016$. Using the Newton–Raphson method, a second approximation is given by

$$R_1 = 0.016 + \frac{-50\left[e^{25(0.016)} - 1 - 30(0.016)\right]}{50\left[25\,e^{25(0.016)} - 30\right]} = 0.014379.$$

Similarly, $R_2 = 0.014171$, which is already very close to the actual value of $R = 0.014168$. The left plot of Figure 4.4 graphs $A(r)$ with the tangent lines determining R_1 and R_2, while the plot on the right gives a more global view.
□

4.3.2   Lundberg's bound on the probability of ruin $\psi(U)$

It will only now become apparent why the adjustment coefficient R for a surplus process is of real interest. The adjustment coefficient R for a Poisson surplus process is in fact very useful in giving an upper bound to the probability of ruin $\psi(U)$ due to a classic inequality of Lundberg [40].

THEOREM 4.1

*If R is the adjustment coefficient in a Poisson surplus process with initial
reserves U, then an upper bound on the probability of ruin is given by e^{-RU}.*

This upper bound e^{-RU} for the probability of ruin is often referred to as
Lundberg's bound. The proof of this result is a nice exercise in using the
principle of induction.

PROOF   Now

$$\psi(U) = \lim_{n \to +\infty} {}_n\psi(U),$$

where ${}_n\psi(U)$ for $n = 1, \ldots$, is the probability of ruin for the process on
or before the occurrence of the nth claim. By the principle of induction, it
therefore suffices to show that

- ${}_1\psi(U) \le e^{-RU}$ for all $U > 0$, and

- for any $n \ge 1$, ${}_n\psi(U) \le e^{-RU}$ for all $U > 0$ implies that

$${}_{n+1}\psi(U) \le e^{-RU} \text{ for all } U > 0.$$

Now

$$
\begin{aligned}
{}_1\psi(U) &= \int_0^\infty P[X_1 > U + ct \mid T_1 = t]\, \lambda e^{-\lambda t} dt \\
&\le \int_0^{+\infty} \left[\int_{U+ct}^{+\infty} e^{-R(U+ct-x)} f_X(x)dx \right] \lambda e^{-\lambda t} dt
\end{aligned}
$$

$$[\text{ since } x > U + ct \Rightarrow R(U + ct - x) < 0]$$

$$
\begin{aligned}
&\le e^{-RU} \int_0^{+\infty} \left[\int_0^{+\infty} e^{-R(ct-x)} f_X(x)dx \right] \lambda e^{-\lambda t} dt \\
&= e^{-RU} \int_0^{+\infty} \left[\int_0^{+\infty} e^{Rx} f_X(x)dx \right] \lambda e^{-(\lambda+cR)t} dt \\
&= e^{-RU} \int_0^{+\infty} \lambda M_X(R)\, e^{-\lambda M_X(R)\, t}\, dt
\end{aligned}
$$

$$[\text{ since } \lambda + cR = \lambda M_X(R)]$$

$$= e^{-RU}.$$

Next, let us assume that ${}_n\psi(U) \le e^{-RU}$ for all $U > 0$. Then

$${}_{n+1}\psi(U) = P(\text{ruin on or before } (n+1)^{\text{st}} \text{ claim})$$

$$= P(\text{ruin on } 1^{\text{st}} \text{ claim}) +$$
$$P(\text{ruin on or before } (n+1)^{\text{st}}, \text{ but not on } 1^{\text{st}})$$

$$= \int_0^{+\infty} [P(X_1 > U + ct \mid T_1 = t)$$
$$+ P(X_1 < U + ct, \text{ruin on or before } (n+1)^{\text{st}} \mid T_1 = t)] \lambda e^{-\lambda t} dt$$

[time T_1 to the 1st claim is exponential with parameter λ]

$$= \int_0^{+\infty} \left[\int_{U+ct}^{\infty} f_X(x)\, dx \right] \lambda e^{-\lambda t} dt$$
$$+ \int_0^{+\infty} \left[\int_0^{U+ct} {}_n\psi(U + ct - x) f_X(x)\, dx \right] \lambda e^{-\lambda t} dt$$

[by induction ${}_n\psi(U + ct - x) \le e^{-R(U+ct-x)}$]

$$\le \int_0^{+\infty} \left[\int_{U+ct}^{\infty} f_X(x)\, dx \right] \lambda e^{-\lambda t} dt$$
$$+ \int_0^{+\infty} \left[\int_0^{U+ct} e^{-R(U+ct-x)} f_X(x)\, dx \right] \lambda e^{-\lambda t} dt$$

$$\le \int_0^{+\infty} \left[\int_0^{\infty} e^{-R(U+ct-x)} f_X(x)\, dx \right] \lambda e^{-\lambda t} dt$$

$$= e^{-RU} \int_0^{\infty} \lambda e^{-(\lambda+cR)t} \left[\int_0^{\infty} e^{Rx} f_X(x)\, dx \right] dt$$

$$= e^{-RU} \int_0^{\infty} \lambda M_X(R)\, e^{-\lambda M_X(R)\, t} dt$$

$$= e^{-RU}.$$

4.3.3 The probability of ruin when claims are exponentially distributed

In general, it is difficult to obtain an explicit and useful expression for the probability of ultimate ruin $\psi(U)$ for a surplus process with initial reserves U. However, in the case of a Poisson surplus process with exponentially distributed claims, one may show that the probability of ruin has the form

$$\psi(U) = \frac{1}{1+\theta} e^{-\beta\theta U/(1+\theta)}. \tag{4.6}$$

Here λ is the rate of claims, θ is the premium loading, the initial reserves are U, and the claim size random variable X is exponential with mean $1/\beta$. A derivation of this result is given in Subsection 4.3.3.2. The following observations about this probability of ruin should be noted:

- The probability of ruin clearly does not depend on λ, the rate at which claims are made per unit time. This initially may seem surprising. In order to gain some insight into why this is the case, consider two Poisson surplus processes $\{U_1(t)\}_t$ and $\{U_2(t)\}_t$ with exponential claims that are identical except that the claim rate λ_1 for the first process is ten times the claim rate λ_2 of the second ($\lambda_1 = 10\lambda_2$). In theory, the only difference between the two processes is that things are happening ten times faster in the first process. There is a natural $1-1$ correspondence between realizations in the two processes, where corresponding to any realization in the second process is the (telescoped) realization in the first, which is identical except that it proceeds at 10 times the rate. In particular, any realization in the second that results in ruin at time T corresponds naturally to a realization in the first where ruin occurs at $T/10$.

- When the security loading θ on premiums is 0, then $\psi(U) = 1$ and ruin is certain. This is not totally unexpected, since in this case we are only collecting in premiums what we expect to pay in claims, and no matter how much we are holding in reserves U, random fluctuations in claims will inevitably lead to ruin.

- $\psi(U)$ is a decreasing function of β. Therefore as the mean ($1/\beta$) of the exponential claim distribution increases (that is, β decreases), the probability of ruin increases when other parameters are held fixed.

- When claims are exponential the adjustment coefficient takes the form $R = \beta\theta/(1+\theta)$, and therefore

$$\psi(U) = \frac{1}{1+\theta}\, e^{-RU} \leq e^{-RU} \quad (= \text{Lundberg upper bound}),$$

and the probability of ruin (as a function of U) is proportional to the Lundberg upper bound.

- The probability of ruin $\psi(U)$ is a decreasing function of U since

$$\frac{\partial}{\partial U}\psi(U) = \frac{-\beta\theta}{1+\theta}\,\psi(U) < 0,$$

and therefore when other parameters are held fixed, the probability of ruin decreases with increasing initial reserves.

- The probability of ruin $\psi(U)$ is a decreasing function of θ since

$$\frac{\partial}{\partial\theta}\psi(U) = -\,\frac{1+\theta+\beta U}{(1+\theta)^2}\,\psi(U) < 0,$$

and therefore when other parameters are held fixed, the probability of ruin decreases as the loading which is put on premiums increases.

   In the following example of three Poisson surplus processes, only the claim size distribution varies, and for each process the adjustment coefficient R is calculated or estimated.

Example 4.4

The surplus process for a risk is modeled by a Poisson surplus process where the security loading for premiums is 0.2 and the Poisson parameter is 50. We determine the adjustment coefficient, R, for the surplus process in each of the following situations (where X denotes the claim random variable).

1. X_1 is exponential with mean $5 = 1/\beta$. Then the adjustment coefficient is $R = \beta\theta/(1+\theta) = 0.2/6 = 1/30 = 0.033333$.

2. $X_2 \sim \Gamma(\beta = 2, \delta = 0.4)$. The adjustment equation takes the form

$$\lambda \left[\left(\frac{0.4}{0.4 - r} \right)^2 - 1 - 6r \right] = 0.$$

   Solving the quadratic equation $150r^2 - 95r + 4 = 0$, one finds that R is either 0.045353 or 0.587980. It must be the former since the adjustment equation (and the moment generating function of X_2) is only defined for $r < 0.4$.

3. $X_3 \sim N(5, 1^2)$. Here the adjustment coefficient R is the unique positive solution to
$$g(r) = A(r)/\lambda = e^{5r + r^2/2} - 1 - 6r = 0.$$

   We know that an upper bound for R is given by

$$R_0 = \frac{2\theta E(X_3)}{E(X_3^2)} = \frac{2(0.2)5}{1 + 5^2} = 0.076923.$$

   Using this as an initial approximation to R and applying the Newton–Raphson method one obtains

$$R_1 = R_0 - \frac{g(R_0)}{g'(R_0)} = 0.068909$$

   and ultimately, that $R = 0.067824$.

The adjustment coefficient R is a measure of risk, and since $\psi(U) \le e^{-RU}$, larger values of R correspond to smaller values of the Lundberg upper bound. Note that although the mean claim size is equal to 5 in each case, one has that $Var(X_1) > Var(X_2) > Var(X_3)$. Hence it is not surprising that more volatility in claim size leads to more risk and corresponding lower values of R. ▯

4.3.3.1 Probability of ruin in finite time

Although Equation (4.6) provides a neat expression for the probability of (eventual) ruin in a Poisson surplus process when claims are exponentially distributed, there is no such expression for the probability of ruin $\psi(U, t)$ in finite time in this situation. However, in this case (and indeed in many such processes) simulation can be useful and informative in estimating $\psi(U, t)$. Consider a Poisson surplus process where $\lambda = 1, \theta = 0.1$, X is exponential with mean 10, and initial reserves are either $U = 50$ or 100. Figure 4.5 gives the result of a simulation exercise carried out in R to evaluate both $\psi(50, t)$ and $\psi(100, t)$ for this process. In each case, 5000 realizations of the process were simulated where the time to ruin T (which in many cases would be in excess of some cutoff point – in this situation, the cutoff was chosen to be 1000) was determined. Then using the procedure (ecdf) for the empirical distribution function of a random variable, the results for $\psi(U, t)$ were plotted on the interval $[0, 400]$. For each plot the upper dotted lines give the Lundberg upper bounds for the probabilities of ruin $(e^{-R\,50} = 0.634736$ and $e^{-R\,100} = 0.402890)$, while the lower dashed lines give the respective probabilities of eventual ruin $\psi(50) = 0.577033$ and $\psi(100) = 0.366264$.

The following R code was used (5000 times) to obtain realizations of the time to ruin (RT) in the process when $U = 50$.

```
EX<-1/0.1 U0<-50 theta<-0.1 lambda<-1 R<-rep(1,5000) for (i
in1:5000) { wait<-rexp(1000,rate=1) T<-rep(0,1000) for (j in 1:1000)
T[j]=sum(wait[1:j]) claim<-rexp(1000,rate=1/EX) U<-rep(0,1000) for
(j in 1:1000) U[j]=U0+1*1.1*10*T[j]-sum(claim[1:j]) a<-1:length(U)
if (length(U[U<0])>0) RT<-T[min(a[U<0])] else RT<-1000 R[i]<-RT }
```

4.3.3.2 Derivation of the probability of ruin when claims are exponentially distributed

We derive the explicit expression for the probability of ruin given by (4.6) using a classic integral-differential equation approach. Another interesting development using Laplace transforms is given in Dickson [20], which also provides an extensive treatment of risk and ruin in general.

We use $\bar{\psi}(U) = 1 - \psi(U)$ to denote the surplus process *survival probability* (the probability that ruin does not occur) with initial reserves U. Assume that we have a Poisson surplus process whereby the time T to the first claim is exponentially distributed with parameter λ. In deriving an expression for $\bar{\psi}(U)$ we will condition on (or average over) the time of the first claim $T = t$ and assume that this first claim X does not cause ruin (hence $X = x < ct+U$). If this happens, then the probability of the process surviving after time t is $\bar{\psi}(U + ct - x))$, and hence we may write

$$\bar{\psi}(U) = \int_0^\infty \left[\int_0^{U+ct} f_X(x)\,\bar{\psi}(U + ct - x)\,dx \right] \lambda\,e^{-\lambda t}dt$$

(letting $w = U + ct$)

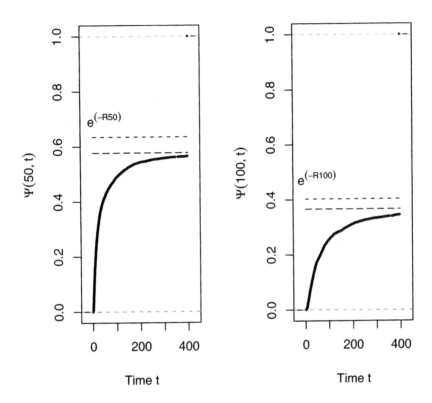

FIGURE 4.5
Probability of ruin in finite time when X is exponential with mean 10, $\lambda = 1$, and $\theta = 0.1$.

$$= \int_U^\infty \left[\int_0^w f_X(x) \, \bar{\psi}(w - x) \, dx \right] \frac{\lambda}{c} e^{-\lambda(w-U)/c} \, dw$$

$$= \frac{\lambda}{c} e^{\lambda U/c} \int_U^\infty \left[\int_0^w f_X(x) \bar{\psi}(w - x) dx \right] e^{-\lambda w/c} \, dw. \qquad (4.7)$$

Using $\bar{\psi}'(U) = \partial \bar{\psi}(U)/\partial U$ to denote the partial derivative with respect to U, one has on differentiating expression (4.7)

$$\bar{\psi}'(U) = \left(\frac{\lambda}{c} \right)^2 e^{\lambda U/c} \int_U^\infty \left[\int_0^w f_X(x) \bar{\psi}(w - x) \, dx \right] e^{-\lambda w/c} \, dw$$

$$+ \frac{\lambda}{c} e^{\lambda U/c} \left(-e^{-\lambda U/c} \right) \int_0^U f_X(x) \, \bar{\psi}(U - x) \, dx$$

$$= \frac{\lambda}{c} \bar{\psi}(U) - \frac{\lambda}{c} \int_0^U f_X(x) \, \bar{\psi}(U - x) \, dx. \qquad (4.8)$$

Since $\bar{\psi}(U) = 1 - \psi(U)$, it follows from Equation (4.8) that

$$\psi'(U) = - \left[\frac{\lambda}{c} (1 - \psi(U)) - \frac{\lambda}{c} \int_0^U f_X(x) \left[1 - \psi(U - x) \right] dx \right]$$

$$= \frac{\lambda}{c} \psi(U) - \frac{\lambda}{c} \int_0^U f_X(x) \, \psi(U - x) \, dx - \frac{\lambda}{c} \bar{F}_X(U). \qquad (4.9)$$

Lundberg's inequality ensures that $\lim_{U \to \infty} \psi(U) = 0$, and hence we may integrate expression (4.9) as a function of U yielding

$$\int_0^\infty \psi'(U) \, dU = \frac{\lambda}{c} \left[\int_0^\infty \psi(U) - \int_0^\infty \int_0^U f_X(x) \psi(U - x) \, dx \right] dU \quad (4.10)$$

$$- \frac{\lambda}{c} \int_0^\infty \bar{F}_X(U) \, dU.$$

The two expressions on the right-hand side of Equation (4.10) are equal since by changing the order of integration

$$\int_0^\infty \left[\int_0^U f_X(x) \psi(U - x) \, dx \right] dU = \int_0^\infty \left[\int_x^\infty \psi(U - x) \, dU \right] f_X(x) \, dx$$

$$\text{(letting } w = U - x)$$

$$= \int_0^\infty \left[\int_0^\infty \psi(w) \, dw \right] f_X(x) \, dx$$

$$\text{(letting } U = w)$$

$$= \int_0^\infty \left[\int_0^\infty \psi(U) \, dU \right] f_X(x) \, dx$$

$$= \int_0^\infty \psi(U) \, dU.$$

Therefore (again using Lundberg's inequality) one may conclude that

$$\psi(0) = - \int_0^\infty \psi'(U) \, dU$$

$$= \frac{\lambda}{c} \int_0^\infty \bar{F}_X(x) \, dx$$

$$= \frac{\lambda E(X)}{c}$$

$$= \frac{1}{1+\theta} \quad (\text{when } c = (1+\theta)\,\lambda\,E(X)).$$

Suppose now that the claim size distribution for X is exponential with mean $1/\beta$. The expression for $\psi'(U)$ given in (4.9) becomes

$$\psi'(U) = \frac{\lambda}{c}\,\psi(U) - \frac{\beta\lambda}{c}\int_0^U e^{-\beta x}\,\psi(U-x)\,dx - \frac{\lambda}{c}e^{-\beta U}$$

$$= \frac{\lambda}{c}\,\psi(U) - \frac{\beta\lambda}{c}e^{-\beta U}\int_0^U e^{\beta x}\,\psi(x)\,dx - \frac{\lambda}{c}e^{-\beta U} \qquad (4.11)$$

and hence

$$\psi''(U) = \frac{\lambda}{c}\,\psi'(U) + \frac{\beta^2\lambda}{c}e^{-\beta U}\int_0^U e^{\beta x}\,\psi(x)\,dx - \frac{\beta\lambda}{c}\psi(U) + \frac{\beta\lambda}{c}e^{-\beta U}.(4.12)$$

Adding expression (4.12) and β times expression (4.11), one finds that

$$\psi''(U) + \beta\,\psi'(U) = \frac{\lambda}{c}\,\psi'(U),$$

or that $\psi(U)$ satisfies the second order homogenous differential equation

$$\psi''(U) + (\beta - \frac{\lambda}{c})\,\psi'(U) = 0.$$

Therefore $\psi(U)$ must be of the form

$$\psi(U) = k_0 + k_1\,e^{-(\beta - \lambda/c)\,U}$$

for some constants k_0 and k_1. Clearly, $k_0 = 0$ since by Lundberg's inequality $\lim_{U\to\infty}\psi(U) = 0$. Evaluating at $U = 0$, it follows that $\psi(0) = 1/(1+\theta) = k_1$, and therefore one obtains the classic expression for probability of ruin in a Poisson surplus process when claims are exponential:

$$\psi(U) = \frac{1}{1+\theta}\,e^{-\beta\theta U/(1+\theta)}.$$

4.4   Reinsurance and the probability of ruin

In sharing a risk with a reinsurance company, a ceding company hopes to reduce its risk by passing on some of the responsibility to the reinsurer. Of course, this will come at a cost, which must be weighed up against any increase in security obtained. What risk-based criteria should one use in deciding on an appropriate level of reinsurance? One could make a decision on the basis of maximizing expected profits. Given that the loading factor ξ a reinsurance company will put on its premium charges to the ceding company is normally greater than that (θ) used by the ceding company on the policyholders, the criteria of maximizing expected profits will usually result in no reinsurance. Another criterion to consider is that of minimizing the probability of ruin. Given that this is often difficult to calculate exactly, we may use the adjustment coefficient R (or equivalently, the Lundberg upper bound) as a proxy for this measure of security. Hence we would be looking for reinsurance agreements yielding large values of R.

In a claim-by-claim based reinsurance agreement, each individual claim X can be split into two components,

$$X = Y + Z = h_I(X) + h_R(X),$$

which are, respectively, handled by the insurance $(Y = h_I(X))$ and reinsurance $(Z = h_R(X))$ companies. Assuming a Poisson process for claim incidence, the expected net profit at time t for the insurer is therefore

$$\lambda t\,[(1+\theta)E(X) - (1+\xi)E(Z) - E(Y)] = \lambda t\,[\theta E(Y) + (\theta - \xi)E(Z)]$$
$$= \lambda t\,[\theta E(X) - \xi E(Z)]$$
$$\geq 0$$
$$\Leftrightarrow \quad \theta/\xi \geq E(Z)/E(X) \quad \text{or equivalently,} \quad E(Y)/E(X) \geq 1 - \theta/\xi.$$

In this section, we investigate ruin for both proportional and excess of loss reinsurance arrangements.

4.4.1 Adjustment coefficients and proportional reinsurance

In proportional claim-by-claim reinsurance, let α be the proportion of any claim retained by the insurer. Hence each individual claim can be represented as $X = \alpha X + (1-\alpha)X = h_I(X) + h_R(X)$. If our objective is to maximize the adjustment coefficient R as a measure of security, what value of α should we select?

Remember that in order for net profit to be nonnegative, we must have $\theta/\xi \geq E(Z)/E(X) = 1 - \alpha$, or equivalently $\alpha \geq 1 - \theta/\xi$. The expected net monetary gain (or profit) per unit time for the insurer is $[(1+\theta) - (1+\xi)(1-\alpha) - \alpha]\lambda E(X)$. This is an increasing linear function of α (assuming that $\theta \leq \xi$) and hence is maximized when $\alpha = 1$ (i.e., there is no reinsurance). Therefore if maximizing net gain is the objective, one would not make use of reinsurance. In practice, however, we would want to achieve a satisfactory degree of security, and the adjustment coefficient R might serve as a proxy for this.

For a Poisson surplus process where a proportion α of each claim is retained by the insurer (and the reinsurer handles the rest), the surplus process (for the insurer) takes the form

$$U(t) = U + c \cdot t - S_I(t),$$

where $c = [(1+\theta) - (1+\xi)(1-\alpha)]\,\lambda E(X)$ is the net premium income per unit time, and $S_I = \alpha(X_1 + \ldots + X_{N(t)}) = \sum_1^{N(t)} Y_i$. Therefore the adjustment equation for the insurer is now

$$A(r) = \lambda M_{\alpha X}(r) - \lambda - \lambda[(1+\theta) - (1+\xi)(1-\alpha)]\,E(X)\ r = 0.$$

In the special case where X is exponential with mean $1/\beta$, the insurer's typical claim Y is also exponential but with mean α/β, hence (from Equation (4.4)) the adjustment coefficient is given by

$$R = \frac{\beta}{\alpha} - \frac{\lambda}{c} = \frac{\beta}{\alpha} - \frac{\beta}{[(1+\theta) - (1+\xi)(1-\alpha)]} = \frac{\beta[\theta - \xi + \xi\alpha]}{\alpha[\theta - \xi + \alpha(1+\xi)]}. \quad (4.13)$$

In the unlikely situation where $\theta = \xi$, then $R = \beta\theta/\alpha(1+\theta)$ which is a decreasing function of the retention level α. Here any value of α is permissable (since $1 - \theta/\xi = 0$) and R would be maximized by selecting $\alpha = 0$. However in this case (where reinsurance is very cheap) one would be passing on all of the business to the reinsurer, including the possibility for the insurer to make a profit!

Example 4.5

Consider the situation where the aggregate claims for an insurance company are modeled by a compound Poisson process where the rate parameter is λ, claims are exponentially distributed with mean 50, and a security loading of $\theta = 0.15$ is used to determine premiums. Proportional reinsurance is available from different companies (which use varying security loadings to determine their pricing structures for ceding insurance companies). Let us investigate how the optimal level of retention for the insurance company varies with different values of ξ, when using the criteria of maximizing the adjustment coefficient R.

When $\theta = 0.15$ and $\xi = 0.20$, the adjustment coefficient as a function of the retention level α in the interval $(1 - \theta/\xi = 0.25, 1)$ takes the form:

$$R(\alpha) = \frac{1}{50\alpha}\left[\frac{-0.05 + 0.20\alpha}{-0.05 + 1.20\alpha}\right] = \frac{4\alpha - 1}{50(24\alpha^2 - \alpha)}.$$

Differentiating with respect to α, one finds that

$$R'(\alpha) = \frac{1}{50}\frac{-96\alpha^2 + 48\alpha - 1}{(24\alpha^2 - \alpha)^2}$$

and hence $R(\alpha)$ has a maximum occurring at $\alpha^* = 0.478218$. The optimal decision for the ceding insurance company in this case is therefore to retain $\alpha = 0.478218$ of each claim and obtain an adjustment coefficient of $R = 0.003644$. This compares with an adjustment coefficient of $R = 0.002609$ if the company decides not to purchase this reinsurance (i.e., use $\alpha = 1$). Note that in theory any retention level α in the range $(0.478218, 1)$ is preferable to $\alpha = 1$ since R as a function of α is decreasing in this interval. □

Table 4.1 gives the optimal values of the retention level α^* as well as the corresponding adjustment coefficients $R(\alpha^*)$ for the different loadings $\xi = (0.20, 0.25, 0.30, 0.35)$. Note in this case that $R(\alpha^*)$ is a decreasing function of ξ. Note also that if $\xi = 0.35$, then the optimal decision is to go with $\alpha = 1$ since reinsurance is too expensive at any level of retention. Figure 4.6 plots $R(\alpha)$ as a function of the proportion α retained for both $\xi = 0.20$ and $\xi = 0.35$.

TABLE 4.1
Optimal retention levels α^* and $R(\alpha^*)$
for varying ξ.

ξ	α^*	$R(\alpha = \alpha^*)$	$R(\alpha = 1)$
0.20	0.478218	0.003644	0.002609
0.25	0.757771	0.002786	0.002609
0.30	0.938529	0.002620	0.002609
0.35	1	0.002609	0.002609

4.4.2 Adjustment coefficients and excess of loss reinsurance

In an individual claim-based excess of loss reinsurance arrangement, there is an excess level M such that for any claim X the insurance company handles $Y = h_I(X) = min(X, M)$ and the reinsurer the remainder $Z = h_R(X) = max(0, X - M)$. Again, using θ and ξ for the respective loadings of the insurer and reinsurer, in order that net premium income for the insurer exceeds expected claim costs, the excess level M must satisfy

$$\frac{E[Y]}{E[X]} \equiv g_{Y|X}(M) = \frac{\int_0^M x f_X(x)dx + M \bar{F}_X(M)}{\int_0^\infty x f_X(x)dx} \geq 1 - \frac{\theta}{\xi}.$$

When X is continuous (with $-\bar{F}'_X(x) = f_X(x)$), then $\partial/\partial M(g_{Y|X}(M)) = \bar{F}_X(M) \geq 0$, and hence $E[Y]/E[X]$ is an increasing function of M. Therefore the ceding company would only consider excess levels M that are greater than the value M_0, satisfying $g_{Y|X}(M_0) = 1 - \theta/\xi$.

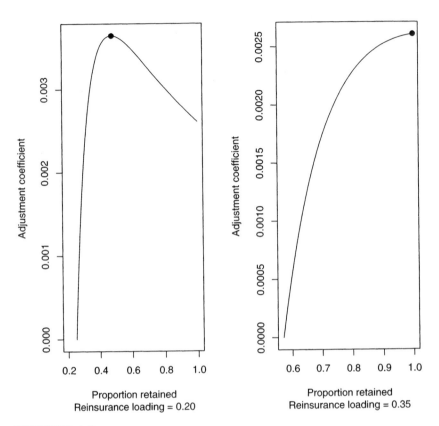

FIGURE 4.6
Adjustment coefficients as a function of the proportion α retained in proportional reinsurance, for reinsurance loadings $\xi = 0.20$ and $\xi = 0.35$.

If X is exponential with mean $1/\beta$, then

$$\frac{E[Y]}{E[X]} \equiv g_{Y|X}(M) = \frac{\int_0^M x\beta e^{-\beta x}dx + Me^{-\beta M}}{1/\beta} = 1 - e^{-\beta M},$$

and the insurer should only consider excess levels M larger than $M_0 = -\log(\theta/\xi)/\beta = -\log(\theta/\xi)\,E(X)$. Therefore if the surplus process has parameters $\theta = 0.2, \xi = 0.3$ and X is exponential with mean 300, then the insurer should only consider excess of loss arrangements where $M \geq M_0 = 121.6395$. Similarly, one may show that when X is Pareto with parameters α and δ, then the minimum excess level M_0 to consider satisfies $[\delta/(\delta + M_0)]^{\alpha-1} = \theta/\xi$.

If $X = Y + Z$ is exponential with mean $1/\beta$, then $E(Z) = e^{-\beta M}/\beta$,

$$M_Y(r) = \int_0^M e^{rx}\beta e^{-\beta x}\, dx + e^{rM}\bar{F}_X(M)$$

$$= \frac{\beta}{\beta - r}\left(1 - e^{-M(\beta-r)}\right) + e^{-M(\beta-r)},$$

and the adjustment equation takes the form

$$\lambda[M_Y(r) - 1 - [(1+\theta)E(X) - (1+\xi)E(Z)]\,r] =$$

$$\lambda\left[\frac{\beta}{\beta-r}\left(1 - e^{-M(\beta-r)}\right) + e^{-M(\beta-r)} - 1 - \frac{1}{\beta}\left[(1+\theta) - (1+\xi)e^{-\beta M}\right]\right] = 0.$$

Example 4.6

Consider again the Poisson surplus process in Example 4.5, but where now excess of loss reinsurance with excess level M is being considered. If $\theta = 0.15$, then the adjustment coefficient R as a function of the excess level M and ξ satisfies

$$\frac{0.02}{0.02 - r}\left[1 - e^{-M(0.02-r)}\right] + e^{-M(0.02-r)} - 1 - \frac{1}{0.02}[(1.15) - (1+\xi)e^{-0.02M}]r = 0.$$

Figure 4.7 plots the adjustment coefficient R as a function of the excess level M for $\xi = 0.20$ and $\xi = 0.35$. When $\xi = 0.20$, then the optimal decision (when trying to maximize R) for the ceding insurance company is to use $M^* = 29.2841$, thereby obtaining an adjustment coefficient of $R = 0.006221$. This compares with an adjustment coefficient of $R = 0.002609$ if the company decides not to purchase this reinsurance (i.e., use $M = \infty$). Note that in theory any excess level M in the range $(29.2841, 100)$ is preferable to $M = 100$ since R as a function of M is decreasing in this interval.

Table 4.2 gives the optimal values of the excess level M, as well as the corresponding optimal adjustment coefficients $R(M^*)$ for the different loadings $\xi = (0.20, 0.25, 0.35, 0.50)$. Note that, $R(M^*)$ is a decreasing function of ξ. The net annual (or other unit of time) gain $P_\$^M$ for the surplus process with excess level M has expected value

$$E(P_\$^M) = (1+\theta)E(X) - (1+\xi)E(Z) - E(Y) = \frac{\theta - \xi e^{-\beta M}}{\beta},$$

which is an increasing function of M.

$\square$

TABLE 4.2
Optimal excess levels M^* and corresponding
$R(M^*)$ for $\theta = 0.15$ and varying ξ.

ξ	M^*	$R(M = M^*)$	$R(M = \infty)$
0.20	29.2841	0.006221	0.002609
0.25	53.9413	0.004140	0.002609
0.35	94.7649	0.003168	0.002609
0.50	143.999	0.002815	0.002609

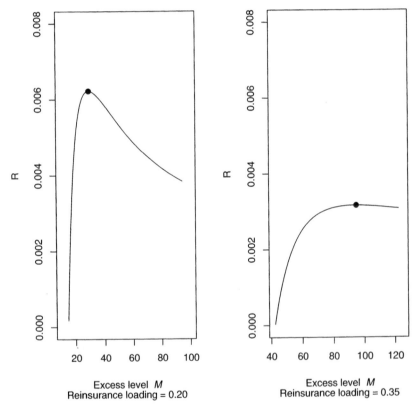

Excess level M
Reinsurance loading = 0.20

Excess level M
Reinsurance loading = 0.35

FIGURE 4.7
Adjustment coefficients as a function of excess of loss reinsurance level M for
reinsurance loadings $\xi = 0.20$ and $\xi = 0.35$.

4.5   Problems

1. The net balance in a surplus process is only checked annually. The
   aggregate claims process is compound Poisson with an annual claims

rate of $\lambda = 0.5$, the loading factor for premiums is $\theta = 0.2$, initial reserves are $U = 0.5$, and the typical claim is uniformly distributed on $\{1, 2, 3\}$. Determine the probability that the aggregate surplus is nonnegative for the first two years. What would the answer be if the claim rate were doubled and the loading factor halved?

2. In a large company, two separate Poisson surplus processes are being monitored. In process A, reserves of $U_A = 3000$ are available, approximately $\lambda_A = 25$ claims are expected annually, $\theta_A = 0.20$, and the typical claim X_A is a $(0.75, 0.25)$ mixture of an exponential random variable with mean $\mu = 60$ and an exponential with mean $\mu = 20$. For process B, $U_B = 1000$, $\lambda_B = 30$, $\theta_B = 0.20$ and $X_B \equiv 50$. For both of these processes, find the adjustment coefficient of the process and comment on the differences. Using Lundberg's inequality, determine upper bounds on the probabilities of ruin for the two processes.

3. Claims made by males in a portfolio of policies can be modeled by a compound Poisson process with Poisson parameter $\lambda_M = 40$ where the typical claim is exponential with mean 20. Similarly, but independently, claims made by females can be modeled as a compound Poisson process but where $\lambda_F = 40$ and a typical claim is exponential with mean 10. Security loadings of 0.4 and 0.2 are used, respectively, for males and females in determining premiums. Find the adjustment coefficient for this surplus process and give an upper bound on the probability of ruin if initial reserves are $U = 40$.

What would the answers be if in each case (for males and females) half as many claims are expected, the sizes of the claims are only half as much (that is, respectively, exponential with means 10 and 5), and reserves are $U = 200$?

4. In a Poisson surplus process suppose that the typical claim X is uniformly distributed on $(0, 1)$ and that $\theta = 0.1$. Use the Newton–Raphson method (with initial estimate $2\theta E(X)/E(X^2)$) to find the adjustment coefficient R. If initial reserves are $U = 10$ what would you estimate for an upper bound on the probability of ruin?

5. In a compound Poisson surplus process the typical claim $X \sim \Gamma(2, 0.5)$. Find the adjustment coefficient R and give a bound on the probability of ruin if initial reserves are $U = 25$ and $\theta = 0.5$.

6. Determine an upper bound for the adjustment coefficient for a Poisson surplus process where the typical claim $X \sim W(c = 0.000001, \gamma = 2)$ is Weibull, and the loading factor on premiums is given by $\theta = 0.25$.

7. Within a large insurance company, three separate Poisson surplus processes are being managed where the respective parameters are given in Table 4.3. For each process, find the adjustment coefficient and the

Lundberg upper bound on the probability of ruin if in each case initial reserves are $U = 150$. Rank the processes by this measure of risk.

TABLE 4.3
Poisson surplus process
parameters (insurance).

Process	λ	θ	Claim size
a	15	0.3	$X \equiv 10$
b	30	0.2	$X \sim \Gamma(1, 0.1)$
c	20	0.1	$X \sim \Gamma(2, 0.2)$

8. In a reinsurance company, three separate Poisson surplus processes are being managed, where the respective parameters are given in Table 4.4. Here for any process, $U =$ initial reserves, $\lambda =$ Poisson parameter, $\theta =$ security loading and X is typical claim size. For each process find the adjustment coefficient. Using the Lundberg bound as an approximation for the probability of ruin, rank the processes relative to their probabilities of ruin.

TABLE 4.4
Poisson surplus process parameters (reinsurance).

Process	U	λ	θ	Claim size X
A	200	10	0.2	$X \equiv 20$
B	130	20	0.4	$X \sim$ exponential $(\mu = 12)$
C	250	25	0.1	$X \sim \Gamma(2, 0.2)$

9. A company manages a Poisson surplus process where the claim random variable X has the following distribution

x	$P(X = x)$
1,000	0.9
10,000	0.1

If the company uses a 20% loading factor, find an upper bound R_0 for the adjustment coefficient R of the form $R_0 = 2\theta\, E(X)/E(X^2)$. With R_0 as an initial estimate, use the Newton–Raphson method to find R. If U, the current reserves of the company is 100,000, give an upper bound for the probability of ruin.

10. The surplus process for company A is Poisson where the typical claim is exponential with mean 1.5, the Poisson parameter $\lambda_A = 200$, the

loading factor on premiums is $\theta_A = 1$, and initial reserves are $U_A = 18$. Find both the exact probability of ruin and the Lundberg upper bound on ruin for this process.

In company B the Poisson surplus process is also Poisson with parameter $\lambda_B = 200$, and $\theta_B = 1$ $U_B = 18$. Here, however, the typical claim is equally likely to come from a single person or a couple. If it comes from a single person it is exponential with mean 1, and if from a couple it is the sum of two independent exponentials each with mean 1. Find the Lundberg upper bound on ruin for company B.

11. The aggregate claims S for a risk is a compound Poisson process with Poisson parameter λ and typical claim X which is exponential with mean $1/\alpha$. The premium for this risk is $(1+\theta)\lambda E(X)$, and initial reserves are U. If $_n\psi(U)$ denotes the probability of ruin at or before the nth claim, for $n = 1, 2, \ldots$, then show that $_1\psi(U) = e^{-\alpha U}/(2 + \theta)$, and use this to find an expression for $_2\psi(U)$.

12. In a Poisson surplus process for aggregate claims, initial reserves are $U = 500$, the annual claim rate is $\lambda = 40$ and the typical claim X is exponential with mean 100. A premium loading of $\theta = 0.3$ is used on policyholders.

   (a) Determine the mean and variance of $U(2)$, the net value of the process after two years.

   (b) What is the adjustment coefficient for the process described above, and what is the probability of ruin for this process? How does this compare with the Lundberg bound for the probability of ruin?

   (c) Proportional reinsurance is available whereby a proportion α of each claim is retained by the insurer (and the reinsurer handles the remaining proportion $1 - \alpha$). The reinsurance loading is $\xi = 0.4$. Show that the adjustment coefficient for the process with this type of reinsurance as a function of α is given by

$$R(\alpha) = \frac{1}{100} \left[\frac{4\alpha - 1}{\alpha(14\alpha - 1)} \right].$$

   What is the maximum value of $R(\alpha)$, and for what values of α is $R(\alpha) \geq R(1)$?

13. Aggregate claims for a risk can be modeled by a compound Poisson process with Poisson parameter λ and where claims are exponential with mean 80. The insurance company uses a security loading of $\theta = 0.3$, while reinsurance is available with a security loading of $\xi = 0.4$.

   (a) The reinsurance company is offering proportional reinsurance where the insurance company pays α of any claim X (and the remainder is taken up by the reinsurance company). What is the minimum value of α which the insurance company should consider if it wishes net premiums to exceed expected claims (net of reinsurance)? Suppose in fact that the insurance company does decide to go ahead with proportional reinsurance, keeping a proportion $\alpha = 0.40$ of every claim. Find the adjustment coefficient and an upper bound on the probability of ruin if initial reserves are $U = 450$.

(b) Suppose that the reinsurance company is also offering an excess of loss type of reinsurance with excess level M (and $\xi = 0.4$). If the insurance company wants net premiums to exceed expected claims, show that the minimum value of M that it should consider is $M = 23.01$.

14. The aggregate claims process in a company is well approximated by a compound Poisson distribution with Poisson parameter 200 and individual claim size density given by $f(x) = e^{-(x-5)}$, $x > 5$. The premium charged by the company to cover this risk includes a security loading of $\theta = 0.25$. Derive a formula for the adjustment coefficient for this claims process and give a good upper bound for R in this situation.

Suppose excess of loss reinsurance is available from a reinsurer with a relative security loading of $\xi = 0.50$. The table below shows for various values the retention limits M and the expected profit for the insurer (net of reinsurance costs) with missing values indicated by *. Complete the table by filling in the missing *. Comment on the relationship between retention limit and profit.

Retention limit M	Expected annual profit
5	200.00
6	⋆
⋆	286.47
∞	⋆

15. The aggregate claims process in a company is being modeled by a compound Poisson surplus process where the Poisson parameter is $\lambda = 20$, initial reserves are $U = 1200$, a typical claim is exponentially distributed with mean 50, and the security or safety loading is $\theta = 0.2$.

The insurance company is considering buying proportional reinsurance where a portion α of every claim is retained (and the reinsurance company handles $1 - \alpha$ of each claim). If the reinsurance loading is $\xi = 0.3$, what is the minimum value of α that the insurance company should consider? Show that the adjustment coefficient for this process when α is the proportion of each claim retained is given by

$$R(\alpha) = \frac{1}{50\alpha} \left[\frac{3\alpha - 1}{13\alpha - 1} \right].$$

For what values of α is the adjustment coefficient (a measure of security) as great as it is when no reinsurance is used?

16. Consider a Poisson Surplus process where claims are exponential with mean $1/\beta$, the unit claim rate parameter is λ, and where an excess of loss reinsurance arrangement is in place with excess level M. Loadings of θ and ξ are used, respectively, by the insurer and reinsurer. Show that an upper bound on the adjustment coefficient R for the insurer is given by

$$\frac{\beta(\theta - \xi e^{-\beta M})}{1 - e^{-\beta M} - M\beta e^{-\beta M}}.$$

17. The process used to model a risk is a Poisson surplus process with Poisson parameter λ and individual claim size X which is exponential with mean 50. The insurance company uses a security loading of $\theta = 0.4$ and reinsurance is available whereby a security loading of $\xi = 0.5$ is used.

   (a) Proportional reinsurance whereby the reinsurance company pays $(1 - \alpha)$ of any claim X is available. Find the minimum value of α which the insurance company should consider if it wants net premiums to exceed expected claims (net of reinsurance). Suppose now the insurance company decides to go ahead with proportional reinsurance with $\alpha = 0.5$. Find the adjustment coefficient and an upper bound on the probability of ruin if initial reserves are $U = 300$.

   (b) Suppose now the same reinsurance company offers excess of loss insurance with excess level M (and $\xi = 0.5$). If the insurance company wants net premiums to exceed expected claims, show that the minimum value of M it should consider is $M = 11.16$.

18. The typical claim X in a (Poisson) surplus process is uniformly distributed on the interval $[0, 50]$. If excess of loss reinsurance is being considered with excess level M, show that the adjustment coefficient for the ceding company satisfies

$$\frac{1}{50r}[e^{rM} - 1] + e^{rM}[1 - M/50] - \{1 + [(1 + \theta)25 - (1 + \xi)E(Z)]r\} = 0.$$

If the insurer's loading $\theta = 0.2$ and the reinsurer's loading $\xi = 0.3$, find the minimum value of M which should be considered.

19. Individual claims in a Poisson surplus process are modeled by a random variable X with density given by

$$f_X(x) = (a + bx)e^{-x} \text{ for } x > 0, \text{ where } a + b = 1, a > 0.$$

A loading factor of θ is used to determine premiums. Determine the mean, variance and moment generating function of X in terms of a and b.

(a) Write down the adjustment equation for this process in terms of a, b and θ, and find an upper bound for the adjustment coefficient R in terms of these parameters.

(b) Now suppose that $a = b = 1/2$ and that a loading factor of $\theta = 1/4$ is used to determine premiums. Find the adjustment coefficient R and give an upper bound on the probability of ruin for this process when initial reserves are $U = 50$.

(c) The direct insurer in this case is considering proportional reinsurance where 40% of each claim is ceded to the reinsurer and the reinsurer's loading factor is $\xi = 0.5$. Write down the adjustment equation for the direct insurer in this case.

20. Consider a Poisson surplus process where $\lambda = 1, \theta = 0.1$, X is Pareto (α, δ) with $E(X) = \delta/(\alpha - 1) = 20/(3 - 1) = 10$, and initial reserves are either $U = 50$ or 100. Using simulation, estimate $\psi(U, t)$ in the interval $[0, 400]$.

5

Credibility Theory

5.1 Introduction to credibility estimates

Credibility theory in general insurance is essentially a form of experience-rating that attempts to use the data in hand as well as the experience of others in determining rates and premiums. An interesting and early historical example of the use of credibility theory deals with the setting of premium rates for employers to cover for workers compensation in the early 20th century [51]. The challenge (to a casualty actuary) was often to balance the claims experience of a particular employer with that of all employers having similar working practices and conditions in determining premiums for insurance coverage.

In this chapter, we address the challenge of trying to estimate expected future claim numbers and/or total aggregate claims for a portfolio of policies on the basis of rather limited sample or current information $\mathbf{x}$, but where other collateral (and possibly useful) information is also at hand. Let us assume there is crucial parameter of interest denoted by θ, which, for example, may be the annual claim rate or a related expected aggregate claims total. Often there is other (in some cases considerable) collateral or prior information from business or portfolios of a somewhat similar nature, which might be useful in estimating θ. Let us denote by $\hat{\theta}_s$ an estimate of θ based on the sample information $\mathbf{x}$, and by $\hat{\theta}_c$ an estimate of θ based on the available collateral information. In the situation where θ is a mean, then $\hat{\theta}_s$ might be the sample mean $\bar{x}$ and $\hat{\theta}_c$ some *prior* estimate (say μ_0) of this mean. In this type of situation, a key question is often "How might we combine the two (sample and collateral) sources of information to get a good estimate of θ, and in particular how much weight or credibility $\mathbf{Z}$ should our estimate put on the sample estimator $\hat{\theta}_s$?" Surely the value of $\mathbf{Z}$ should both be an increasing function of the amount of sample information which we might acquire over time, and also take account of the *relative* values of the sample and collateral information available.

A credibility estimate of θ is a linear combination of the sample estimator $\hat{\theta}_s$ and the collateral estimate $\hat{\theta}_c$ of the form

$$\mathbf{Z}\,\hat{\theta}_s + (1 - \mathbf{Z})\,\hat{\theta}_c \quad \text{(for example } \mathbf{Z}\,\bar{x} + (1 - \mathbf{Z})\,\mu_0\text{)}, \tag{5.1}$$

where $\mathbf{Z}$ is the *credibility* we put on the sample estimator $\hat{\theta}_s$. The general expression given by (5.1) is often called the *credibility premium formula*. Traditionally, there has been an emphasis on only using estimates $\hat{\theta}_s$ of the form $\sum_1^n a_j x_j$ (i.e., linear in the observations) in the credibility premium formula, and although such estimates have considerable appeal there is no theoretical reason why other sample estimates cannot be used.

Consider the situation where the annual claims (for theft and malicious damage) to four rural churches have over the past three years averaged at 24,000, and we are asked to quote a pure premium θ (representing expected annual claims) to insure them next year. Also available is very extensive data (both in terms of past years and numbers of buildings) on claims data from churches in the capital city of the country where the recent average annual amount of claims per church was 8000. Relying exclusively on (the directly relevant but somewhat limited) information from the rural churches, one could estimate θ by 24,000, while using only the collateral information from the extensive city data one might estimate θ to be $4 \cdot 8000 = 32,000$. A credibility approach would use both sources of information and estimate θ by some weighted average of the form

$$\mathbf{Z}\ 24{,}000 + (1 - \mathbf{Z})\ 32{,}000.$$

We shall investigate several approaches to the credibility problem. In the classical approach to credibility theory, concern centers on establishing when there is enough data to use $\mathbf{Z} = 1$ (that is, give full credibility to the data), and if this is not the case how does one determine a reasonable value for $\mathbf{Z}$? This is basically a frequentist approach to credibility, and is often called either the *limited fluctuation* or *American approach to credibility theory*. Whitney [60] suggested that the credibility factor $\mathbf{Z}$ should be of the form $\mathbf{Z} = P/(P + K)$, where P represents *earned premiums* and K is to be *determined by judgement*.

In the *Bayesian approach* to credibility theory, the credibility estimate is based on a posterior distribution of θ where the credibility factor $\mathbf{Z}$ is the weight our (Bayesian) posterior estimate puts on the sample information and $1 - \mathbf{Z}$ is the weight put on the prior estimate (often the prior mean μ_0) of θ. Another approach which was popularized by Bühlmann [11], and Bühlmann and Straub [12], is essentially an empirical Bayes approach (and sometimes called the greatest accuracy or least squares approach to credibility [28]). Here the data available is used to estimate prior parameters for the model, and then estimates for θ are determined. Bühlmann [11] suggested using $\mathbf{Z} = n/(n+K)$ as the credibility factor where the Bühlmann credibility parameter K is the ratio of the *mean of the (so-called) process variance* to the *variance of the hypothetical means*. The Bühlmann and Straub [12] model generalizes this model to the situation where the exposure units (for example, annual number of policyholders) may vary with time.

5.2 Classical credibility theory

In the classical approach to credibility theory, one would ask how much sample information do I need before I can rely completely on the sample in estimating the parameter θ of interest? In such a case, one would use $\mathbf{Z} = 1$ and say that there is *full credibility* in the sample information (and consequently, that the collateral information is not necessary in estimating the parameter). When there is not sufficient sample information for full credibility, the question arises as to what partial credibility $\mathbf{Z}$ one should assign to the sample information (with the remainder or complimentary $1 - \mathbf{Z}$ weighting being assigned to the collateral information).

This approach to credibility theory is clearly frequentist in nature, and because of its extensive use in earlier times by American actuaries it is often called *American credibility theory*. The classical method of credibility can be useful in establishing full credibility estimates for claim frequencies and pure premiums, but some doubt (see, for example, [59]) the basis on which the concepts of partial credibility are used. Although the classical approach is infrequently used today, it is of historical interest and raises some interesting statistical questions.

5.2.1 Full credibility

If θ is the parameter of interest, what does it really mean to say that the sample data is fully credible? For example, the sample mean $\bar{x}$ is, after all, a random variable, and if we are to rely on it as a good estimator of θ we would hope that with high probability it is relatively close to θ. In most applications, the parameter of interest θ is either an annual claims rate λ or a pure premium $E(S)$ representing expected annual aggregated claims. Let k and $0 < \alpha < 1$ be two constants. We say that the estimator $\hat{\theta}_s$ based on the sample data $\mathbf{x}$ is *fully credible* (k, α) *for* θ, if

$$Prob[\, |\, \hat{\theta}_s - \theta \,| \leq k\theta \,] = Prob[\, \hat{\theta}_s - k\theta \,\leq\, \theta \leq \hat{\theta}_s + k\theta \,] \geq 1 - \alpha. \qquad (5.2)$$

Usually, k and α are small, with $k = 0.05$ and $\alpha = 0.10$ being a common choice (see [38]). In a situation where there is sufficient data to claim full credibility (k, α), one can say that with probability $1 - \alpha$ the relative fluctuations of $\hat{\theta}_s$ from θ are *limited* to $k\theta$, and hence this approach to credibility theory is often called *limited fluctuation credibility theory*. Some texts use the notation full credibility (k, p) instead of (k, α) where $p = 1 - \alpha$, and our preference for using (k, α) stems from the similarity of Equation (5.2) to the classical frequentist approach to confidence interval estimation of parameters.

Example 5.1

Aggregate annual claims are modeled by a compound Poisson distribution S with Poisson parameter λ and typical claim size X. The parameter of interest is the pure premium $\theta = E(S) = \lambda E(X)$. Aggregate annual claims data $S_1, \ldots, S_r$ is collected over r years. For given values of (k, α), approximately how many claims does one need to put full credibility (k, p) on $\bar{S} = \sum S_i/r$ as an estimator of $E(S)$?

$\bar{S}$ is approximately normal for large $r\lambda$, and since $Var(\bar{S}) = \lambda E(X^2)/r$, it follows that

$$P(|\bar{S} - \theta| \le k\theta) = P\left(|\frac{\bar{S} - \theta}{\sqrt{\lambda E(X^2)/r}}| \le \frac{k\theta}{\sqrt{\lambda E(X^2)/r}}\right)$$

$$\doteq 2\Phi\left(\frac{k\theta}{\sqrt{\lambda E(X^2)/r}}\right) - 1$$

$$\ge 1 - \alpha$$

$$\Leftrightarrow \quad r\lambda \ge \frac{z_{1-\alpha/2}^2}{k^2} \frac{E(X^2)}{E^2(X)} = \frac{z_{1-\alpha/2}^2}{k^2}[1 + cv^2(X)], \qquad (5.3)$$

where $\Phi(z_{1-\alpha/2}) = 1 - \alpha/2$ and $cv(X) = \sqrt{Var(X)/E^2(X)} = \sigma_X/\mu_X$ is the coefficient of variation of the random claim size X.

Suppose, for example, that $(k, \alpha) = (0.10, 0.05)$ and claims are exponential. Then $E(X^2) = 2E^2(X)$ and for full credibility $(0.10, 0.05)$ one should have at least $r\lambda \ge (1.96/0.10)^2(2) \doteq 769$ claims. In particular, if the annual rate of claims was of the order of $\lambda \doteq 160$, then one would have full credibility $(0.10, 0.05)$ for the pure premium after $r = 5$ years of data. One can say that in theory the estimator $\bar{S} = (S_1 + \cdots + S_5)/5$ differs from $\theta = E(S)$ by at most $0.10\,E(S)$ with probability at least 0.95.

Now suppose instead that the parameter of interest is the annual claims rate λ as opposed to the pure premium $E(S) = \lambda E(X)$. If we want to use the observed average annual rate $\bar{N} = \sum N_i/r$ to estimate λ with full credibility (k, α), then arguing as above it is clear that one has

$$P(|\bar{N} - \lambda| \le k\lambda) \ge 1 - \alpha \quad \Leftrightarrow \quad r\lambda \ge \frac{z_{1-\alpha/2}^2}{k^2}. \qquad (5.4)$$

〔

Unless X is constant, one has that $E(X^2)/E^2(X) > 1$, and it follows that one always needs more claims experience to estimate well a pure premium than a claim rate. This is not surprising since for a pure premium one has to allow for variability in severity (claim size) as well as variability in frequency (claim rate). For an exponentially distributed claim the ratio $E(X^2)/E^2(X) = 2$, while for the lognormal distribution (with parameters μ and σ^2) it takes the value e^{σ^2}. Longley–Cook [38] states that for many types of insurance claims, $2 \leq E(X^2)/E^2(X) \leq 5$.

Consider the situation where annual aggregate claims $S = X_1 + \ldots + X_N$ are modeled by a compound distribution but where N is not necessarily Poisson. Using again a normal approximation for S and arguing as above when $r = 1$, it is clear that one can achieve full credibility (k, α) by using S to estimate θ if $E^2(S)/Var(S) \geq z^2_{1-\alpha/2}/k^2$. For example, suppose that S is compound binomial where the number of claims N is $B(m, q)$. Then $E^2(S)/Var(S) = mqE^2(X)/[E(X^2) - qE^2(X)]$, and hence full credibility (k, α) for $\theta = E(S)$ is achieved if

$$ mq \geq \frac{z^2_{1-\alpha/2}}{k^2} \left[\frac{E(X^2)}{E^2(X)} - q \right]. $$

In particular, suppose that total claims S for a year result from an exposure of m units with claim frequency rate $q = 0.02$ per exposure, and that a typical claim X is exponential. Then S is a fully credible (k, α) estimate for $E(S) = \theta$ if the expected number of claims $mq \geq (z^2_{1-\alpha/2}/k^2)(1.98)$. On the other hand, if the interest is only in the parameter $E(N) = mq$, then the expected number of claims necessary for a fully credible (k, α) estimate is $mq \geq (z^2_{1-\alpha/2}/k^2)(0.98)$.

5.2.2   Partial credibility

Often one does not have enough data to justify using full credibility to estimate a parameter, and the question therefore arises as to how much credibility or weight $\mathbf{Z}$ to put on the sample information? The most commonly used method for assigning partial credibility is the *square root rule*, whereby the credibility factor $\mathbf{Z}$ is given by $\sqrt{n/n_{F(k,\alpha)}}$ where n is the expected (or, in practice, the observed) number of claims for the data and $n_{F(k,\alpha)}$ is the number required for full credibility. The justification for this formula is that with this value of $\mathbf{Z}$ one limits the fluctuation (or variance) of $\mathbf{Z}\bar{X}$ around $\mathbf{Z}\theta$ to that which would be obtained if the data were fully credible for θ.

Suppose that we want to estimate $\theta = E(S) = \lambda E(X)$ for a compound Poisson distribution, and that we have r years of aggregate claim data $S_1, \ldots, S_r$. We know from Equation (5.3) that for full credibility (k, α), we need approximately $n_{F(k,\alpha)} = (z_{1-\alpha/2}/k)^2 [E(X^2)/E^2(X)]$ claims. If we do not have this amount of data, then we will use $\mathbf{Z}\bar{S} + (1 - \mathbf{Z})\mu_0$ to estimate $\theta = E(S)$ where $0 < \mathbf{Z} < 1$ and μ_0 is a collateral estimate of θ. Suppose we decide to use a value of $\mathbf{Z}$ so that $\mathbf{Z}\bar{S}$ as an estimator of $\mathbf{Z}\theta$ has the same credibility (k, α) that we would require of full credibility, or in other words that

$$ P(|\, \mathbf{Z}\bar{S} - \mathbf{Z}\theta \,| \leq k\theta) = 1 - \alpha. $$

Arguing as in Equation (5.3), one sees that this is equivalent to

$$r\lambda = \mathbf{Z}^2 \frac{z_{1-\alpha/2}^2 E(X^2)}{k^2 E^2(X)} = \mathbf{Z}^2 \, n_{F(k,\alpha)} \text{ or } \mathbf{Z} = \sqrt{\frac{r\lambda}{n_{F(k,\alpha)}}}.$$

Example 5.2

Assume that expected total losses for a company last year were 40,000,000, but in fact the total observed was 47,000,000. This total was based on 12,000 observed claims, while in fact a fully credible estimate of the company's annual total losses would require about 18,500 claims. Using the square root rule for partial credibility, one might estimate total claims for next year (in millions) by

$$\sqrt{\frac{12,000}{18,500}} \, 47 + \left(1 - \sqrt{\frac{12,000}{18,500}}\right) 40 = 45.638.$$

□

Historically, some questions have been raised about this (and other similar) approaches to partial credibility. Note that, for example, in using the above (square root) approach to determine the partial credibility factor $\mathbf{Z}$, one does limit in a probabilistic sense the variation between $\mathbf{Z}\bar{X}$ and $\mathbf{Z}\theta$. However, this does not give a probabilistic statement of how reliable $\mathbf{Z}\hat{\theta}_s + (1-\mathbf{Z})\hat{\theta}_c$ is as an estimator of $\theta = \mathbf{Z}\theta + (1-\mathbf{Z})\theta$, and in fact the reliability of $\hat{\theta}_c$ as an estimator of θ seems to be completely ignored!

5.3   The Bayesian approach to credibility theory

If θ is an unknown characteristic or parameter related to a population or random variable, then the frequentist statistician will attempt to make inferences about θ on the basis of the sample information $\mathbf{x}$. The Bayesian statistician, however, would always believe that there is additional prior information available about θ which should be combined with the sample information $\mathbf{x}$ in order to make inferences about θ. The reader might find it useful at this point to consult the appendix to this book entitled *An Introduction to Bayesian Statistics* in order to review the Bayesian approach to statistics and the use of loss functions in estimation.

5.3.1   Bayesian credibility

In the Bayesian approach to credibility theory, one would summarize collateral knowledge about the unknown parameter θ by means of a prior distribution.

As such, one is really saying that the unknown parameter can be viewed as a random variable denoted by Θ. The choice of a prior distribution for Θ is often a subjective decision. Informative and accurate information is usually reflected in a prior distribution which is centered near the unknown parameter and/or is precise (has small variability). Given sample information $\mathbf{x}$, one would determine the posterior distribution for $\Theta \mid \mathbf{x}$ and use this to estimate the true value of θ (or some function of it). The most common estimator is the posterior mean (used with quadratic loss), but one may on occasion also use the posterior median or mode. If $\mu_0 = E(\Theta)$ is the prior mean for Θ, then in some (but not all) cases one may express the posterior mean $E(\Theta \mid \mathbf{X} = \mathbf{x})$ in the form of credibility estimate for θ, that is, as a linear combination of a statistic θ_s (depending on $\mathbf{x}$) and μ_0 of the form

$$E(\Theta \mid \mathbf{X} = \mathbf{x}) = \mathbf{Z}\,\theta_s + (1 - \mathbf{Z})\,\mu_0.$$

The following classic example is useful in estimating expected aggregate claims when they are normally distributed.

Example 5.3 Normal | Normal model.
In the normal | normal model, we assume that the sampling distribution for $X \mid \theta$ is $N(\theta, \sigma^2)$ where σ^2 is known, and that prior information for $\theta \equiv \mu$ can be summarized by the normal distribution $N(\mu_0, \sigma_0^2)$. With a little bit of algebra one may show that the sampling distribution for $\mathbf{X} \mid \theta$ is

$$f_{\mathbf{X}\mid\Theta}(\mathbf{x} \mid \theta) = \prod_{j=1}^{n} \frac{1}{\sqrt{2\pi}\sigma} e^{-(x_j - \theta)^2/2\sigma^2}$$

$$\propto e^{-[n\theta^2 - 2\theta \sum_{j=1}^{n} x_j]/2\sigma^2}$$

$$\propto e^{-\frac{n}{2\sigma^2}(\theta - \bar{x})^2},$$

where $\propto$ means "is proportional to" as a function of θ. Therefore the posterior for $\Theta \mid \mathbf{x}$ is a normal distribution of with density of the form

$$f_{\Theta\mid\mathbf{x}}(\theta \mid \mathbf{x}) \propto e^{-\frac{n}{2\sigma^2}(\theta - \bar{x})^2} \frac{1}{\sqrt{2\pi}\sigma_0} e^{-\frac{(\theta - \mu_0)^2}{2\sigma_0^2}}$$

$$\propto e^{-\frac{(\theta - \theta^*)^2}{2\sigma^{*2}}},$$

where

$$\theta^* = \left(\frac{n\bar{x}}{\sigma^2} + \frac{\mu_0}{\sigma_0^2} \right) \Big/ \left(\frac{n}{\sigma^2} + \frac{1}{\sigma_0^2} \right) \quad \text{and} \quad \sigma^{*2} = \left(\frac{1}{\sigma^2/n} + \frac{1}{\sigma_0^2} \right)^{-1} \tag{5.5}$$

are, respectively, the mean and the variance. Hence the credibility estimate for $\theta \mid \mathbf{x}$ may be written as

$$\theta^* = \frac{\sigma_0^2}{\sigma_0^2 + \sigma^2/n}\,\bar{x} + \frac{\sigma^2/n}{\sigma_0^2 + \sigma^2/n}\,\mu_0. \tag{5.6}$$

with credibility factor

$$Z = \frac{\sigma_0^2}{\sigma_0^2 + \sigma^2/n} = \frac{n/\sigma^2}{n/\sigma^2 + 1/\sigma_0^2} = \frac{n}{n + \sigma^2/\sigma_0^2}. \tag{5.7}$$

It is both reasonable and clear from expression (5.7) that in this model the credibility factor Z is an increasing function of n for fixed σ_0^2 and σ^2, and also an increasing function of σ_0^2 for fixed $\sigma^2/n = Var(\bar{X})$. Furthermore, Z is a decreasing function of the ratio $K = \sigma^2/\sigma_0^2$ (the ratio of the *process variance* to the prior variance of the *hypothetical means* μ) for fixed sample size n.

Note also that the posterior variance is actually the reciprocal of the reciprocals of σ^2/n (the variance of $\bar{X}$ which is the maximum likelihood estimator of θ) and σ_0^2 (the variance for the prior). Furthermore, it is easy to see from (5.5) that

$$\sigma^{*2} \leq min(\sigma^2/n, \sigma_0^2),$$

and hence as one might expect the posterior distribution is less variable than the prior as a distribution for θ. Since $\sigma^{*2} \leq \sigma^2/n$, the variance of the posterior can be made as small as desirable by taking a large enough sample size n. □

Example 5.4

In past years a lognormal distribution has been used to model the claim size X in a portfolio of household contents policies. If $Y = \log X$, then up to now in the company one has assumed that $Y \sim N(6.9, \sigma^2 = 0.1^2)$. Changes have been recently made in the procedure for making claims, and as a result it is expected that the mean μ for Y in the future will be reduced, but that σ^2 will stay approximately the same. Initial feelings about the new value for μ are that it will be approximately 6.4, and that with 90% certainty it is within the interval $[6.35, 6.45]$. If a random sample of 40 new claims yields $\bar{y} = 6.51$, what is the Bayes estimator (with quadratic loss) for μ? What would be the Bayes estimate for $E(Y)$ if an absolute value loss function were used? What sample size (of claims) would be necessary in order to obtain a posterior distribution for μ with a standard deviation ≤ 0.01?

We assume a normal prior for μ and since

$$0.9 = P(6.4 - 1.645\sigma_0 \leq \mu \leq 6.4 + 1.645\sigma_0),$$

it follows that $\sigma_0 = 0.05/1.645 = 0.0304$. Thus the prior distribution for μ is $N(\mu_0 = 6.4, \sigma_0^2 = 0.0304^2)$. The posterior will also be normal with mean given by

$$\frac{\sigma_0^2}{\sigma_0^2 + \sigma^2/n} \bar{y} + \frac{\sigma^2/n}{\sigma_0^2 + \sigma^2/n} \mu_0 = \frac{0.0304^2}{0.0304^2 + 0.1^2/40} 6.51$$

$$+ \frac{0.1^2/40}{0.0304^2 + 0.1^2/40} 6.4$$

$$= 0.7870\,(6.51) + 0.2130\,(6.4) = 6.4866.$$

To obtain a posterior with standard deviation ≤ 0.01 in this situation, we need a sample of size n such that

$$\left(\frac{1}{0.1^2/n} + \frac{1}{0.0304^2}\right)^{-1/2} \leq 0.01,$$

or equivalently, that $n \geq 90$. ▯

Example 5.5 Poisson | Gamma model.
The Poisson model is frequently used to model claim incidence. Let us suppose that annual claim numbers on a portfolio of policies are modeled by a Poisson distribution with parameter λ, and that prior information about λ may be summarized by a gamma distribution $\Gamma(\alpha, \beta)$. (We might view λ as the realization of a random variable $\Lambda \sim \Gamma(\alpha, \beta)$.) After collecting sample information about claim numbers $\mathbf{x} = (x_1, \ldots, x_n)$ over n years where x_j is the number of claims from year j, the posterior density for λ takes the form

$$f_{\Lambda|\mathbf{x}}(\lambda \mid \mathbf{x}) \propto \frac{\prod_{j=1}^{n} \lambda^{x_j} e^{-\lambda}}{\prod_{j=1}^{n} x_j!} \; \frac{\beta^{\alpha} \lambda^{\alpha-1} e^{-\beta\lambda}}{\Gamma(\alpha)}$$

$$\propto \lambda^{\sum x_j + \alpha - 1} e^{-\lambda(n+\beta)}.$$

Therefore the posterior distribution for $\lambda \mid \mathbf{x}$ is the gamma distribution $\Gamma(\sum x_j + \alpha, n + \beta)$, and the posterior mean (Bayesian estimator of λ with quadratic loss) can be expressed as a credibility estimate in the form

$$E(\Lambda|\mathbf{x}) = \frac{\sum x_j + \alpha}{n + \beta}$$

$$= \frac{n}{n+\beta} \frac{\sum x_j}{n} + \frac{\beta}{\beta+n} \frac{\alpha}{\beta}$$

$$= \mathbf{Z}\,\bar{x} + (1 - \mathbf{Z})\,\mu_0.$$

The credibility we put on the data itself is $\mathbf{Z} = n/(n + \beta)$, and $\mu_0 = \alpha/\beta$ is the prior mean. Note that for a fixed prior distribution on λ, the credibility factor $\mathbf{Z} = n/(n+\beta)$ is an increasing function of the number of years of data observed, and moreover

$$\beta = \frac{\alpha/\beta}{\alpha/\beta^2} = \frac{E[\Lambda]}{Var[\Lambda]} = \frac{E[V(X \mid \Lambda)]}{Var[E(X \mid \Lambda)]} \equiv K.$$

For example, suppose that the $\Gamma(3, 1)$ distribution is used as a prior for the annual claims rate λ, and that the numbers of annual claims observed in a

5-year period are given by $\mathbf{x} = (2, 3, 6, 0, 3)$. Then the posterior distribution
for $\lambda \,|\, \mathbf{x}$ is $\Gamma(14 + 3, 5 + 1) = \Gamma(17, 6)$ and a credibility estimate for λ is

$$\frac{17}{6} = \frac{5}{6}\frac{14}{5} + \frac{1}{6}\frac{3}{1}.$$

⬜

Example 5.6

Suppose that the annual number of claims resulting from water damage in
a small city is a Poisson random variable with $\lambda = 160$, but that this is not
actually known. Instead, an initial feeling about the possible values for λ is
expressed by a $\Gamma(120, 1)$ distribution, which clearly underestimates the reality
of the situation. After n years of data have been collected ($n = 1, \ldots, 12$),
the credibility estimate (posterior mean) for λ can be expressed as

$$\frac{\sum_1^n x_j + \alpha}{n + \beta} = \frac{\sum_1^n x_j + 120}{n + 1} = \frac{n}{n + 1}\bar{x} + \frac{1}{n + 1}\frac{120}{1},$$

where the credibility factor $\mathbf{Z} = n/(n + 1)$ converges rather quickly to 1.
In this instance, this is crucial since the prior information on λ gives little
credence to values as large as the actual value 160. Note, however, that had
one used a more precise prior for λ with the same mean, then convergence
of the credibility factor to 1 would be slower. For example, if the prior were
the $\Gamma(960, 8)$ distribution with the same mean of 120, then the corresponding
credibility factor after observing n years of data would be $n/(n+8) < n/(n+1)$.

A sample of actual claims arising over 12 successive years is given in Table
5.1, together with the corresponding credibility estimates for the two different
prior distributions for λ. In Figure 5.1, these estimates are plotted over years
to illustrate different rates of convergence to the true value of 160.   ⬜

TABLE 5.1
Annual water damage claims and credibility estimates.

Year	1	2	3	4	5	6	7	8	9	10	11	12
Claims	156	150	157	150	167	134	157	157	155	156	161	178
Prior					Credibility estimates							
$\Gamma(120, 1)$	138	142	146	147	150	148	149	150	150	151	152	154
$\Gamma(960, 8)$	124	127	129	131	134	134	135	137	138	139	140	142

In the Poisson | gamma model for claim numbers, we have assumed our
knowledge about the unknown parameter (annual claim rate) λ is summarized

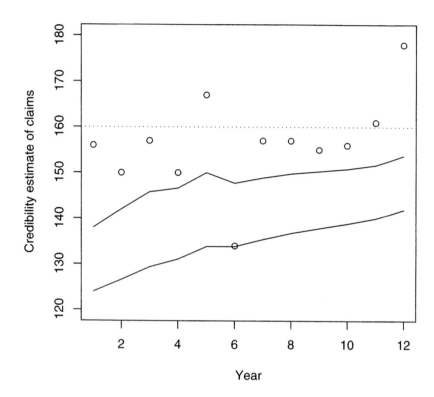

FIGURE 5.1

Posterior estimates for λ (claim rate on water damage) with different priors.

by a $\Gamma(\alpha, \beta)$ distribution. If we let $\mathbf{X} = (X_1, \ldots, X_n)$ represent the numbers of claims to be observed in n subsequent years, then we should be careful to distinguish between the theoretical distributions of $\mathbf{X}$ *with* and *without* knowledge of λ.

Given a particular value of λ, we assume that the random variables $(X_1 \mid \lambda, \ldots, X_n \mid \lambda)$ are independent identically distributed Poisson random variables with parameter λ. Furthermore, $X_{n+1} \mid \lambda$ is Poisson with mean λ, and our posterior Bayesian estimate of its mean (having actually observed $\mathbf{x} = (x_1, \ldots, x_n)$) is $E(X_{n+1} \mid \mathbf{x}) = (\alpha + \sum_1^n x_j)/(n + \beta)$. However, if we do not know the value of λ (but assume that our prior feeling about its possibilities are summarized by a $\Gamma(\alpha, \beta)$ distribution), then the components of $\mathbf{X} = (X_1, \ldots, X_n)$ are neither Poisson nor independent (although they are identically distributed). In fact, we can essentially view any X_j as a $\Gamma(\alpha, \beta)$

mixture of Poisson random variables since

$$P(X_j \leq k) = \int_0^\infty P[X_j \leq k \mid \Lambda = \lambda] \, f_\Lambda(\lambda) \, d\lambda$$

$$= \int_0^\infty \sum_{i=0}^k \frac{\lambda^i e^{-\lambda}}{i!} \, \frac{\beta^\alpha \lambda^{\alpha-1} e^{-\lambda\beta}}{\Gamma(\beta)} \, d\lambda \,.$$

Once we condition on (or assume a value for) λ, the observations in $\mathbf{X}$ are independent. However, before conditioning, for any $i \neq j$,

$$E(X_i X_j) = E_\Lambda(E(X_i X_j \mid \lambda)) = E_\Lambda[E(X_i \mid \lambda) \, E(X_j \mid \lambda)]$$

$$= E_\Lambda(\lambda^2) = \frac{\alpha}{\beta^2} + \left(\frac{\alpha}{\beta}\right)^2$$

$$\neq \left(\frac{\alpha}{\beta}\right)^2 = E(X_i) \, E(X_j),$$

and hence in particular the components of $\mathbf{X}$ are dependent!

5.4 Greatest accuracy credibility theory

One of the most useful situations for a theory of credibility arises when there are n years of claims experience $\mathbf{X} = (X_1, \ldots, X_n)$, and one is interested in X_{n+1} (the number or aggregate amount of claims that will be observed next year). In most cases, this will be done by estimating its conditional mean (or what might be termed the pure premium) given $\mathbf{X} = \mathbf{x}$.

Let us assume that there is some unknown *risk* parameter θ determining the distribution of our observations. Even if we feel we have very little information about θ, the Bayesian statistician (and there are many such people who can argue their case very strongly) would always say that we can (by asking ourselves and learning from the experience of so-called experts) always construct a prior distribution for this parameter! Such prior information allows us to view the true risk parameter as (the realization of) a random variable Θ. The parameter θ might be something as straightforward as a mean, but it might also be a more complex indicator of the type of insurance (say personal liability) and/or characteristics of the group insured (say middle aged drivers of large vehicles).

In the most basic claims model of this type, we assume that the yearly observations of $(X_1, \ldots, X_n)$ which we make are of equal value in making inferences about the mean of X_{n+1}. In particular, we assume that conditional on knowing that $\Theta = \theta$, the random observations $X_j \mid \theta$, $j = 1, \ldots, n+1$ are independent and identically distributed. We shall use $m(\theta)$ to denote the mean of any of these observations given knowledge of θ, that is, $m(\theta) = E(X_j \mid \theta)$ for $j = 1, \ldots, n + 1$. In this context, having observed $\mathbf{X} = \mathbf{x}$, it is actually $m(\theta) = E(X_{n+1})$ which is usually of more interest than θ itself (although in some of the more classic situations like the normal $|$ normal or Poisson $|$ gamma models, $m(\theta)$ is actually equal to θ).

Given the risk parameter θ, there will be variation in the observations $X_1, \ldots, X_{n+1}$ which we denote by $s^2(\theta)$, that is, $Var(X_j \mid \theta) \equiv s^2(\theta)$. Note that in this basic model both $m(\theta) = E(X_j \mid \theta)$ and $s^2(\theta) = Var(X_j \mid \theta)$ are independent of $j = 1, \ldots, n, n + 1$.

Of course, as a function of the unknown risk parameter $\Theta = \theta$, the mean $m(\theta)$ varies between risks and hence can be viewed as a random variable $m(\Theta)$. The mean and variance of $m(\Theta)$ will be denoted, respectively, by $E[m(\Theta)]$ and $Var[m(\Theta)]$ (remember the Bayesian always believes there exists a prior distribution for $m(\theta)$, even if it is not one of the more *classic* ones). Furthermore, the annual within risk variation $s^2(\theta)$ for $X_j \mid \theta$ can be viewed as a random quantity (with respect to Θ), and its mean value will be denoted by $E[s^2(\Theta)]$.

It is very important to realize that although for a given $\Theta = \theta$ the random variables $X_1 \mid \theta, \ldots, X_{n+1} \mid \theta$ are independent, the (unconditional) random variables X_j, $j = 1, \ldots, n + 1$ are not (although they are identically distributed). In fact, one might view the observation of the data $\mathbf{X} = \mathbf{x}$ as a two-step random operation. In the first step, a selection of the risk parameter $\Theta = \theta$ takes place, and in the second the n values $(X_1 = x_1, \ldots, X_n = x_n \mid \Theta = \theta)$ are made conditional on a value for $\Theta = \theta$. The following basic properties will be useful (established using the classic double expectation theorem) in estimating $E(X_{n+1} \mid \mathbf{X})$, $E[m(\Theta) \mid \mathbf{X}]$, or X_{n+1}:

$$E(X_j) = E_\Theta[E(X_j \mid \theta)] = E[m(\Theta)], \tag{5.8}$$

$$\begin{aligned} E(X_i X_j) = E_\Theta(X_i X_j \mid \theta)) &= E_\Theta[\, E(X_i \mid \theta) \, E(X_j \mid \theta)\,] \\ &= E[m^2(\Theta)] \quad \text{for } i \neq j \quad \text{and} \end{aligned} \tag{5.9}$$

$$\begin{aligned} E(X_j^2) = E_\Theta[E(X_j^2 \mid \theta)] &= E_\Theta[s^2(\theta) + m^2(\theta)] \\ &= E[s^2(\Theta)] + E[m^2(\Theta)]. \end{aligned} \tag{5.10}$$

Observing sample information $\mathbf{X} = \mathbf{x} = (x_1, \ldots, x_n)$ allows us to update our information on θ, giving rise to a posterior distribution for θ which we will conveniently denote with its distribution function $F_{\Theta|\mathbf{X}}$. The pure premium is the expected value of X_{n+1} with the updated information $F_{\Theta|\mathbf{X}}$ on θ, and this can be expressed in equivalent ways as

$$\begin{aligned} E(X_{n+1} \mid \mathbf{X} = \mathbf{x}) &= E_{\Theta|\mathbf{X}}\left[E(X_{n+1} \mid \Theta = \theta, \mathbf{X} = \mathbf{x})\right] \\ &= \int m(\theta) f_{\Theta|\mathbf{X}}(\theta \mid \mathbf{x}) \, d\theta \end{aligned}$$

$$= \int \left[\int x f_{X|\Theta}(x \mid \theta) \, dx \right] f_{\Theta|\mathbf{X}}(\theta \mid \mathbf{x}) \, d\theta$$

$$= E[m(\Theta) \mid \mathbf{X} = \mathbf{x}],$$

and hence is the posterior mean of $m(\Theta)$ given $\mathbf{X} = \mathbf{x}$.

We shall use $E_{\Theta|\mathbf{X}}[m(\Theta)]$ to denote the random variable which on observing $\mathbf{X} = \mathbf{x}$ takes the value $E[m(\Theta) \mid \mathbf{X} = \mathbf{x}]$. Bühlmann [11] states that experience rating is "a sequence of estimates for $E[m(\Theta)]$ based on the observations of $X_1, \ldots, X_n$."

5.4.1 Bayes and linear estimates of the posterior mean

In some cases, the posterior mean (or Bayes estimate when using quadratic loss) $E[m(\Theta) \mid \mathbf{X} = \mathbf{x}]$ is conveniently expressible as a linear function of the sample observations $\mathbf{x}$ and what one might call the prior mean $\mu_0 = E[m(\Theta)]$. For example, in the basic normal|normal model of Example 5.3, $m(\theta) = \theta$ and hence the pure premium $E(X_{n+1} \mid \mathbf{X} = \mathbf{x}) = E(\Theta \mid \mathbf{X} = \mathbf{x})$ can be expressed as a linear function of the observations $\mathbf{x}$ given by (see Equation (5.6))

$$E(X_{n+1} \mid \mathbf{X} = \mathbf{x}) = \frac{\sigma_0^2}{\sigma_0^2 + \sigma^2/n} \, \bar{x} + \frac{\sigma^2/n}{\sigma_0^2 + \sigma^2/n} \, \mu_0.$$

In the Poisson | gamma model for the number of annual claims, the pure premium takes the form

$$E(X_{n+1} \mid \mathbf{X} = \mathbf{x}) = E(\Lambda \mid \mathbf{X} = \mathbf{x})$$

$$= \frac{n}{n+\beta} \, \frac{\sum x_j}{n} + \frac{\beta}{n+\beta} \, \frac{\alpha}{\beta}$$

$$= \mathbf{Z} \, \bar{x} + (1 - \mathbf{Z}) \, \mu_0.$$

In other situations of interest, however, the posterior distribution for $m(\Theta)$ may be difficult to determine, and the posterior mean $E(m(\Theta) \mid \mathbf{X} = \mathbf{x})$ may not be conveniently expressible as a linear combination of the observations $\mathbf{x}$ and μ_0. This led Hans Bühlmann [11] to develop an elegant model for credibility theory where the credibility formula is a *linear combination of the observations in the sample data* $\mathbf{X} = (X_1, \ldots, X_n)$. His stated objective was to find the linear coefficients $a_0, a_1, \ldots, a_n$ which minimize $E(X_{n+1} - [a_0 + \sum_1^n a_j X_j])^2$ with respect to the posterior distribution of $\Theta \mid \mathbf{X}$. One may show that this is equivalent to minimizing either

$$E(E[m(\Theta) \mid \mathbf{X}] - [a_0 + \sum_1^n a_j X_j])^2 \quad \text{or} \quad E(m(\Theta) - [a_0 + \sum_1^n a_j X_j])^2.$$

In what follows we give some insight into how one can determine (optimal) values for $a_0, a_1, \ldots, a_n$ in order to minimize

$$E(X_{n+1} - [a_0 + \sum_{j=1}^n a_j X_j])^2. \tag{5.11}$$

Remember that the random observations $X_j \mid \theta$, $j = 1, \ldots, n+1$ are independent and identically distributed. We will therefore make the reasonable assumption (which may be justified) that in minimizing (5.11), we may restrict consideration to the case where $a_i = a$ for all $i = 1, \ldots, n$. Hence we want to find a_0 and a which minimize $E(X_{n+1} - [a_0 + a \sum_1^n X_j])^2$. The following technical result is both insightful and of considerable use.

THEOREM 5.1

$$E(X_{n+1} - [a_0 + a \sum_{j=1}^n X_j])^2 = (na^2 + 1) E[s^2(\Theta)] + E[(na - 1)m(\Theta) + a_0]^2.$$

(5.12)

PROOF   Conditioning on $\Theta = \theta$,

$$E\left([X_{n+1} - a_0 - a \sum_{j=1}^n X_j]^2 \mid \theta\right)$$

$$= E([X_{n+1} - m(\theta)]^2 \mid \theta) + E([m(\theta) - a_0 - a \sum_{j=1}^n X_j]^2 \mid \theta)$$

$$\{\text{since } E([X_{n+1} - m(\theta)] [m(\theta) - a_0 - a \sum_{j=1}^n X_j] \mid \theta) = 0\}$$

$$= s^2(\theta) + E([-(na - 1)m(\theta) - a_0 - a \sum_{j=1}^n [X_j - m(\theta)]]^2 \mid \theta)$$

$$= s^2(\theta) + a^2 \sum_{j=1}^n E([X_j - m(\theta)]^2 \mid \theta) + [(na - 1)m(\theta) + a_0]^2$$

$$= (na^2 + 1) s^2(\theta) + [(na - 1)m(\theta) + a_0]^2.$$

Theorem 5.1 follows on taking expectation with respect to Θ.   ∎

Note that the first term on the right-hand side of the expression (5.12) does not involve a_0, and thus $E(X_{n+1} - [a_0 + a \sum_1^n X_j])^2$ is clearly minimized if $a_0 = -(na - 1)E[m(\Theta)]$. Finding the optimal value for a therefore reduces to minimizing

$$(na^2 + 1) E[s^2(\Theta)] + (na - 1)^2 E(m(\Theta) - E[m(\Theta)])^2$$
$$= (na^2 + 1) E[s^2(\Theta)] + (na - 1)^2 Var[m(\Theta)],$$

and therefore the optimal value for a is

$$a = \frac{1}{n + K} = \frac{1}{n + E[s^2(\Theta)]/Var[m(\Theta)]},$$

where $K = E[s^2(\Theta)]/Var[m(\Theta)]$. Since $s^2(\theta)$ is the random variation (or variance) of $X_j \mid \theta$, the numerator in the expression for K is the average value of this variation with respect to the prior distribution of Θ, and hence is often called the *expected value of the process variance*. Analogously, the denominator is called the *variance of the hypothetical means*.

In summary, the best (in terms of minimizing mean square deviation) linear estimator of X_{n+1} of the form $a_0 + \sum_1^n a_j X_j$, is given by

$$E(X_{n+1} \mid \mathbf{X} = \mathbf{x}) = \sum_{j=1}^{n} \frac{1}{n + E[s^2(\Theta)]/Var[m(\Theta)]} X_j$$

$$+ \left(1 - \frac{n}{n + E[s^2(\Theta)]/Var[m(\Theta)]}\right) E[m(\Theta)]$$

$$= \frac{n}{n + E[s^2(\Theta)]/Var[m(\Theta)]} \bar{\mathbf{X}}$$

$$+ \left(1 - \frac{n}{n + E[s^2(\Theta)]/Var[m(\Theta)]}\right) E[m(\Theta)]$$

$$= \mathbf{Z} \, \bar{\mathbf{X}} + (1 - \mathbf{Z}) \, E[m(\Theta)],$$

where $\mathbf{Z} = n/(n + K)$.

Credibility estimators of this form are often called *Bühlmann credibility estimators*, and this approach to credibility theory is sometimes called *Greatest Accuracy Credibility Theory*. Jewell [30] provides a good analysis of Bühlmann credibility estimators and Bayesian estimators. Note that the Bühlmann credibility factor $\mathbf{Z} = n/(n + K)$ approaches 1 as the sample size n increases. It is also a decreasing function of $E[s^2(\Theta)]$, that is, the larger the average volatility (or process variance) of observations between different risk parameters is, the less weight we put on the sample mean $\bar{\mathbf{X}}$. On the other hand, it is an increasing function of $Var[m(\Theta)]$ since the larger the variation is between the hypothetical means $m(\theta)$, the more weight we put on our actual sample data.

Example 5.7
Small companies in an industrial estate are insured for fire and theft. Prior claims experience with such companies has led to modeling annual aggregate claims by compound Poisson distributions. Generally speaking, companies are divided into two types, where the different assumptions made on the compound Poisson distributions are given in Table 5.2. Approximately 60% of companies are classified as type θ_1. The claims experience over the past three years from a relatively new company is given by

$$\mathbf{S} = (S_1, S_2, S_3) = (0, \, 591 + 790, \, 740 + 846).$$

Here, for example, in year 3 there were 2 claims of respective sizes 740 and 846. Given this information, what would you quote as a pure premium (for next year) for this company?

TABLE 5.2
Risk groups for fire and theft insurance.

Company type θ	Prior on θ	Poisson rate λ	Claim X_j
θ_1	0.6	1	$X_1 \sim$ Pareto (3,1200)
θ_2	0.4	1.5	$X_2 \sim$ Pareto (3,1600)

The pure premium of interest is $E(S_4 \mid \mathbf{S} = \mathbf{s})$, and although we do not know which type of company this came from, our information (s_1, s_2, s_3) allows us to update the prior information. Now $m(\theta_1) = E(S \mid \theta_1) = 1 \cdot (1200/2) = 600$, and similarly $m(\theta_2) = 1200$. Therefore the prior mean of $m(\Theta)$ is

$$\mu_0 = E[m(\Theta)] = 600\,(0.6) + 1200\,(0.4) = 840.$$

Now

$$s^2(\theta_1) = \lambda_1 E(X_1^2) = 1[600^2 + 600^2(3)] = 4[600]^2,$$

and similarly, $s^2(\theta_2) = 6[800]^2$. It follows that $E[s^2(\Theta)] = 2{,}400{,}000$. Finally,

$$Var[m(\Theta)] = 600^2\,(0.6) + 1200^2\,(0.4) - 840^2 = 86{,}400\,,$$

and hence the factor $K = 2{,}400{,}000/86{,}400 = 27.7778$. This gives a Bühlmann credibility factor of $\mathbf{Z} = 3/(3 + K) = 0.0975$. Therefore the pure premium is estimated as

$$
\begin{aligned}
E(S_4 \mid \mathbf{s}) &= (0.0975)\,\bar{s} + (1 - 0.0975)\,\mu_0 \\
&= (0.0975)\,989 + (0.9025)\,840 \\
&= 854.52.
\end{aligned}
$$

□

5.4.2   Predictive distribution for X_{n+1}

The distribution of X_{n+1} given the posterior distribution $\Theta \mid \mathbf{X}$ is useful for making predictions about the next claims value, and hence is called the *predictive distribution*. This is actually a mixture distribution of $X \mid \theta$ with respect to $\Theta \mid \mathbf{X}$, and is given by

$$f_{X_{n+1}}(x) = \int f_{X \mid \Theta}(x \mid \theta) f_{\Theta \mid \mathbf{X}}(\theta \mid \mathbf{x})\, d\theta.$$

Example 5.8 Poisson | Gamma model.

Let $X \mid \lambda$ be Poisson with mean λ, where prior information about λ is given by a $\Gamma(\alpha, \beta)$ distribution. On observing numbers of claims $\mathbf{X} = (x_1, \ldots, x_n)$ over n years, the predictive distribution for X_{n+1} (the random number of claims which will be made next year) is the negative binomial distribution $NB(r = \sum x_j + \alpha, p = (n+\beta)/(n+\beta+1))$. If $\sum x_j + \alpha$ is an integer, then the probability that there will be k claims next year is the same as the probability that it will take $\sum x_j + \alpha + k$ Bernoulli trials to observe the $(\sum x_j + \alpha)^{\text{th}}$ success, where the probability of success is $p = (n+\beta)/(n+\beta+1)$. Hence

$$
\begin{aligned}
P[X_{n+1} = k] &= \int_0^\infty P[X_{n+1} = k \mid \lambda] \, f_{\Lambda \mid \mathbf{X}}(\lambda \mid \mathbf{x}) \, d\lambda \\
&= \int_0^\infty \frac{\lambda^k e^{-\lambda}}{k!} \frac{(n+\beta)^{\sum x_j + \alpha} \lambda^{\sum x_j + \alpha - 1} e^{-\lambda(n+\beta)}}{\Gamma(\sum x_j + \alpha)} \, d\lambda \\
&= \frac{1}{\Gamma(\sum x_j + \alpha) \, \Gamma(k+1)} \cdot \\
&\qquad \frac{(n+\beta)^{\sum x_j + \alpha}}{(n+\beta+1)^{\sum x_j + \alpha + k}} \int_0^\infty w^{\sum x_j + \alpha + k - 1} e^{-w} \, dw \\
&= \frac{\Gamma(\sum x_j + \alpha + k)}{\Gamma(\sum x_j + \alpha) \, \Gamma(k+1)} \left(\frac{n+\beta}{n+\beta+1} \right)^{\sum x_j + \alpha} \left(\frac{1}{n+\beta+1} \right)^k
\end{aligned}
$$

In particular, continuing with Example 5.5, it follows that

$$
P[X_6 = 4] = \frac{\Gamma(14 + 3 + 4)}{\Gamma(14 + 3) \, \Gamma(4 + 1)} \left(\frac{5+1}{5+1+1} \right)^{14+3} \left(\frac{1}{5+1+1} \right)^4
$$

$$
= 0.1468.
$$

$\Box$

5.5   Empirical Bayes approach to credibility theory

Bühlmann or greatest accuracy credibility estimates for pure premiums are theoretically appealing, but in practice it may be difficult to actually determine the *prior* quantities $E[m(\Theta)]$, $Var[m(\Theta)]$ and $E[s^2(\Theta)]$ necessary to calculate these estimates. In the (nonparametric) *empirical Bayes* approach to credibility theory, the data itself is used to estimate these quantities.

In this section we will discuss two models where one possesses knowledge and data on N risks over n years, with the objective of estimating a pure premium for the next year for one or all of the risks. Model 1 is the simpler of the two models; however, it is more restrictive than Model 2 in that it does not account for the varying annual volume of business that may contribute to the observed amount or number of claims. We denote the unknown risk parameter by θ, while $m(\theta)$ will be the quantity of interest. This quantity may be θ itself, but there are many other possibilities like standardized versions of pure premiums or claim rates. In Model 1 we use the available data to estimate the quantities $E[m(\Theta)], Var[m(\Theta)]$ and $E[s^2(\Theta)]$, and hence obtain a Bühlmann-type credibility estimate for a particular risk. In Model 2 we follow the approach of Bühlmann and Straub [12] allowing for varying annual volumes in the risks, and then after standardizing annual aggregate claims by these volumes, we obtain empirically based credibility estimates for pure premiums.

5.5.1 Empirical Bayes credibility – Model 1

The data (random observations) are of the form $\{\{X_{ij}\}_{i=1}^N\}_{j=1}^n$, where X_{ij} represents the aggregate claims in the jth year from the ith risk. Assume that there is an unknown risk parameter θ_i for the ith risk, which for $i = 1, \ldots, N$ represents a realization from a random variable Θ. We denote by $m(\theta_i)$ and $s^2(\theta_i)$, respectively, the mean and variance of $X_{ij} \mid \theta_i$ for $j = 1, \ldots, n$. Although we assume that $X_{ij} \mid \theta_i$ for $j = 1, \ldots, n$ are independent and identically distributed for any given risk i, the (unconditional) random variables $X_{ij}, j = 1, \ldots, n$ are not necessarily independent, even though they are identically distributed.

We do not assume any particular form for the distribution of the random variable $m(\Theta)$, but we shall denote its mean and variance by $E[m(\Theta)]$ and $Var[m(\Theta)]$, respectively. The mean value of the variability within a risk, or the expected value of the process variance, is denoted by $E[s^2(\Theta)]$. As we have already noted, in the normal | normal model, $m(\theta) = \theta$, $s^2(\theta) = \sigma^2$ (which is constant in θ), $E[m(\Theta)] = \mu_0$, $Var[m(\Theta)] = \sigma_0^2$ and $E[s^2(\Theta)] = \sigma^2$.

From the previous section, it follows that the greatest accuracy credibility estimate of the pure premium for risk i (the predicted mean aggregate claim amount $E(X_{i,n+1} \mid \mathbf{X} = \mathbf{x})$ for next year in risk i) given the data takes the form

$$\frac{n}{n + E[s^2(\Theta)]/Var[m(\Theta)]}\, \bar{X}_i + \left(1 - \frac{n}{n + E[s^2(\Theta)]/Var[m(\Theta)]}\right) E[m(\Theta)].$$

In the empirical Bayes approach to credibility theory, one actually uses the data to estimate the three parameters $E[m(\Theta)]$, $E[s^2(\Theta)]$ and $Var[m(\Theta)]$. Letting X_{ij} represent the aggregate claims in the jth year from the ith risk, $\bar{X}_i = \sum_{j=1}^n X_{ij}/n$ and $\bar{X} = \sum_{i=1}^N \bar{X}_i/N$, one may show that the estimators given in Table 5.3 are unbiased.

The estimators for $E[m(\Theta)]$ and $E[s^2(\Theta)]$ are intuitively what one might expect (that for $E[s^2(\Theta)]$ is simply the average of the sample variances in each risk). Naively, one might initially expect that $\sum_{i=1}^N (\bar{X}_i - \bar{X})^2/(N-1)$ would

TABLE 5.3
Model 1: Unbiased estimators for $E[m(\Theta)]$, $E[s^2(\Theta)]$ and $Var[m(\Theta)]$.

Estimator	
$\widehat{E[m(\Theta)]}$	$= \bar{X}$
$\widehat{E[s^2(\Theta)]}$	$= \sum_{i=1}^{N} \sum_{j=1}^{n} (X_{ij} - \bar{X}_i)^2 / (N(n-1))$
$\widehat{Var[m(\Theta)]}$	$= \sum_{i=1}^{N} (\bar{X}_i - \bar{X})^2 / (N-1) - \sum_{i=1}^{N} \sum_{j=1}^{n} (X_{ij} - \bar{X}_i)^2 / (nN(n-1))$

be a good estimator for $Var[m(\Theta)]$, but a correction factor (namely, $\widehat{E[s^2(\Theta)]}$) divided by n) is needed to make it unbiased.

Example 5.9
Table 5.4 gives data on aggregate motor claims ($000's$) in each of three regions of a country for a national insurer over five years.

TABLE 5.4
Aggregate motor claims over five years.

Region	Year				
	1	2	3	4	5
1	5,841	7,782	5,373	7,020	7,773
2	5,910	4,491	6,102	5,373	6,651
3	7,011	8,045	7,078	7,266	9,027

Using Model 1 of Empirical Bayes credibility we calculate the credibility factor and credibility premium for next year for each of the three regions. For each of the regions we calculate (sample) means and variances, yielding

$$(\bar{X}_1, \bar{X}_2, \bar{X}_3) = (6{,}757.8, \ 5{,}705.4, \ 7{,}685.4), \ \text{and}$$
$$(s^2(\theta_1), s^2(\theta_2), s^2(\theta_3)) = (1{,}226{,}639.7, \ 669{,}642.3, \ 732{,}212.3).$$

Hence we estimate $E[m(\Theta)]$ and $E[s^2(\Theta)]$ by 6,716.2 and 876,164.8, respectively. Since the sample variance of $(\bar{X}_1, \bar{X}_2, \bar{X}_3) = (990.6553)^2$, it follows that $Var[m(\Theta)]$ is estimated by $(990.6553)^2 - 876{,}164.8/5 = 806{,}165$. Therefore the credibility factor (which is the same for each region) is

$$Z = \frac{n}{n + \widehat{E[s^2(\Theta)]}/\widehat{Var[m(\Theta)]}} = 5/(5 + 1.0868) = 0.8214.$$

The credibility premium for region 1 is therefore

$$Z\bar{X}_1 + (1 - Z)(6716.2) = 0.8214(6757.8) + 0.1786(6716.2) = 6{,}750.37,$$

and similarly, those for regions 2 and 3 are, respectively, 5,885.88 and 7,512.35.

▯

Example 5.10

A general insurance company has four separate groups of policyholders for residential household contents insurance. Total claim amounts for five successive years have been calculated for each of the separate groups, and the figures in Table 5.5 have been adjusted to take account of inflation. The objective is to estimate the pure premium for group A for the following year. Two methods (empirical Bayes and Bayes) are being proposed by two different analysts.

TABLE 5.5
Aggregate claims for 5 groups.

Year →	1	2	3	4	5
Group A	58	42	98	130	64
Group B	204	186	246	222	186
Group C	183	153	215	171	147
Group D	78	104	77	116	118

Analyst 1 would like to estimate the pure premium for Group A using empirical Bayes credibility theory. Analyst 2 does not feel information from Groups B, C and D is relevant and decides to ignore it completely. She does, however, feel that she has prior information for the pure premium $m(\theta_A)$ which is $N(90, 10^2)$, and she also feels that the five years of data from Group A can be viewed as a random sample from an $N(\theta_A, 35^2)$ distribution.

Let us denote by $m(\hat{\theta}_A^1)$ and $m(\hat{\theta}_A^2)$ the respective estimates of $m(\theta_A)$ for Analysts 1 and 2. Summary statistics on aggregate claims for each of four risks over five years are given in Table 5.6. Here X_{ij} denotes aggregate claims in year j for risk i. These give estimates as follows:

TABLE 5.6
Summary statistics on aggregate claims.

Risk	$\overline{X}_i$	$\sum_{j=1}^{5}(X_{ij} - \overline{X}_i)^2/4$
A	78.4	1248.8
B	208.8	655.2
C	173.8	735.2
D	98.6	399.8

$$\widehat{E[m(\Theta)]} = 139.9,$$
$$\widehat{E[s^2(\Theta)]} = 759.7, \text{ and}$$
$$\widehat{Var[m(\Theta)]} = \sum (\bar{X}_i - \bar{X})^2/3 - E[s^2(\Theta)]/5$$

$$= (61.6018)^2 - 151.9487 = 3{,}642.83.$$

Hence

$$Z = \frac{5}{5 + E\widehat{[s^2(\Theta)]}/\widehat{Var[m(\Theta)]}} = \frac{5}{5 + 0.2086} \doteq 0.96.$$

Therefore the pure premium for Analyst 1 is estimated to be

$$m(\hat{\theta}_A^1) = 0.96\,(78.4) + 0.04\,(139.9) = 75.26 + 5.60 = 80.86.$$

The posterior distribution for Analyst 2 is

$$N\left(\frac{(5/35^2)78.4 + 90/10^2}{5/35^2 + 1/10^2} = 86.64, \quad \left(\frac{1}{35^2/5} + \frac{1}{10^2} \right)^{-1} \right),$$

and hence her estimate is $m(\hat{\theta}_A^2) = 86.64.$ □

5.5.2   Empirical Bayes credibility – Model 2

In many practical situations one has not only information on annual aggregate claims, but also on other relevant information like premium income or number of policies. The data might be given as in Table 5.7.

TABLE 5.7

Year	1	2	,	,	,	n
Aggregate claims	Y_1	Y_2	,	,	,	Y_n
Volume of business	P_1	P_2	,	,	,	P_n

Usually, P_j will denote a volume or amount of business for the jth year, and knowing P_{n+1} one would like to estimate aggregate claims Y_{n+1} for next year. Let us define the random variable $X_j = Y_j/P_j$ for $j = 1, \ldots, n, n+1$, where the risk parameter is θ. $E(X_j \mid \theta) = m(\theta)$ may be viewed (or interpreted) as the average amount of claims per unit of risk. One might then estimate $m(\theta)$ with $\widehat{m(\theta)}$, and then predict Y_{n+1} by $P_{n+1}\,\widehat{m(\theta)}$. Our motivation for introducing X_j (instead of using just Y_j) is to obtain some form of standardized measure of claim size when the amount of business P_j varies in different years.

Somewhat similar to our assumptions in Model 1, we shall assume in Model 2 that given the risk parameter θ for a company, then $X_1, \ldots, X_n, X_{n+1}$ are independent random variables – however they are not necessarily *identically* distributed as they were in Model 1. We do assume, however, that they have the same mean (essentially, that the average size of a claim per unit of risk is the same in each year), and this is denoted by $m(\theta) = E(X_j \mid \theta)$. We also assume that $s^2(\theta) = P_j\,Var(X_j \mid \theta)$ is independent of j, and hence that $Var(X_j \mid \theta) = s^2(\theta)/P_j$ only varies from year to year because of the varying volume of business. In Model 1 the annual aggregate claim amounts with risk factor θ are treated equally in estimating $E[m(\theta) \mid \mathbf{Y}]$, while in Model 2 we weight the aggregate claims by the (reciprocal of the) amount of business they represent.

In practice, there will be an unknown value of $m(\theta)$ for a particular company, but there will also be collateral information on other companies. We will use both sources of information to estimate $m(\theta)$ for a given risk factor θ. Generally speaking, we shall have data on N risks over n years, where the data is of the form $\{\{Y_{ij}\}_{i=1}^{N}\}_{j=1}^{n}, \{\{P_{ij}\}_{i=1}^{N}\}_{j=1}^{n}$. Here Y_{ij} represents the aggregate claims in the jth year from the ith risk (or company) and P_{ij} is the corresponding risk volume. We use the following notation:

$$\bar{P}_i = \sum_{j=1}^{n} P_{ij}, \quad \bar{P} = \sum_{i=1}^{N} \bar{P}_i$$

$$P^* = (Nn-1)^{-1} \sum_{i=1}^{N} \bar{P}_i(1 - \bar{P}_i/\bar{P})$$

$$X_{ij} = Y_{ij}/P_{ij}, \quad \bar{X}_i = \sum_{j=1}^{n} P_{ij}X_{ij}/\bar{P}_i \quad \text{and} \quad \bar{X} = \sum_{i=1}^{N} \sum_{j=1}^{n} P_{ij}X_{ij}/\bar{P}.$$

The best (Bühlmann and Straub [12]) linear estimator for next year's total (as well as expected) aggregate claims $Y_{i\,n+1}$ for company (or risk) i is of the form $P_{i,n+1} X_{i\,n+1}$, where $X_{i\,n+1}$ is estimated by

$$\left(\mathbf{Z}_i\,\bar{X}_i + (1 - \mathbf{Z}_i)\,E[m(\Theta)]\right) =$$

$$\frac{\sum_j P_{ij}}{\sum_j P_{ij} + E[s^2(\Theta)]/Var[m(\Theta)]}\,\bar{X}_i + \frac{E[s^2(\Theta)]/Var[m(\Theta)]}{\sum_j P_{ij} + E[s^2(\Theta)]/Var[m(\Theta)]}\,E[m(\Theta)].$$

Note that the credibility factors $\mathbf{Z}_i$ would be the same for all risks if the total volumes of business $(\sum_j P_{ij})$ were the same for all risks. As in Model 1, we need estimates of the parameters $E[m(\Theta)], E[s^2(\Theta)], Var[m(\Theta)]$, and in the empirical Bayes approach unbiased estimators based on the data obtained are given in Table 5.8.

Example 5.11
The data in Table 5.9 was collected on aggregate claims resulting from smoke damage ($\$000's$) for three risks over the years $2002 - 2006$. Here Y_{ij} denotes total claims for risk i in the jth year. Using Model 1 of empirical Bayes credibility, we calculate the credibility factor and premium for next year for each of these risks. Now

$$\widehat{E[m(\theta)]} = 5{,}512\,,$$

$$\widehat{E[s^2(\theta)]} = (891{,}136 + 272{,}484 + 320{,}356)/12 = 123{,}664.67 \text{ and}$$

$$\widehat{Var[m(\theta)]} = 3{,}823{,}804 - (123{,}664.67)/5 = 3{,}799{,}071.2.$$

TABLE 5.8
Model 2: Unbiased estimators for $E[m(\Theta)]$, $E[s^2(\Theta)]$ and $Var[m(\Theta)]$.

Estimator	
$\widehat{E[m(\Theta)]}$	$= \bar{X}$
$\widehat{E[s^2(\Theta)]}$	$= N^{-1}\sum_{i=1}^{N}(n-1)^{-1}\sum_{j=1}^{n}P_{ij}(X_{ij}-\bar{X}_i)^2$
$\widehat{Var[m(\Theta)]}$	$= P^{*-1}((Nn-1)^{-1}\sum_{i=1}^{N}\sum_{j=1}^{n}P_{ij}(\bar{X}_{ij}-\bar{X})^2$
	$\quad -N^{-1}\sum_{i=1}^{N}(n-1)^{-1}\sum_{j=1}^{n}P_{ij}(X_{ij}-\bar{X}_i)^2\ )$

TABLE 5.9
Aggregate claims from smoke damage.

Year	2002	2003	2004	2005	2006	$\bar{Y}_i$	$\sum_1^5(Y_{ij}-\bar{Y}_i)^2$
Risk i							
1	4,560	4,825	4,965	5,325	5,775	5,090	891,136
2	3,425	3,700	3,825	3,940	4,120	3,802	272,484
3	7,200	7,540	7,760	7,810	7,910	7,644	320,356

Hence the credibility factor $\mathbf{Z}$ takes the form

$$\mathbf{Z} = \frac{n}{n+K} = \frac{5}{5+(123,664.67/3,799,071.2)} = 0.9935.$$

Note that in using Model 1 here, we have almost full credibility ($\mathbf{Z} = 0.9935$) in the data from any particular risk. This is essentially due to the very small estimate for $K = E[s^2(\Theta)]/Var[m(\Theta)]$ of 0.03255 (since the variance $Var[m(\Theta)]$ between the hypothetical means is so much more than the mean $E[s^2(\Theta)]$ of the process variance). The risk premiums for risks 1, 2 and 3 are therefore, respectively, 5,092.74, 3,813.12 and 7,630.14. Note that they are all very close to the values $\bar{Y}_1, \bar{Y}_2$ and $\bar{Y}_3$.

Now suppose that we also have a measure of risk volume P_{ij} corresponding to each Y_{ij} above. Suppose also that by using Model 2 for empirical Bayes credibility, the credibility premium per unit of risk volume for next year for risk 1 has already been calculated to be 7.10. What would the corresponding credibility premiums be for risks 2 and 3? Some summary statistics for the risk volumes are:

Risk i	$\sum_{j=1}^{5}P_{ij}$
1	3,760
2	2,448
3	4,624

We have that $\bar{X} = \sum 5\bar{Y}_i/\sum P_{ij} = 7.6329$. Therefore

$$7.10 = \mathbf{Z}_1(6.7686) + (1-\mathbf{Z}_1)(7.6329) \quad \Rightarrow \quad \mathbf{Z}_1 = 0.5329/0.8643 = 0.6166.$$

Hence

$$0.6166 = \frac{3760}{3760 + \widehat{E[s^2(\theta)]}/\widehat{Var[m(\Theta)]}} \quad \Rightarrow \quad \widehat{E[s^2(\theta)]}/\widehat{Var[m(\Theta)]} = 2337.98.$$

Therefore, $\mathbf{Z}_2 = 0.5115$, and the credibility premium per unit risk for risk 2 is estimated to be

$$0.5115\,(5)\,(3802/2448) + 0.4885\,(7.6329) = 7.70.$$

Similarly for risk 3, $\mathbf{Z}_3 = 0.6642$ with risk premium 8.05.   ∎

5.6 Problems

1. Aggregate annual claims are modeled by a compound Poisson distribution where a typical claim is Pareto with mean 600 and variance 720,000. For the past few years the annual claim rate λ has been of the order of 800. How many years of claims experience r would be necessary to ensure that $|\,\bar{S} - E(S)\,| \leq 0.05\,E(S)$ with probability at least 0.98? If the parameter of interest was actually only the annual claim rate, how many years r would be necessary to estimate λ with full credibility $(0.05, 0.01)$ when using the average annual claim rate?

2. In Chapter 2 on loss distributions (Example 2.3), a lognormal distribution with parameters $\tilde{\mu} = 8.74927$ and $\tilde{\sigma}^2 = 0.36353$ was fitted to data on automobile damage claims in ($\$000's$) during the year 2002 for a fleet of rental cars. If the annual aggregate claims S are modeled by a compound Poisson distribution with $\lambda = 500$, how many years r of data do we need to be able to say that $\bar{S}$ will be a fully credible $(0.10, 0.05)$ estimator of $E(S)$?

3. A group of individuals with similar age have a mortality rate of 0.02 per annum (based on previous experience). It is desired to quote an annual pure premium for a group policy for these individuals on three bases:

   (a) Each life is insured for $16,000.
   (b) Each life is insured for one year's salary, and the distribution of salary is approximately $N(16{,}000, (1000\sqrt{3})^2)$.
   (c) Each life is insured for an amount determined by both salary and experience. It is reasonable to assume that this variable is distributed uniformly on $[13{,}000,\ 19{,}000]$ for the group.

   How many lives would you require in each of the above situations to obtain full credibility $(k = 0.05, p = 0.95)$ (or, equivalently, full credibility $(k = 0.05, \alpha = 0.05)$) for your estimate?

4. Aggregate claims S are modeled by a compound distribution of the form $S = X_1 + \cdots + X_N$. The parameter of interest is the pure premium $\theta = E(N)E(X)$.

   (a) Show that the number of claims $n_{(k,\alpha)}$ necessary for full credibility (k, α) is of the form

$$n_{(k,\alpha)} = \frac{z_{1-\alpha/2}^2}{k^2} \left(\frac{Var(N)}{E(N)} + \frac{Var(X)}{E^2(X)} \right)$$

$$= \frac{z_{1-\alpha/2}^2}{k^2} \left(\frac{Var(N)}{E(N)} + \frac{E(X^2)}{E^2(X)} - 1 \right). \qquad (5.13)$$

   (b) Show that if N is Poisson, then Equation (5.13) reduces to

$$n_{(k,\alpha)} = (z_{1-\alpha/2}/k)^2 (E(X^2)/E^2(X)),$$

   while if X is constant it reduces to

$$n_{(k,\alpha)} = (z_{1-\alpha/2}/k)^2 (Var(N)/E(N)).$$

   The pure premium is $\theta = \lambda E(X)$. Assume the mean claim frequency per policy is 0.015, while the mean and variance of a claim are, respectively, 4,000 and 2,000,000. With this information, what would you estimate the minimum number n of *policies* for full credibility $(k = 0.05, \alpha = 0.02)$ to be?

5. Annual aggregate claims are modeled by a compound Poisson distribution where the claim distribution is well approximated by a gamma distribution with shape parameter 2 $(X \sim \Gamma(2, \beta))$. Full credibility in estimating the pure premium $\theta = E(S)$ is defined by having a 95% chance of the estimator differing from the true value by at most $0.04\,\theta$. What partial credibility (using the square root rule) would be assigned to experience from 1200 claims?

6. The probability that an insured individual will give rise to no claims next year is $e^{-\theta}$, where θ varies by individuals according to the density function $f_\Theta(\theta) = 25\,\theta\,e^{-5\theta}$. What is the probability that a randomly selected individual will give rise to no claims next year?

7. The number of annual claims for malicious damage to a local school are modeled by a Poisson distribution with parameter λ, where λ is initially assumed to be one of the values $(1, 2, 3, 4, 5)$ with prior distribution $f_\Lambda(\lambda) = (0.3, 0.3, 0.2, 0.1, 0.1)$. Over a three-year period, the annual claims experience was $(3, 2, 7)$. Update the distribution for λ (find the posterior) and find the Bayesian estimates of λ when using the quadratic, absolute value, and zero-one loss functions.

8. A group scheme of term life insurance covers 200 lives, each of which is independent with respect to mortality for the coming year. Assume that the probability of death q for the next 12 months is the same for all lives, and although q is unknown previous experience suggests that a prior density of the form $(\beta+1)\,\beta\,q\,(1-q)^{\beta-1}$ for $0 < q < 1$ is appropriate where the prior mean is 0.1. Calculate the parameter β for this prior. If 25 people in this scheme die in the coming year, determine the posterior estimate for q and write it as a credibility estimate.

9. Claims in an automobile portfolio are modeled by a Burr distribution with density function given by

$$f(x \mid \theta) = \begin{cases} \dfrac{2\theta x}{(1+x^2)^{\theta+1}} & \text{for } x > 0 \\ 0 & \text{otherwise,} \end{cases}$$

where θ is an unknown parameter. Prior information suggests that a gamma distribution $\Gamma(42, 20)$ is appropriate for θ. If a sample of $n = 200$ claims is taken, determine the form of the Bayesian estimator for θ, assuming a quadratic loss function is used. Give an expression for a 90% Bayesian belief interval for θ.

10. A random variable Y modeling claim size has density function of the form

$$f_Y(y) = 2\,c\,y\,e^{-cy^2} \qquad \text{for } y > 0,$$

where c is an unknown parameter. Prior information on c suggests the gamma distribution $\Gamma(\alpha = 15, \lambda = 24)$ is suitable. A sample of 40 values of Y yields $\sum_1^{40} y_i^2 = 32.5$. Determine the Bayesian estimator of c using a quadratic loss function.

11. Claims in a portfolio are made according to a Poisson process at the rate of λ per year. Based on collateral data and experience, it is felt that a gamma distribution $\Gamma(r, \alpha)$ with mean 50 and variance 25 is appropriate for λ.

   (a) Determine the parameters r and α of the prior distribution.

   (b) Suppose that data collected over eight years gave rise to 448 claims. Determine the posterior distribution for λ and show that the posterior mean may be written as a credibility estimate. Find an (approximate) 95% Bayesian interval for λ.

12. Claim amounts handled by a general insurer follow a probability distribution with density function

$$f(x \mid \lambda) = \begin{cases} \frac{1}{2}\lambda^3 x^2 e^{-\lambda x} & x > 0 \\ 0 & \text{otherwise.} \end{cases}$$

Prior information on λ can be summarized by an exponential distribution with mean 0.002. A sample of $n = 100$ such claims is taken. Determine the form of the Bayesian estimate for λ (using a quadratic loss function). Would the estimate for λ using an "all or nothing (or zero-one) loss" be higher or lower than using quadratic loss? Justify your answer.

13. Claims in an insurance portfolio are uniformly distributed on the interval $(0, \theta)$ where θ is unknown. A prior distribution for θ is given by the density

$$f(\theta) = \frac{5 \cdot (500)^5}{\theta^6} \quad \text{for } \theta \geq 500 \text{ and } 0 \text{ otherwise.}$$

   (a) Determine an estimate of θ based on the mean of the prior distribution.

   (b) The claims record for one year yields the following 10 values:

   $$350, \ 410, \ 200, \ 520, \ 135, \ 175, \ 600, \ 450, \ 800, \ 250.$$

   Determine the posterior distribution for θ based on this data and obtain the Bayes estimate of θ using a quadratic loss function.

14. Claims are uniformly distributed on the interval $(0, \theta)$, although the upper limit θ on the size of a claim itself is not actually known. It is felt, however, that prior knowledge on θ can be summarized by a distribution with density function

$$f(\theta) = \begin{cases} \alpha\, l^\alpha / \theta^{\alpha+1} & \text{for } \theta \geq l, \text{ and} \\ 0 & \text{otherwise} \end{cases}$$

   where α and l (lower limit on θ) are known positive parameters. Show that the posterior distribution on observing a sample of n claims has the same general form, and find the resulting Bayesian estimate for θ.

15. Let Y represent the proportion of a 20-year mortgage that has been repaid at the end of 16 years. Assume that Y has a density function of the form $\alpha y^{\alpha-1}$ for $0 < y < 1$, and that prior knowledge about the parameter α is summarized by a $\Gamma(\alpha_1, \beta_1)$ distribution. Find the posterior mean and variance for α resulting from a sample of proportional repayments $\mathbf{y} = (y_1, \ldots, y_n)$.

16. Annual claim numbers X_i are modeled by a Poisson distribution with unknown parameter λ. Prior knowledge about λ can be summarized by a uniform distribution on the interval $[60, 120]$.

   (a) Determine the form of the Bühlmann credibility estimator of the pure premium $E(X_{n+1})$ based on a random sample $\mathbf{x}$ of claim numbers over n years.

(b) What would the estimate be if $\mathbf{X} = (106, 105, 110, 98, 101, 113)$?

(c) Generate for yourself six observations from a Poisson distribution with mean 80, and determine the Bühlmann credibility estimate.

17. Small companies in an industrial estate are insured for building damage. Prior claims experience with such companies has led to modeling annual aggregate claims by a compound Poisson distribution. Generally speaking, companies are divided into three types – those of type θ_1, θ_2 and θ_3, where the different assumptions on the compound Poisson distributions are made as given in Table 5.10. Approximately 50% of companies are classified as type θ_1, while 30% are of type θ_2. If the claims experience over the past four years from a relatively new company is given by

$$S = (s_1, s_2, s_3, s_4) = (300 + 705 + 400, 520 + 635, 475, 702 + 235 + 528),$$

then what would you quote as a pure premium (for next year) for this company?

TABLE 5.10
Types of building insurance in industrial estate.

Company type θ	Probability (θ)	Rate λ	Severity X
θ_1	0.5	1	Pareto (4,1200)
θ_2	0.3	1.5	Pareto (4,1500)
θ_3	0.2	2	Pareto (4,1800)

18. Continuing with Examples 5.5 and 5.8, suppose that X_6 is actually equal to 4. Then what is the predictive probability distribution for X_7, and what is $P(X_7 = 4)$?

19. A group of n policies gives rise to a random number N of claims in a given year. The incidence rate λ for claims is the same for each policy, and a prior density for λ which is $\Gamma(\alpha, \beta)$ is appropriate – i.e., the prior density takes the form

$$f(\lambda) = \beta^\alpha \lambda^{\alpha-1} e^{-\beta\lambda}/\Gamma(\alpha), \quad \lambda > 0.$$

Assume that, conditional on a given λ, the probability distribution for N is given by

$$P(N = k|\lambda) = e^{-n\lambda}\frac{(n\lambda)^k}{k!} \quad k = 0, 1, 2, \ldots.$$

(a) Having observed $N = k$, show that the posterior density for λ is a gamma density and express the (posterior mean) estimate of λ as a credibility estimate. How does the credibility factor depend on n, α and β?

(b) Suppose now it is felt that an appropriate prior mean and variance for λ are 0.30 and 0.006, respectively. If there were 40 claims out of 100 policies last year, what would be the credibility estimate of λ? How many claims would you predict for next year if business doubles?

20. A life insurance portfolio consists of 500 policies, each independent of the others with respect to mortality. Suppose the common rate of mortality is q. Although q is unknown, it is felt that a prior density of the form $f(q) = \beta(1 - q)^{\beta-1}$, $0 < q < 1$ is appropriate, where the prior mean is 0.1. Calculate the parameter β for this prior. If 40 policyholders die within the following year, determine the posterior estimate of q and write it as a credibility estimate.

21. Total aggregate claims in a particular company are modeled with a $N(\mu, 12,000^2)$ random variable, where μ is unknown. Prior information about μ suggests that it is "95% likely" to lie in the interval [130,000, 170,000]. Aggregate claims from the last five years were not incorporated in the prior information and are as follows:

$$146,000, \quad 142,000, \quad 153,000, \quad 127,000, \quad 132,000$$

(a) Convert the prior information given into a suitable prior distribution for μ.

(b) Using your assumed prior from (a), determine the Bayesian credibility estimate for μ using quadratic loss. Also find a 95% Bayesian interval for μ.

(c) Determine the standard 95% confidence interval for μ based on the claims data. Contrast briefly the difference in interpretation between the standard confidence interval and the Bayesian interval.

22. Claims are made in an insurance portfolio according to a Poisson distribution with rate parameter λ per year. A prior gamma distribution $\Gamma(\alpha, \beta)$ with mean 10 and variance 20 is used for λ. Calculate the parameters α and β for the prior distribution of λ. Suppose that over the past six years a total of 72 claims have been made on this portfolio. Determine the posterior distribution for λ, and show that the posterior mean for λ may be written as a credibility estimate.

23. The data in Table 5.11 give the aggregate claims for household damage insurance in six successive years by four separate (regional) groups of policyholders. Assume that the claim amounts have been adjusted to remove any effect of inflation and that the unit of money is millions of dollars. Using empirical Bayes credibility Model 1, calculate estimates of the pure premiums for the coming year for each of the regions.

TABLE 5.11

Household damage claims.

Year →	1	2	3	4	5	6
Region A	206	146	271	178	136	162
B	144	284	310	218	266	301
C	64	57	43	97	132	110
D	204	186	248	222	188	204

24. Summary statistics on aggregate claims for each of four car rental company risks over five years are given in Table 5.12 below. Here X_{ij} denotes aggregate claims in year j for risk i. Using Model 1 of empirical Bayes credibility, calculate the credibility factor and credibility premium for risk 1. In your opinion, is Model 1 suitable to calculate credibility premiums for these risks?

TABLE 5.12

Car rental company risks.

Risk	$\bar{X}_i$	$\sum_{j=1}^{5}(X_{ij} - \bar{X}_i)^2$
1	6,132	5,321,643
2	7,465	5,974,212
3	4,927	4,321,615
4	23,416	41,271,314

25. Data on annual aggregate claims ($000's$) for each of three employer liability risks over five years is given in Table 5.13. Here Y_{ij} denotes claims in year j for risk i.

TABLE 5.13

Annual aggregate claims for employer liability risks.

			Year				
Risk	1	2	3	4	5	$\bar{Y}_i$	$\sum_{1}^{5}(Y_{ij} - \bar{Y}_i)^2$
1	3,894	5,188	3,582	4,680	5,182	4,505	2,180,690
2	3,940	2,994	3,582	4,068	4,434	3,804	1,190,476
3	4,382	5,028	4,434	4,844	5,642	4,866	1,049,784

(a) Using Model 1 of empirical Bayes credibility, calculate the credibility factor and credibility premium for next year for each of the three risks.

(b) Suppose now that you also have a risk volume denoted P_{ij} corresponding to each Y_{ij} above. Some summary statistics for the risk volumes are as follows:

Risk	$\sum_{1}^{5} P_{ij}$
1	3,560
2	2,276
3	4,012

In using Model 2 for empirical Bayes credibility theory, the credibility premium per unit of risk volume for next year for risk 1 has been calculated to be 6.46. Calculate the credibility premium per unit of risk volume for next year for risks 2 and 3.

26. The annual numbers of fatal traffic motor accidents are monitored in three different geographical regions, with the results given in Table 5.14. Two analysts are asked to predict the number of fatal traffic accidents for Region A for next year. Analyst 1 would like to like to use the empirical Bayes Model 1 to make a prediction. Analyst 2 is going to use a Bayesian approach assuming the number of annual traffic accidents is a Poisson variable with parameter λ, and where she can assume prior knowledge of λ summarized by a gamma $\Gamma(54, 3)$ distribution. Find the predictions which the two analysts would make in this situation!

TABLE 5.14
Annual fatal traffic accidents.

Year $\rightarrow$	1	2	3	4	5	6
Region A	14	17	12	19	18	16
Region B	21	32	16	7	17	33
Region C	11	7	9	8	12	13

6

No Claim Discounting in Motor Insurance

6.1 Introduction to No Claim Discount schemes

If risks are not all equal in an insurance scheme, it seems fair (and perhaps essential) to require insured parties to contribute (in the form of premiums) roughly in proportion to their relative risk. The risk of a motor accident, which often gives rise to an insurance claim, varies from driver to driver. What are the factors that contribute most to high risks of claims in motor insurance?

It would seem that the risk of a motor accident must be correlated with the annual driving distance (mileage) of a driver, but this is often difficult to measure in an economical way. Some attempts have been made to use it directly (Sweden and Holland) or indirectly (using, for example, distance between home and work in the USA), but many of these measures are often underreported. In the USA there are indications (Lemaire [37]) that such underreporting results in more motor insurance fraud than faked accidents.

The ability or skill of a driver is also related to the propensity to give rise to a claim. This encompasses such factors as knowledge and appreciation of the rules of the road, care with respect to speed restrictions and driving conditions, and good judgement about driving when tired or under the influence of factors such as alcohol, tension and aggressiveness. However, these factors are also difficult if not impossible to measure cheaply and effectively. Some variables that are more accessible and commonly used in rating are the age, gender, marital status, occupation, residence and driving experience of the driver. The age, engine size, model, purpose and even color of the vehicle are also used. In his study, Coutts [17] suggested using type of cover, policyholder age, vehicle age and vehicle size as rating factors, while Brockman and Wright [10] suggested adding district and vehicle use to these. Having selected appropriate variables, each set of individuals with common values for these factors (or covariates) constitutes a rating or tariff group. The group is then charged an appropriate premium by the insurance company based on its own (and others') experience. However, such groups can still be quite heterogeneous in their risk factors, and insurers are constantly trying to reduce heterogeneity in risk groups.

Research has shown (e.g., Lemaire [35]) that if insurers are allowed to use only one rating category then it should be based on some form of merit-rating. Evidence suggests that the best indicator of future claims for a driver is the individual's past claims history. In the 1950s the idea emerged to adjust one's initial premium assessment following the actual observation of the individual's claims experience with the use of a No Claim Discount (NCD) system for rating.

No Claim Discount systems (sometimes also called Bonus–Malus systems) are experience rating systems which are commonly used in motor insurance. NCD schemes (or systems) represent an attempt to categorize policyholders into relatively homogeneous risk groups who pay premiums relative to their claims experience. Those who have made few claims in recent years are rewarded with discounts on their initial premium, hence are enticed to stay with the company. Depending on the rules in the scheme, new policyholders may be required to pay the full premium initially and then obtain discounts in the future as a result of claim-free years.

There are a wide variety of NCD schemes. The regulatory systems in the UK and Ireland give insurance companies considerable latitude in designing their own schemes and tariff structures. The number of discount classes, their range, and the rules for moving between them can vary considerably, but typical features of such schemes are the following:

1. The number of classes (or discount levels) in the scheme varies between four and eight, with six being the most common.

2. The discount levels increase in uniform steps (from a 0% discount) and depend on the number of classes, with the maximum discount varying between 40% and 70%.

3. Transition rules vary considerably. A policyholder may go back two levels for the first claim in any year and to the 0% discount level if there is more than one. In the event of a claim-free year, the policyholder may move up one class (to a better discount level) or remain in the maximum discount class.

Bonus–Malus systems are used in some European countries and although they are similar to the NCD systems described above, they have the following differences:

1. There are often many classes (varying from 7 to 22) and a much wider range of premium levels (for the same rating group), with the maximum being up to six times the minimum premium.

2. Drivers usually start in a class in the middle of the system and obtain discounts (bonuses) following claim-free years. Step-backs of several classes result when claims are made, often into classes where the premium is more than that of the initial class (maluses).

Why are NCD schemes in such popular use in motor insurance? Introducing discounts for those with a good claims record should, in theory, reduce the heterogeneity between policyholders within the various rating classes and allow the insurer to charge premiums that are more appropriate to the individual risks. Since the consequence of making a claim is normally an increased premium in the following year, policyholders are naturally discouraged from making small claims. This will presumably reduce both the number of claims made by the policyholders and the overall management costs incurred by the insurer. It is also felt that the penalty associated with making a claim will lead to more caution in driving.

Although NCD schemes do seem to reduce the number of small claims, they do not seem to be sufficiently effective in reducing heterogeneity, and there is no evidence that they have resulted in safer driving. NCD schemes are, however, well established, and the public perception is that they penalize the poorer drivers while rewarding those with good records, consequently making it difficult to introduce radically different systems into the market.

6.2 Transition in a No Claim Discount system

A No Claim Discount system (NCD) is defined by a set of discount classes $\mathcal{E} = \{E_0, E_1, \ldots, E_k\}$ and a transition rule for moving between them.

6.2.1 Discount classes and movement in NCD schemes

Normally we will use E_0 to denote the class with no (0%) discount, and it is usually the class where new entrants (without claims experience) enter the scheme. As i increases from 0 to k, E_i represents a class with a higher premium discount. At the end of a policy year, the insured may move to another class (and premium rate) depending on their claims experience during the year. The specific rule for movement in the system is called the *transition* or *stepback rule*. Most NCD systems penalize on the basis of the number but not the size of claims, implicitly making the assumption that claim frequency and loss severity are independent. (It seems somewhat reasonable to accept that for the most part the cost of an accident is independent of whatever caused it. Someone skidding on a wet road might crash headlong into a car full of passengers, or be more fortunate in landing softly in a green median strip in the road. The same driving mistake can in these two instances result in claims of very different magnitudes.)

Example 6.1
An NCD system has the discount classes: E_0 (no discount), E_1 (20% discount)

and E_2 (40% discount). Movement in the system is determined by the rule whereby one steps back one discount level (or stays in E_0) with one claim in a year, and returns to a level of no discount if more than one claim is made. A claim-free year results in a step up to a higher discount level (or one remains in class E_2 if already there). This elementary example of an NCD system has only three discount classes (see Table 6.1), but can be efficiently used to illustrate many of the general ideas about how NCD systems operate.

TABLE 6.1

NCD classes for Example 6.1.

NCD class	E_0	E_1	E_2
% Discount	0	20	40
% Pure premium	100	80	60

▯

For our second example, we consider a more realistic system with six discount classes ($k = 5$) and a more typical transition rule.

Example 6.2

An NCD system has six discount classes $\mathcal{E} = \{E_0, E_1, \ldots, E_5\}$ as illustrated in Table 6.2. Consider the *normal* or *typical* transition rule whereby one moves up one class as a result of no claims being made in the current year (or stays in the top discount class if that is the present class), drops back two classes as a result of one claim (in the case where one is in class E_0 or E_1 one simply steps back to class E_0), and goes back to paying the full pure premium (class E_0) if two or more claims are made in the current year.

TABLE 6.2

NCD classes for Example 6.2.

NCD Class	E_0	E_1	E_2	E_3	E_4	E_5
% Discount	0	10	20	30	40	50
% Pure premium	100	90	80	70	60	50

Suppose that Richard joined the system (and starts in E_0) as a policyholder on July 1, 1992, paying the full (or pure) premium of 800. If he made claims on May 12, 1996 and September 1, 1999, then the following diagram indicates his path through the classes in the 10 year period up to July 1, 2002, when his premium was 480 (being in class E_4 on that date):

$$E_0 \rightarrow E_1 \rightarrow E_2 \rightarrow E_3 \rightarrow E_1 \rightarrow E_2 \rightarrow E_3 \rightarrow E_4 \rightarrow E_2 \rightarrow E_3 \rightarrow E_4$$
$$(1992) \rightarrow \qquad\qquad\qquad\qquad\qquad\qquad\qquad\qquad \rightarrow (2002)$$

□

A less harsh rule for the insured in Example 6.2 would be where the policyholder only steps back *one* class or discount level consequent to making one or more claims in the past year. We shall refer to this (gentle way of penalizing claimants) as the *soft* transition rule. The harshest rule in this situation would be to send a policyholder back to the beginning (that is, to the class E_0 where the full premium is paid) on making one or more claims, and we will refer to this as the *severe* transition rule. There are of course a plethora of possible transition rules which are intermediate to the soft and severe rules described above. Problem 1 asks one to determine the premium Richard would have paid on July 1, 2002, if either the soft or severe transition rules had been applied.

6.2.2 One-step transition probabilities in NCD schemes

Modeling movement in an NCD system via a stochastic process can give insight into future numbers of policyholders in the various discount classes as well as the annual premium income. Given a set of discount classes $\mathcal{E}$ and a transition rule for an NCD system, let us assume that movement from discount class E_i in one year to discount class E_j in the next is a random event with probability p_{ij}, which is the same for all of the insured in a specific rating group. We are implicitly assuming that movement from any class E_i in one year to another class E_j in the following is independent of how the individual arrived in class E_i to begin with. This property (that the future depends on the present but not the past) is commonly called the *Markov property*.

The square $k + 1$ by $k + 1$ matrix of one-step (one year) transition probabilities is given by $\mathbf{P} = (p_{ij})$, where we write

$$p_{ij} = P(\text{in discount class } E_j \text{ next year} \mid \text{in discount class } E_i \text{ this year}).$$

Observe that if one is presently in class E_i, then one must move to some state E_j (for $j = 0, 1, .., i, .., k$) next year, and hence the probabilities in the ith row of $\mathbf{P}$ must sum to 1 (for each i). Such a matrix is known as a *stochastic matrix*. Note that we are labeling our $k+1$ rows from $i = 0$ to $i = k$, hence in particular, row 0 consists of the probabilities of going from discount class E_0 to each of the other classes in one year.

In Example 6.1 with three discount levels and $k = 2$, let us suppose that for anyone in this scheme the probability of one claim in a year is 0.2 while the probability of two or more claims is 0.1. Then the matrix of one-step transition probabilities for this system is given by:

$$\mathbf{P} = \begin{pmatrix} 0.3 & 0.7 & 0.0 \\ 0.3 & 0.0 & 0.7 \\ 0.1 & 0.2 & 0.7 \end{pmatrix}. \tag{6.1}$$

Transition in any NCD system from state E_i to E_j may also occur over any two-year period by passing through any intermediate state E_l, where $l = 0, 1, \ldots, k$. Summing over the possibilities for l, we see that the probability p_{ij}^2 of moving from state E_i to E_j in any two-year (two-step) period is given by

$$p_{ij}^2 = \sum_{l=0}^{k} p_{il}\, p_{lj}.$$

Note that p_{ij}^2 is **not** the square $(p_{ij})^2$ of the probability p_{ij}. In fact the two-step transition probability p_{ij}^2 corresponds to multiplying (term by term) row i of the transition matrix $\mathbf{P}$ by column j of $\mathbf{P}$. The matrix of two-step transition probabilities is actually the matrix product (denoted by $\cdot$) of $\mathbf{P}$ with itself, which is written as $\mathbf{P}^2 = \mathbf{P} \cdot \mathbf{P}$. For example, if $\mathbf{P}$ is the one-step transition matrix associated with Example 6.1 above, then the matrix $\mathbf{P}^2$ of two-step transition probabilities is given by:

$$\mathbf{P}^2 = \begin{pmatrix} 0.3 & 0.7 & 0.0 \\ 0.3 & 0.0 & 0.7 \\ 0.1 & 0.2 & 0.7 \end{pmatrix} \cdot \begin{pmatrix} 0.3 & 0.7 & 0.0 \\ 0.3 & 0.0 & 0.7 \\ 0.1 & 0.2 & 0.7 \end{pmatrix} = \begin{pmatrix} 0.30 & 0.21 & 0.49 \\ 0.16 & 0.35 & 0.49 \\ 0.16 & 0.21 & 0.63 \end{pmatrix}.$$

In this system, if one is in discount class E_1 (and paying 80% of the full premium) in a given year, then the probability that one is still paying the same premium two years later is 0.35. The probability is 0.63 that an individual who is on the maximum discount retains that discount two years later.

Of course, one may also consider the probabilities of going from one discount class to another in n years for larger values of n. In a fashion similar to the above development, one may show that the matrix of n-step transition probabilities (p_{ij}^n) is given by the n-fold matrix product of $\mathbf{P}$ with itself, that is $\mathbf{P}^n$. This emphasizes the importance of the basic transition matrix $\mathbf{P}$ in studying movement in an NCD system.

The matrix of two-step transition probabilities $\mathbf{P}^2$ is, like $\mathbf{P}$, a stochastic matrix, and it can easily be seen that this is the case for $\mathbf{P}^n$ for any n. Using computer software (like Excel, R, S-Plus or Mathematica) to multiply matrices, one may quickly demonstrate that with the transition matrix $\mathbf{P}$ of Example 6.1,

$$\mathbf{P}^3 = \begin{pmatrix} 0.202 & 0.308 & 0.490 \\ 0.202 & 0.210 & 0.588 \\ 0.174 & 0.238 & 0.588 \end{pmatrix} \quad \text{and} \quad \mathbf{P}^6 = \begin{pmatrix} 0.1883 & 0.2435 & 0.5682 \\ 0.1855 & 0.2463 & 0.5682 \\ 0.1855 & 0.2435 & 0.5710 \end{pmatrix}.$$

Interestingly, we shall see that usually the sequence of matrices $\mathbf{P}^n$ converges as $n \to +\infty$ to a matrix whose rows are constant, any one of which represents the long-run probability (denoted by $\boldsymbol{\pi} = (\pi_0, \pi_1, \ldots, \pi_k)$) that an individual will be in the various states or discount levels. In particular, the entries in each individual column of this limiting matrix will be identical. For example, using the transition matrix $\mathbf{P}$ of (6.1), one may show that

$$\mathbf{P}^{20} = \mathbf{P}^{21} = \begin{pmatrix} 0.1860465 & 0.2441860 & 0.5697674 \\ 0.1860465 & 0.2441860 & 0.5697674 \\ 0.1860465 & 0.2441860 & 0.5697674 \end{pmatrix}.$$

Most companies will place drivers in class E_0 with no discount at the start of their driving career. However, they often recognize bonuses and experience earned with other companies in order to entice customers to transfer into their own system. For an NCD system with discount classes $\mathcal{E} = \{E_0, E_1, \ldots, E_k\}$, we shall use $\mathbf{p}^0 = (p_0^0, p_1^0, \ldots, p_k^0)$ to denote the vector of probabilities with which an individual starts in the various discount levels. The set of classes (or states) $\mathcal{E} = \{E_0, E_1, \ldots, E_k\}$, the matrix of transition probabilities $\mathbf{P}$, and the *initial distribution* $\mathbf{p}^0$ completely define a Markov chain model for an NCD system. Although our models for NCD systems have a finite number of classes or states, Markov chains may have an infinite number of states, and are special cases of Markov processes. There is a well-developed theory for Markov chains (see, for example, Feller [24] or Ross [53]), and the theory of Markov chains is now extensively used as a simulating tool in Markov Chain Monte Carlo (MCMC) methods. Assuming that a person initially starts in category E_0 and that no discount is given initially in year 0, then we have that

$$\mathbf{p}^0 = (p_0^0, p_1^0, \ldots p_k^0) = (1, 0, \ldots 0).$$

Similarly, we use p_j^{n+1} to denote the probability that an individual is in class E_j at time (year) $n+1$. In order to be in class E_j at time $n+1$, an individual has to be in some class E_l in year n and then pass to E_j in the following year. Therefore we must have

$$P(E_j \text{ in year } n+1) = p_j^{n+1} = \sum_{l=0}^{k} p_l^n \, p_{lj}.$$

In other words

$$\mathbf{p}^{n+1} = (p_0^{n+1}, \ldots, p_k^{n+1}) = (p_0^n, \ldots, p_k^n) \cdot \mathbf{P} = \mathbf{p}^n \cdot \mathbf{P} \qquad (6.2)$$

and in particular, that

$$\mathbf{p}^1 = (p_0^1, \ldots, p_k^1) = (p_0^0, \ldots, p_k^0) \cdot \mathbf{P}.$$

Similarly, one may show that for any n and m,

$$\mathbf{p}^{n+m} = (p_0^{n+m}, \ldots, p_k^{n+m}) = (p_0^n, \ldots, p_k^n) \cdot \mathbf{P}^m.$$

Suppose that in Example 6.1 policyholders enter the classes E_0, E_1 and E_2 with respective probabilities $0.5, 0.3$ and 0.2 (that is, the initial distribution is given by $\mathbf{p}^0 = (0.5, 0.3, 0.2)$). Then the probability distribution for year 1 is given by the vector:

$$\mathbf{p}^1 = \mathbf{p}^0 \cdot \mathbf{P} = (0.5, 0.3, 0.2) \cdot \begin{pmatrix} 0.3 & 0.7 & 0.0 \\ 0.3 & 0.0 & 0.7 \\ 0.1 & 0.2 & 0.7 \end{pmatrix} = (0.26, 0.39, 0.35),$$

and for year 2 by

$$\mathbf{p}^2 = \mathbf{p}^0 \cdot \mathbf{P}^2 = (0.5, 0.3, 0.2) \cdot \begin{pmatrix} 0.30 & 0.21 & 0.49 \\ 0.16 & 0.35 & 0.49 \\ 0.16 & 0.21 & 0.63 \end{pmatrix} \quad \text{or}$$

$$= \mathbf{p}^1 \cdot \mathbf{P} = (0.26, 0.39, 0.35) \cdot \begin{pmatrix} 0.3 & 0.7 & 0.0 \\ 0.3 & 0.0 & 0.7 \\ 0.1 & 0.2 & 0.7 \end{pmatrix}$$

$$= (0.230, 0.252, 0.518).$$

Table 6.3 gives an indication of how the probability distributions $\mathbf{p}^n$ stabilize as the years go by. Note the relative stability in these distributions even after four or five years. We will see that for most Markov chain NCD models, no matter what the initial distribution $\mathbf{p}^0$ is, there is a stationary distribution $\boldsymbol{\pi} = (\pi_0, \pi_1, \ldots, \pi_k)$ to which $\mathbf{p}^n$ converges as $n \to \infty$. Moreover, the convergence generally speaking is quite rapid.

Suppose that 2000 policyholders are placed in the discount classes of Example 6.1 in year 0 according to the initial distribution $\mathbf{p}^0 = (0.5, 0.3, 0.2)$, and that the pure premium is \$600. Assuming premiums are paid at the beginning of a year, the initial (or year 0) "expected" premiums would be $2000 [0.5(600) + 0.3(480) + 0.2(360)] = \$1,032,000$. Similarly, expected premiums in year 1 would be $2000 [0.26(600) + 0.39(480) + 0.35(360)] = \$938,400$. Expected premiums for other years are given in Table 6.3, and again we note that these stabilize quickly.

6.2.3   Limiting distributions and stability in NCD models

Let us suppose that for a given Markov chain NCD model, the limiting probabilities $\pi_j = \lim_{n\to\infty} p_j^n$ exist for all j. Then from Equation(6.2), the probability vector $\boldsymbol{\pi} = (\pi_0, \pi_1, \ldots, \pi_k)$ satisfies

$$\boldsymbol{\pi} = \lim_{n\to\infty} \mathbf{p}^n = \lim_{n\to\infty} \mathbf{p}^{n+1}$$

$$= \lim_{n\to\infty} \mathbf{p}^n \cdot \mathbf{P}$$

$$= \boldsymbol{\pi} \cdot \mathbf{P}.$$

A probability distribution $\mathbf{a} = (a_0, a_1, \ldots, a_k)$ is a *stationary* (or *equilibrium*) distribution for the Markov chain with transition matrix $\mathbf{P}$ if $\mathbf{a} = \mathbf{a} \cdot \mathbf{P}$.

TABLE 6.3

Probability distributions and expected premiums for
Example 6.1 when $\mathbf{p}^0 = (0.5, 0.3, 0.2)$.

Discount class	E_0	E_1	E_2	Expected
Year n	p_0^n	p_1^n	p_2^n	premiums
0	0.500	0.300	0.200	1,032,000.0
1	0.260	0.390	0.350	938,400.0
2	0.230	0.252	0.518	890,880.0
3	0.1964	0.2646	0.539	877,776.0
4	0.1922	0.2453	0.5625	871,123.2
5	0.1875	0.2470	0.5655	869,288.6
6	0.1869	0.2443	0.5688	868,357.2
10	0.1861	0.2442	0.5697	867,915.8
15	0.1860	0.2442	0.5698	867,907.0
30	0.1860	0.2442	0.5698	867,907.0

Therefore in our situation π is a stationary distribution. A brief description will now be given of some of the terminology which implies that in the standard NCD Markov chain model there exists a unique stationary distribution. An interesting aspect of such a stationary distribution π is that although it clearly depends on the (one-step) transition matrix $\mathbf{P}$, it is independent of the initial distribution $\mathbf{p}^0$.

In a Markov chain model for movement in NCD systems, we normally assume that starting with no discount in class E_0, it is possible to reach any other discount class in a finite number of years (for any $j > 0$ there exists some n_j such that $p_{0j}^{n_j} > 0$). We also assume that if one is in the maximum discount level E_k, then it is possible (by having a sufficiently large number of claims in the current or subsequent years) to return to the discount class E_0. In the technical jargon of Markov chains, this means that any two states E_i and E_j can *communicate* with each other. In other words, it is possible to go (with positive probability in a finite number of steps) from any one state to any other. More precisely, given i and j, there exist m and n such that $p_{ij}^m > 0$ and $p_{ji}^n > 0$. This establishes what is termed the *irreducible* property of the Markov chain model (that is, it cannot be reduced further to a smaller self-contained collection of classes or states). We also assume that once in the maximum discount level E_k, it is possible to stay in that class for the next year (usually following a claim-free year), that is, $p_{kk} > 0$. This property (together with the irreducible property) enables one to show all discount classes are *aperiodic** (they are not periodic or cyclic in a technical sense).

---

*The discount class E_i has *period* r if $p_{ii}^n = 0$ whenever n is not divisible by r, and r is the greatest integer with this property. For example if E_i has period 2, then return visits to E_i can only occur in an even number of years and $p_{ii}^2 > 0$. A class (or state) with period 1 is called an aperiodic class.

To summarize, our models for NCD systems are *irreducible aperiodic* finite state Markov chains, and moreover the states or discount levels are *ergodic*[†]. This is important, for a classic result in the theory of Markov chains (see, for example, Ross [53], p.175) implies that in this situation[‡] there exists a unique probability distribution $\boldsymbol{\pi} = (\pi_0, \pi_1, \ldots, \pi_k)$ which is stationary for the Markov chain and has the property that

$$\pi_j = \lim_{n \to \infty} p_{ij}^n = \lim_{n \to \infty} p_j^n,$$

independent of the initial distribution $\mathbf{p}^0$.

To find the equilibrium distribution, we solve the $k+1$ equations given by the matrix expression $\boldsymbol{\pi} = \boldsymbol{\pi} \cdot \mathbf{P}$, and they can be written as:

$$\pi_0 = \pi_0 p_{00} + \pi_1 p_{10} + \pi_2 p_{20} + \ldots + \pi_k p_{k0}$$
$$\pi_1 = \pi_0 p_{01} + \pi_1 p_{11} + \pi_2 p_{21} + \ldots + \pi_k p_{k1}$$

$$\vdots \quad \vdots$$

$$\pi_k = \pi_0 p_{0k} + \pi_1 p_{1k} + \pi_2 p_{2k} + \ldots + \pi_k p_{kk}.$$

For Example 6.1 these are the three equations given by the matrix expression

$$\boldsymbol{\pi} = \boldsymbol{\pi} \cdot \mathbf{P} = (\pi_1, \pi_2, \pi_3) \cdot \begin{pmatrix} 0.3 & 0.7 & 0.0 \\ 0.3 & 0.0 & 0.7 \\ 0.1 & 0.2 & 0.7 \end{pmatrix},$$

or equivalently,

$$\pi_0 = 0.3\pi_0 + 0.3\pi_1 + 0.1\pi_2$$
$$\pi_1 = 0.7\pi_0 \qquad\quad + 0.2\pi_2$$
$$\pi_2 = \qquad\qquad 0.7\pi_1 + 0.7\pi_2.$$

From the last of these equations, it follows that

$$\pi_2 = 7/3\pi_1,$$

and therefore using the second equation, one finds

$$\pi_1 = 0.7\pi_0 + 0.2(7/3)\pi_1 \;\Rightarrow\; \pi_1 = (21/16)\pi_0.$$

[†]The eventual or limiting distribution of the states of the system is independent of the initial state.

[‡]Another sufficient condition is that the matrix of transition probabilities $\mathbf{P}$ has one eigenvalue $= 1$, with the rest being less than 1 in absolute value. The equilibrium distribution is the eigenvector (when standardized to be a probability vector) corresponding to the eigenvalue 1.

Since π is a probability distribution,

$$1 = \pi_0 + \pi_1 + \pi_2 \ = \ \pi_0 + (21/16)\pi_0 + (7/3)(21/16)\pi_0, \quad \text{thus}$$

$$\pi_0 = \frac{1}{1 + (21/16) + (7/3)(21/16)} \ = \ 0.1860.$$

The equilibrium distribution for this system is $\pi = (0.1860, 0.2442, 0.5698)$. Note in particular from Table 6.3 that the probability distribution $\mathbf{p}^{15}$ is already (to four decimal places) equal to this equilibrium distribution. In general, $\mathbf{p}^n$ itself depends on the initial distribution $\mathbf{p}^0$, but the *rate* of convergence to the equilibrium distribution π depends on $\mathbf{P}$, and usually it is quite rapid.

Although in Example 6.1 we assumed that the probability distribution for the number of claims N in a year had distribution defined by

$$P(N = 0) = 0.7, \quad P(N = 1) = 0.2, \quad \text{and } P(N \geq 2) = 0.1,$$

it is perhaps more common to model N by a Poisson distribution with *rate* parameter $\lambda = q$. In this situation, the probabilities of $0, 1, 2, 3, \ldots, k-1$, and k or more claims in a year, are, respectively,

$$e^{-q}, \ qe^{-q}, \ q^2 e^{-q}/2!, \ \ldots, \ q^{k-1}e^{-q}/(k-1)!, \ \text{and } 1 - \sum_{i=0}^{k-1} q^i e^{-q}/i!.$$

Consider the NCD system of Example 6.2 with six discount classes where we model the number of claims in a given year for an insured individual by a Poisson random variable with rate parameter $\lambda = 0.1$. Then the one-step probability transition matrix for the normal rule (two steps back for one claim, one step ahead for none, etc.) is given by

$$\mathbf{P} = \begin{pmatrix}
1 - e^{-0.1} & e^{-0.1} & 0 & 0 & 0 & 0 \\
1 - e^{-0.1} & 0 & e^{-0.1} & 0 & 0 & 0 \\
1 - e^{-0.1} & 0 & 0 & e^{-0.1} & 0 & 0 \\
1 - 1.1e^{-0.1} & 0.1e^{-0.1} & 0 & 0 & e^{-0.1} & 0 \\
1 - 1.1e^{-0.1} & 0 & 0.1e^{-0.1} & 0 & 0 & e^{-0.1} \\
1 - 1.1e^{-0.1} & 0 & 0 & 0.1e^{-0.1} & 0 & e^{-0.1}
\end{pmatrix}$$

$$= \begin{pmatrix}
0.0952 & 0.9048 & 0 & 0 & 0 & 0 \\
0.0952 & 0 & 0.9048 & 0 & 0 & 0 \\
0.0952 & 0 & 0 & 0.9048 & 0 & 0 \\
0.0047 & 0.0905 & 0 & 0 & 0.9048 & 0 \\
0.0047 & 0 & 0.0905 & 0 & 0 & 0.9048 \\
0.0047 & 0 & 0 & 0.0905 & 0 & 0.9048
\end{pmatrix}.$$

The equilibrium distributions are given in Table 6.4 for this NCD system using the three transition rules (in the *soft* rule a person steps back one class for

TABLE 6.4

Equilibrium distributions for Example 6.2.

Transition rule	Discount classes					
	E_0	E_1	E_2	E_3	E_4	E_5
Soft	0.000	0.000	0.001	0.010	0.094	0.895
Normal	0.009	0.016	0.022	0.091	0.082	0.780
Severe	0.095	0.086	0.078	0.070	0.064	0.607

any number of claims, and in the *severe* transition rule a person goes all the way back to E_0 for any number of claims) for Example 6.2. Note that with the *severe* transition rule we expect 9.5% of policyholders to be paying the full premium annually once stability has been reached, while with the *soft* transition rule we expect relatively few (less than 0.05% or almost no one) to be paying this. An interesting observation from Table 6.4 is that in the long run, no matter what the transition rule (soft, normal or severe) is, a majority of the drivers are in the maximum discount class E_5.

Of course, it is of interest to know how long it takes to reach the so-called *steady state* in such systems, in addition to what are the factors which influence this? When using an initial distribution where everyone starts in the class E_0 with no discount (that is $\mathbf{p}^0 = (1, 0, \dots,)$), Table 6.5 (Colgan [15]) gives the period of convergence in years to equilibrium. Note that convergence is quickest with the *severe* transition rule. Note also that (in this example) the time to convergence for the soft and normal transition rules is initially increasing and then decreasing as a function of λ.

Here time to equilibrium is defined[§] as the smallest number n such that $| p_i^n - \pi_i | < 0.005$ for all $i = 0, 1, \dots, k$. Rates of convergence clearly depend on the *stopping threshold* (chosen to be 0.005 in this case), but in fact most of these systems settle down to reasonably steady levels before the stated number of years to convergence. Studies have shown that the following factors are influential in determining the convergence period: (a) the claim rate λ, (b) the step-back rule, (c) the number of discount classes and (d) the initial distribution $\mathbf{p}^0$.

6.2.3.1 Equilibrium distributions for soft and severe transition rules

There are compact forms for the stability distributions when the soft or severe transition rules are used and the probability of making no claims in a year is the same (say b) for all policyholders (that is, independent of the discount class). For the soft transition rule the one-step transition matrix P takes the

---

[§]An alternative measure of convergence is given by Bonsdorff [5] who used a measure called the *total variation* defined as $TV_n = \sum_i | p_i^n - \pi_i |$.

TABLE 6.5
Time to equilibrium
convergence for Example 6.2.

λ	Transition rule		
	Soft	Normal	Severe
0.05	10	12	6
0.10	14	15	6
0.20	18	18	6
0.35	26	18	6
0.50	32	13	6
0.75	30	10	6
1.00	22	7	6

form:

$$\mathbf{P} = \begin{pmatrix} 1-b & b & 0 & 0 & . & . & 0 & 0 & 0 \\ 1-b & 0 & b & 0 & . & . & 0 & 0 & 0 \\ 0 & 1-b & 0 & b & . & . & 0 & 0 & 0 \\ . & . & . & . & . & . & . & . \\ . & . & . & . & . & . & . & . \\ 0 & 0 & 0 & 0 & . & . & 0 & b & 0 \\ 0 & 0 & 0 & 0 & . & . & 1-b & 0 & b \\ 0 & 0 & 0 & 0 & . & . & 0 & 1-b & b \end{pmatrix}.$$

Therefore the system of equations $\pi = \pi \cdot \mathbf{P}$ becomes

$$\pi_0 = (1-b)(\pi_0 + \pi_1)$$
$$\pi_i = b\pi_{i-1} + (1-b)\pi_{i+1} \quad \text{for } i = 1, \ldots, k-1, \text{ and}$$
$$\pi_k = (1-b)\pi_{k-1} + b\pi_k.$$

It then follows easily that $\pi_i = [b/(1-b)]^i \pi_0$ for $i = 1, \ldots, k$, and therefore

$$\pi_i = \frac{[b/(1-b)]^i}{1 + [b/(1-b)] + [b/(1-b)]^2 + \ldots + [b/(1-b)]^k} \quad \text{for } i = 0, \ldots, k.$$

When the severe transition rule is used in an NCD system, then the one-step transition matrix P takes the form

$$\mathbf{P} = \begin{pmatrix} 1-b & b & 0 & 0 & . . & 0 & 0 & 0 \\ 1-b & 0 & b & 0 & . . & 0 & 0 & 0 \\ 1-b & 0 & 0 & b & . . & 0 & 0 & 0 \\ . & & . & . & . . . & . & . \\ . & & . & . & . . . & . & . \\ 1-b & 0 & 0 & 0 & . . & 0 & b & 0 \\ 1-b & 0 & 0 & 0 & . . & 0 & 0 & b \\ 1-b & 0 & 0 & 0 & . . & 0 & 0 & b \end{pmatrix}.$$

When there are $k+1$ discount classes it is easy to see that the k-step transition matrix takes the form

$$
\mathbf{P}^k =
\begin{pmatrix}
1-b & b(1-b) & b^2(1-b) & . & . & b^{k-1}(1-b) & b^k \\
1-b & b(1-b) & b^2(1-b) & . & . & b^{k-1}(1-b) & b^k \\
1-b & b(1-b) & b^2(1-b) & . & . & b^{k-1}(1-b) & b^k \\
. & . & . & & . & . & . \\
1-b & b(1-b) & b^2(1-b) & . & . & b^{k-1}(1-b) & b^k \\
1-b & b(1-b) & b^2(1-b) & . & . & b^{k-1}(1-b) & b^k
\end{pmatrix}.
$$

Therefore when the severe rule is in effect, stability occurs in $k+1$ years, and the equilibrium distribution is given by

$$
\pi_{severe} = (1-b,\ b(1-b),\ b^2(1-b),\ b^3(1-b),\ \ldots,\ b^{k-1}(1-b),\ b^k).
$$

Note that at any point in time the probability that any individual will return to class E_0 is $1-b$ (the probability of one or more claims being made) with the severe rule. Consequently, it should come as no surprise that the limiting probability π_0 is equal to $1-b$.

Consider again the NCD system described in Example 6.2, but where now the number of claims N is Poisson with parameter $\lambda = 0.2$. Then $b = e^{-0.2}$ and the equilibrium distribution when the soft rule is in effect is

$$
\pi_{soft} = (0.0004,\ 0.0019,\ 0.0085,\ 0.0382,\ 0.1724,\ 0.7787),
$$

while for the severe rule the equilibrium distribution is

$$
\pi_{severe} = (0.1813,\ 0.1484,\ 0.1215,\ 0.0995,\ 0.0814,\ 0.3679).
$$

Table 6.5 shows that for Example 6.2, the time to convergence for various values of λ (or b) is always 6 years with the severe transition rule.

6.3   Propensity to make a claim in NCD schemes

Following an accident, an insured individual will normally consider whether or not it is actually worthwhile to make a claim due to the increased premium payments (or discounts foregone) that will result by doing so. A decision might be based on a quick calculation of the estimated increase in premium payments. In such calculations, one compares the premiums which one would expect to pay in the coming few years under the two situations where one actually makes a claim now (C) and where one foregoes making a claim (NC) and instead absorbs the loss. One therefore can determine for each discount level a *threshold value*. If the loss due to the accident is greater than the threshold for an individual in a given discount class, then the individual should make a claim, but otherwise forego doing so and absorb the loss. In making these calculations one assumes that there will be no further claims made in the short term (next couple of years), and that Year 0 refers to the current year (when the accident has occurred, and presumably the premium for the year has already been paid), Year 1 to next year, etc.

6.3.1 Thresholds for claims when an accident occurs

Consider Example 6.1 where there are the three discount levels of $0\%, 20\%$ and 40%, and the full premium is $500. A person in E_0 who has just had an accident and now makes a claim (C) would expect to pay $500 + 400 + 300 = \$1200$ in premiums over the next three years, while if no claim (NC) is made this would be only $400 + 300 + 300 = \$1000$. Hence the difference or threshold is $200, and the individual would (usually) only decide to make a claim if the accident damage exceeded this threshold of $200. Similarly, an individual in discount class E_1 (respectively, E_2) would claim if the loss exceeds $300 ($100). These calculations are given in Table 6.6. Note that for this NCD system, the premium effect of making (or not making) a claim in year 0 has worn off by year 3. Essentially, in this system there is a maximum two-year horizon on a decision to make a claim.

TABLE 6.6
Premiums and thresholds (**T**) for making claims in Example 6.1.

Class	E_0 0% Discount		E_1 20% Discount		E_2 40% Discount	
Premium Year 0	500		400		300	
	C	NC	C	NC	C	NC
Year 1	500	400	500	300	400	300
Year 2	400	300	400	300	300	300
Year 3	300	300	300	300	300	300
Total (Years 1-3)	1200	1000	1200	900	1000	900
T	200		300		100	

For Example 6.2 with the same full premium of $500, the six discount levels of $0\%, 10\%, 20\%, 30\%, 40\%$ and 50%, and the transition rule of dropping back two steps for one claim and to paying the full premium for more than one, the effect of a single claim in the NCD system wears off in at most five years. This can be seen from Table 6.7. For example, according to this table, a policyholder in discount class E_1 who has just incurred an accident should make a claim if the loss exceeds $450, otherwise not. However, remember that this table was constructed assuming no new claims are to be made in the near future (for this type of person it would be five years). If such a person is very prone to accidents (and claims) and the chances are considerable that another claim might be made in the next five years, then perhaps this should also be taken into account in any decision. For instance, a person presently in E_1 who suffers a loss of $400 might well decide to make a claim if it is felt that another accident within the next year is very likely.

TABLE 6.7
Premiums and thresholds (**T**) for making claims in Example 6.2.

Class	E_0		E_1		E_2		E_3		E_4		E_5	
Year 0	500		450		400		350		300		250	
	C	NC	C	NC	C	NC	C	NC	C	NC	C	NC
Year 1	500	450	500	400	500	350	450	300	400	250	350	250
Year 2	450	400	450	350	450	300	400	250	350	250	300	250
Year 3	400	350	400	300	400	250	350	250	300	250	250	250
Year 4	350	300	350	250	350	250	300	250	250	250	250	250
Year 5	300	250	300	250	300	250	250	250	250	250	250	250
T	250		450		600		450		300		150	

Example 6.3
An NCD scheme operates with four levels of discount: 0%, 20%, 30% and 40%. The rule for movement between discount levels is the soft rule whereby a person moves up one step (discount level) next year if no claims are made this year, while if one or more claims are made in a year then one drops to the next lower discount level in the following year (or stays at 0% discount). The full premium for an individual is $600 per year. A deductible is in effect where the first $150 of any claim must be paid by the insured. When an accident occurs, the appropriate damage (loss) distribution X is lognormal with parameters $\mu = 6.466769$ and $\sigma^2 = 0.1686227$. We answer the following questions with respect to this scheme:

1. For each discount level determine the size of damage below which it is not worthwhile making a claim, assuming a three-year time horizon. What are the mean and variance of a typical loss?

2. Assume the number of accidents an individual has in a year is a Poisson random variable with parameter 0.2. Find the matrix of one-step transition probabilities for one-year movements in this NCD scheme.

3. If 20,000 policyholders all begin initially in E_0, what would be the expected total premiums in year 2? Find the stationary distribution for this NCD system.

In order to determine the threshold on damage for making a claim in Example 6.3, we must consider the deductible of $150 as well as the difference in future premium payments that will be made if a claim is made. If the damage incurred in an accident is represented by the random variable X, then the insured will make a claim if this exceeds the deductible plus the difference in additional premiums that must be paid in the next few years as a result of making a claim (equivalently, $X - 150$ exceeds the additional cost of premiums on making a claim). In determining such a threshold, we are again assuming that the chance of another loss (in excess of the deductible $150) being incurred in the next few years is negligible. Table 6.8 illustrates the calculation of these thresholds for each of the discount classes.

TABLE 6.8
Premiums and thresholds ($\mathbf{T}$) for making claims in Example 6.3.

Class	E_0 (0%)		E_1 (20%)		E_2 (30%)		E_3 (40%)	
Year 0	600		480		420		360	
	C	NC	C	NC	C	NC	C	NC
Year 1	600	480	600	420	480	360	420	360
Year 2	480	420	480	360	420	360	360	360
Year 3	420	360	420	360	360	360	360	360
Year 4	360	360	360	360	360	360	360	360
$\mathbf{T}$	240+150=390		360+150=510		180+150=330		60+150=210	

Given that the random loss X is lognormal with parameters $\mu = 6.466769$ and $\sigma^2 = 0.1686227$, it follows that

$$E(X) = e^{\mu+\sigma^2/2} = e^{6.466769+0.1686227/2} = 700$$

and

$$Var(X) = e^{2\mu+\sigma^2}(e^{\sigma^2} - 1) = 300^2 = 90,000.$$

For someone in the discount class E_0 (that is, with no discount), the intensity rate for making a *first* claim is the rate for suffering a loss (0.2) multiplied by the conditional probability of making a claim having suffered a loss (or $P(X > 390) = 0.8886037$). Another way of thinking about this is that the number of accidents an individual has which result in damage in excess of 390 is a Poisson process with rate parameter $0.2(0.8886037) = 0.1777207$. It follows that the transition probabilities from this class are given by $p_{00} = 1 - e^{-0.1777207} = 0.1628238$, $p_{01} = 0.8371762$, and $p_{02} = p_{03} = 0$. Similarly, one may show that the *first* claim rates for individuals in classes E_1, E_2 and E_3 are, respectively, $0.1428503, 0.1896040$ and 0.1993602. Therefore the one-step transition matrix for this system is given by

$$\mathbf{P} = \begin{pmatrix} 0.1628238 & 0.8371762 & 0 & 0 \\ 0.1331162 & 0 & 0.8668838 & 0 \\ 0 & 0.1727134 & 0 & 0.8272866 \\ 0 & 0 & 0.1807452 & 0.8192548 \end{pmatrix}.$$

Hence the probability distribution for the classes in year 2 is

$$\mathbf{p}^2 = (1,0,0,0) \cdot \mathbf{P}^2 = (0.1379533, 0.1363122, 0.7257345, 0.0000000)$$

and the expected premiums in year 2 are

$$E(\text{year 2 premiums}) = 20{,}000[(0.1379533)600 + (0.1363122)480$$
$$+ (0.7257345)420 + (0)360]$$
$$= 9{,}060{,}207.$$

Solving the Equation $\pi = \pi \cdot \mathbf{P}$ (and remembering that the components of π add to one), we find that

$$\pi = (0.005454467, 0.03430349, 0.1721763, 0.7880658).$$

Expected premiums when stability has been reached are therefore equal to

$$E(\text{premiums at stability}) = 20{,}000\,[(0.005454467)600 + (0.03430349)480$$
$$+ (0.1721763)420 + (0.7880658)360]$$
$$= 7{,}515{,}122.$$

Note that this is considerably less than the expected premiums for year 2, but this is not surprising as most of the policyholders will (in the limit) be in the top discount level.   []

6.3.2   The claims rate process in an NCD system

Suppose that in an NCD system, accidents for individuals (in a given discount level) occur as a Poisson process with rate parameter λ, and that a first claim is made following an accident only if the loss X exceeds a threshold M. If $p = P(X > M)$, then it is correct to say that losses in excess of M are occurring as a Poisson process with rate $p\lambda$, but usually not correct to say that the *claims process* itself is Poisson with parameter $p\lambda$. This is because once a first claim is made, then (depending on the transition rule) the rate for the next claim may very well increase. For example, suppose the soft rule is being used, and that there are no deductibles in effect. Once an individual makes a claim in a given year (because of a loss in excess of M), then any subsequent loss may as well be reported as a claim since the individual has nothing further to lose. In such a situation, the rate for the first claim is $p\lambda$ (and the time to such a claim is [a censored] exponential distribution with parameter $p\lambda$), but after such a first claim other claims will occur at the increased rate of λ. The following example may illustrate some of these points.

Example 6.4
An NCD scheme for motorists operates with three discount levels of $0\%, 20\%$ and 30%. The full normal premium is $\$500$, and the soft operating rule is

used whereby if no claims are made in a given year the policyholder moves up one discount category in the following year, while in the event of one or more claims being made the policyholder moves back one discount level. A deductible of $100 is also in effect, whereby the policyholder pays the first $100 of any claim.

- What is the size of damage below which it is not worthwhile making a claim at each level of discount, assuming policyholders have a two-year time horizon?

- Assume the loss distribution for an accident is Pareto with parameters $\alpha = 3$, $\lambda = 3200$. For each discount level find the probability that, given an accident occurs, a claim will be made (using again a two-year time horizon).

- Assume all policyholders have the same underlying rate of 0.2 for sustaining accidents. What is the underlying transition matrix $\mathbf{P}$ appropriate for (one-year) movement between discount levels in this scheme? Determine the stationary distribution for this NCD system, and estimate the annual net profit (expected premiums – expected claims payable) once stability has been reached if there are 10,000 policyholders.

Proceeding as before, one may determine the appropriate thresholds for making a claim once a loss has been incurred, and these are given in Table 6.9. Hence the probability that a person in class E_0 who has suffered a loss $X > 0$ will make a claim is

$$P(X > 250 \mid X > 0, E_0) = \left(\frac{3200}{3200 + 250}\right)^3 = 0.7979812.$$

In a similar way, one determines

$$P(X > 300 \mid X > 0, E_1) = 0.7642682 \quad \text{and}$$
$$P(X > 150 \mid X > 0, E_2) = 0.8715966.$$

The Poisson rate parameter for an individual in class E_0 to make a claim in a given year is therefore $0.7979812(0.2) = 0.1595962$, and similarly, they are 0.1528536 and 0.1743193 for individuals in classes E_1 and E_2, respectively. Therefore the one-step transition matrix for this NCD system is given by

$$\mathbf{P} = \begin{pmatrix} 0.1475121 & 0.8524879 & 0 \\ 0.1417447 & 0 & 0.8582553 \\ 0 & 0.1599714 & 0.8400286 \end{pmatrix}$$

and the stationary distribution is $\pi = (0.02545758, 0.15310824, 0.82143419)$.

Assume now that stability has been reached, and let $S = S_0 + S_1 + S_2$ be the total amount of claims in a year, where S_j is the total claims from those

TABLE 6.9
Premiums and thresholds (**T**) for making claims in
Example 6.4.

Class	E_0 (0%)		E_1 (20%)		E_2 (30%)	
Year 0	500		400		350	
	C	NC	C	NC	C	NC
Year 1	500	400	500	350	400	350
Year 2	400	350	400	350	350	350
Year 3	350	350	350	350	350	350
T	150+100=250		200+100=300		50+100=150	

who are in discount class E_j for $j = 0, 1, 2$. Here $S_0 = Y_1^0 + \cdots + Y_{N_0}^0$, where N_0 is the (random) number of individuals in E_0 and Y_i^0 is the amount payable to individual i. In the long run we expect about

$$E(N_0) = 10,000\,(0.02545758) \doteq 255$$

in discount class E_0 in a given year, and similarly, approximately 1531 and 8214 in classes E_1 and E_2, respectively.

Note that Y_i^0 will be positive only if the time T_i^0 until the first loss in excess of \$250 encountered by individual i (in class E_0) occurs in the given year (that is, $T_i^0 < 1$), which happens with probability 0.1475121. If, for example, $T_i^0 = t < 1$, then Y_i^0 will be the excess over \$100 of the first loss above \$250, plus the excess above \$100 for each loss suffered in the interval $(t, 1]$. Losses encountered in the time period $(t, 1]$ which do not exceed \$100 are of course totally absorbed by the insured.

A Pareto random variable X with parameters α and λ has the property that $X - M \mid [X > M] \sim \text{Pareto}(\alpha, \lambda + M)$. Hence given that $T_i^0 = t < 1$, Y_i^0 will be composed of two parts, the first of which is the sum of 150 plus a Pareto $(3, 3200 + 250)$ random variable. The second part is a compound Poisson random variable with Poisson parameter $(1-t)(0.2)(3200/(3200+100))^3$ and typical component a Pareto $(3, 3200 + 100)$ random variable. Therefore

$$E(Y_i^0 \mid T_i^0 = t < 1) = 150 + 3450/(3 - 1)$$
$$+ (0.2)\,(1 - t)\,(3200/3300)^3\,[3300/(3 - 1)]$$
$$= 1875 + 300.8999(1 - t)$$

and hence using $\lambda_0 = 0.7979812(0.2) = 0.1595962$,

$$E(Y_i^0) = \int_0^1 E(Y_i^0 \mid T_i^0 = t)\, f_{T_i^0}(t)\, dt$$

$$= \int_0^1 [1875 + 300.8999(1 - t)]\,\lambda_0\, e^{-\lambda_0 t} dt$$

$$= (1875 + 300.8999)\,(1 - e^{-\lambda_0}) - 300.8999 \int_0^1 \lambda_0 t e^{-\lambda_0 t} dt$$

$$= 320.9715 - 300.8999\,[1 - e^{-\lambda_0} - \lambda_0 e^{-\lambda_0}]/\lambda_0$$
$$= 320.9715 - 300.8999(0.07179501)$$
$$= \$299.3684.$$

Therefore $E(S_0) = 10{,}000\,(0.02545758)\,299.3684 = \$76{,}212.$

Similarly, using $\lambda_1 = 0.2(3200/3500)^3 = 0.1528536$, one finds

$$E(Y_i^1) = \int_0^1 [1950 + 300.8999(1 - t)]\,\lambda_1 e^{-\lambda_1 t} dt$$
$$= 319.0531 - 300.8999[1 - e^{-\lambda_1} - \lambda_1 e^{-\lambda_1}]/\lambda_1$$
$$= \$298.2707, \quad \text{and}$$
$$E(S_1) = 10{,}000\,(0.15310824)\,298.2707 = \$456{,}677.$$

Likewise, using $\lambda_2 = 0.2(3200/3350)^3 = 0.1743193$, one finds

$$E(Y_i^2) = 300.7172 \quad \text{and} \quad E(S_2) = \$2{,}470{,}194.$$

Therefore in the long run, expected annual claims are

$$E(S) = E(S_0) + E(S_1) + E(S_2) = 76{,}212 + 456{,}677 + 2{,}470{,}194 = \$3{,}003{,}083.$$

Since expected annual premiums (in the long run) are

$$10{,}000\,[500\,(0.02545758 + 0.15310824(0.8) + 0.82143419(0.7))] = 3{,}614{,}740,$$

the expected annual net profit once stability has been reached is

$$3{,}614{,}740 - E(S) = 3{,}614{,}740 - 3{,}003{,}083 = \$611{,}658$$

and hence premiums exceed expected claims by approximately 20%. ⬚

6.3.2.1 The number of claims made by an individual

Assume that accidents to a policyholder occur according to a Poisson process with rate parameter λ, and that the first claim in a year is only made when a loss suffered in an accident exceeds a threshold M (which occurs with probability p). Furthermore, assume that the soft rule for transition between classes is in effect and hence any loss suffered during the rest of the year is reported to the insurance company. The time T_1 to the first claim therefore has an exponential distribution with parameter $p\lambda$. Letting C_N be the random variable representing the number of claims made by an individual policyholder in a one-year time period, then clearly $P(C_N = 0) = e^{-p\lambda}$. We derive the expression for $P(C_N = 1)$ as follows:

$$P(C_N = 1) = \int_0^1 p\lambda e^{-p\lambda t}\,[P(\text{no further accidents in } [t, 1])]\,dt$$

$$= \int_0^1 p\lambda e^{-p\lambda t} \left[\frac{(\lambda(1-t))^0}{0!} e^{-\lambda(1-t)} \right] dt$$

$$= \frac{p}{1-p} \left[e^{-p\lambda} - e^{-\lambda} \right]$$

$$= \frac{p}{1-p} \left[P(C_N = 0) - e^{-\lambda} \right].$$

Similarly,

$$P(C_N = 2) = \int_0^1 p\lambda e^{-p\lambda t} \left[\frac{(\lambda(1-t))^1}{1!} e^{-\lambda(1-t)} \right] dt$$

$$= p\lambda e^{-\lambda} \int_0^1 \lambda(1-t) e^{-\lambda t(p-1)} dt$$

$$= \frac{1}{1-p} \left[P(C_N = 1) - p\lambda e^{-\lambda} \right]$$

and more generally for $k \geq 1$

$$P(C_N = k+1) = \int_0^1 p\lambda e^{-p\lambda t} \left[\frac{(\lambda(1-t))^k}{k!} e^{-\lambda(1-t)} \right] dt$$

$$= \frac{1}{1-p} \left[P(C_N = k) - p\frac{\lambda^k}{k!} e^{-\lambda} \right].$$

6.4   Reducing heterogeneity with NCD schemes

One of the objectives of using NCD systems is to make those with high claim rates pay appropriately in the form of higher premiums. Although NCD systems do punish individuals who make claims in the form of reduced premium discounts, they are not as effective as one might expect or like them to be. In the following example we compare the premium income from two groups of (relatively speaking) *good* and *bad* drivers.

Example 6.5

Again consider the NCD system introduced in Example 6.1, where the discount levels are E_0 (no discount), E_1 (20% discount) and E_2 (40% discount), and the full premium is \$500. The transition rule is to drop back one discount level if one claim is made, and to go back to paying the full premium if more than one claim is made. Let us assume that we have 10,000 relatively good drivers in this scheme who have the one-step transition matrix $\mathbf{P}_G$ given below (and in Equation(6.1)). Assume also however that we have another group of 10,000 relatively bad drivers who are (in some sense) twice as likely to make claims. More precisely, let us assume that for one of these drivers, the probability of one claim in a year is 0.4 while the probability of two or more claims is 0.2. It follows that the one-step matrix $\mathbf{P}_B$ of transition probabilities for these *bad* drivers and that for the *good* drivers are given by:

$$\mathbf{P}_G = \begin{pmatrix} 0.3 & 0.7 & 0.0 \\ 0.3 & 0.0 & 0.7 \\ 0.1 & 0.2 & 0.7 \end{pmatrix} \quad \text{and} \quad \mathbf{P}_B = \begin{pmatrix} 0.6 & 0.4 & 0.0 \\ 0.6 & 0.0 & 0.4 \\ 0.2 & 0.4 & 0.4 \end{pmatrix}.$$

TABLE 6.10
Expected premium income from two
groups of *good* and *bad* drivers.

Year	Good drivers	Bad drivers
0	5,000,000	5,000,000
1	4,300,000	4,600,000
2	3,810,000	4,440,000
3	3,712,000	4,376,000
4	3,643,400	4,350,400
8	3,616,811	4,333,770
16	3,616,279	4,333,334
32	3,616,279	4,333,334
∞	3,616,279	4,333,333

If we assume that all drivers start in class E_0 in year 0, then Table 6.10 gives the annual expected premium income from the two groups as the numbers in the different classes stabilize. The stationary distributions for the two groups are, respectively,

$$\pi_G = (0.1860465, \ 0.2441860, \ 0.5697674) \text{ and}$$
$$\pi_B = (0.5238095, \ 0.2857143, \ 0.1904762).$$

The results about premium incomes for the two groups are disappointing in that even after numbers stabilize there is relatively little difference in premium income between the *good* and *bad* drivers. In fact, the limiting ratio of premium income from the two groups is 1.2, and consequently the bad drivers are only paying 20% more than the good drivers in the long run! □

6.5   Problems

1. If Richard had joined the NCD system of Example 6.2 where the soft transition rule was in force, what premium would he have paid on the first day of July 2002? What would the answer be if the severe rule had been used? In each case, draw a diagram to illustrate the transition through the discount classes.

2. Using the one-step transition matrix for the NCD system of Example 6.1, what is the probability p_{02}^3 of a person who is presently on no discount getting the maximum discount three years from now? If one is presently paying 80% of the full premium, what is the probability that one will be doing the same six years hence?

3. Use a basic text editor to make a text file as follows, and "source it" or bring it into R. (Note: Any line in a text file beginning with # is ignored by R.) The text file below will create the matrix of transition probabilities $\mathbf{P}$ in Example 6.1. What is $\mathbf{P}^8$?

```
# NCD Example 1
ex1<-c(0.3,0.7,0,  0.3,  0,0.7,  0.1,0.2,0.7)
P1<-matrix(ex1,ncol=3,byrow=T)
P2<-P1%*%P1
P4<-P2%*%P2
P8<-P4%*%P4
```

4. Calculate a table similar to Table 6.3 for Example 6.1 when all 2000 policyholders start with no discount with a (pure) premium of $600.

5. Let us assume for the NCD system of Example 6.2 that for an insured individual, the probabilities of making $0, 1$, and > 1 claims in a year are, respectively, $0.7, 0.2$ and 0.1. Then the one-step transition matrix for this system (two steps back with one claim, and back to E_0 with more than one claim) is:

$$\mathbf{P} = \begin{pmatrix} 0.3 & 0.7 & 0 & 0 & 0 & 0 \\ 0.3 & 0 & 0.7 & 0 & 0 & 0 \\ 0.3 & 0 & 0 & 0.7 & 0 & 0 \\ 0.1 & 0.2 & 0 & 0 & 0.7 & 0 \\ 0.1 & 0 & 0.2 & 0 & 0 & 0.7 \\ 0.1 & 0 & 0 & 0.2 & 0 & 0.7 \end{pmatrix}.$$

Determine the equilibrium distribution for this NCD system.

6. The NCD scheme in Company B has three levels – 0%, 15% and 30%. The rules for moving between discount levels are: (a) in a year with no claims one moves up a discount level (or stays at 30%), (b) if one claim is made in a year, one steps back one discount level in the following year (or stays at 0%) and otherwise, (c) one goes back to having no discount. It is believed that policyholders at the various discount levels have different Poisson claim rates which are, respectively, $(0.5, 0.4, 0.3)$ for the respective discount levels $(0\%, 15\%, 30\%)$. Determine the transition matrix for movement between the discount levels, and find the steady state probabilities for being in the respective discount states.

7. Verify in Example 6.4 that $E(S_2) = \$2,470,194$.

8. An NCD scheme for motorists operates with three levels of discount: $E_0(0\%)$, $E_1(30\%)$ and $E_2(50\%)$. The normal full premium is $800. The soft transition rule applies whereby an individual moves to the next higher discount level (or stays at 50% discount) following a year with no claim, and when one or more claims are made in a year the individual moves back to the next lower discount level (or stays in E_0). The loss distribution appropriate for losses suffered by individuals in this scheme is a Weibull random variable with parameters $\gamma = 2$ and $c = 1/600,000$.

   (a) For each discount level, determine the size of a loss below which it is not worthwhile making a claim (given a two-year time horizon during which no further loss is suffered).

   (b) Losses are suffered by individuals at a Poisson rate of $\lambda = 0.3$ per year. For each discount level determine the Poisson rate of making a claim.

   (c) Determine the basic transition probability matrix for this scheme, and find the long-run probabilities of being in the three states.

   (d) What is the expected annual premium income for a group of 20,000 policyholders once stability has been reached?

9. Company A uses an NCD scheme with the four discount levels of $E_0 = 0\%$, $E_{20} = 20\%$, $E_{40} = 40\%$ and $E_{50} = 50\%$. The rules for movement between discount levels are: (a) in a year with no claims, one moves up a discount level (or stays at 50%), and (b) on making one or more claims in a year if one is in discount level E_0 or E_{20} then next year the individual will be in E_0, while if in E_{40} or E_{50} then one goes to discount level E_{20} next year. The loss suffered by an insured when an accident occurs is modeled by an exponential distribution with mean 2000. When an accident occurs an insured will only make a claim if the size of the loss exceeds the total extra premium payment that would be paid over the next three years (assuming no further accidents). The normal full premium is 900. For each discount level calculate the smallest loss an

insured will incur before making a claim, and the probability that an insured will make a claim given that a (first) loss is suffered.

10. An NCD scheme for motorists operates with three discount levels of $0\%, 20\%$ and 30%. The full normal premium is $500, and the operating rule is that if no claims are made in a given year the policyholder moves up one discount category in the following year, while in the event of one or more claims being made the policyholder moves back one discount level. A deductible of $100 is also in effect, whereby the policyholder pays the first $100 of any claim.

   (a) Calculate the size of damage below which it is not worthwhile making a claim at each level of discount, assuming policyholders have a two-year time horizon.

   (b) Assume the damage (loss) distribution for an accident is Pareto with parameters $\alpha = 3$, $\lambda = 800$. For each discount level find the probability that given an accident occurs a claim will be made (using again a two-year time horizon).

   (c) Assume all policyholders have the same underlying rate of 0.3 for sustaining accidents. Find the underlying transition matrix $\mathbf{P}$ appropriate for (one-year) movement between discount levels for this portfolio, taking into account calculations in (b).

11. A car insurance company operates a no claims discount scheme with three levels of discount: $0\%, 20\%$ and 40%. A policyholder who makes no claim in a given year moves to the next (higher) level of discount in the following year – or remains at the top level of discount. If one or more claims are made in a given year, the policyholder drops one level of discount in the following year (or remains with 0% discount). Let us assume that the accident rate is 0.1 for each individual holding a policy. The full normal premium is $400 per annum and the random damage X on a car in an accident is uniformly distributed on $[0, 1000]$.

   Suppose that a policyholder who has an accident only decides to make a claim if the damage incurred exceeds the subsequent increase in premium payments that he will pay over the next two years as a result of making the claim. His calculations are made assuming that no more accidents will occur over the next few years. For each level of discount determine the probability that a policyholder will make a claim in a given year. Estimate the proportion of policyholders in each of the discount levels once stability has been reached.

12. An NCD motor insurance scheme operates with the four discount levels (states) of $0\%, 20\%, 40\%$ and 60%. If one claim is made in a given year, the individual drops two levels of discount (or stays at 0% discount) in the following year. If no claim is made then the policyholder moves up

one level of discount for the following year (or stays at 60% discount). Assume the probability of one accident in a year is 0.1 for any individual, and that we can ignore the possibility of two or more accidents by an individual in a year. The random loss on occurrence of an accident is modeled by a Pareto random variable with parameters $\lambda = 2400$ and $\alpha = 3$. Assume a claim is made following an accident only when the damage X exceeds the additional premiums the individual would have to pay (assuming a three-year horizon without additional claims). The normal full premium is $800.

(a) For each level of discount, determine the minimum damage that must occur before a claim is made.

(b) Determine the matrix of one-step transition probabilities for this NCD Markov chain.

(c) If 10,000 policyholders have just paid the full premium for the year 2007, how much premium income would one expect from this group in the year 2009?

(d) How much premium income would be expected from those on the maximum discount level in the year 2010?

13. An NCD scheme for automobile insurance has three levels of discount: 0%, 30% and 50%. A policyholder who makes no claim in a given year moves to the next higher level of discount (if possible). When a claim is made in a given year the policyholder drops one level of discount in the following year (or stays at 0% discount). Assume that the chances of one accident in a year is 0.3 for any policyholder, and that we can ignore the possibility of more than one accident. The full premium is $500. The cost of an accident X is a random variable with density function

$$f(x) = 2(400)^2/(400 + x)^3, \quad \text{for } x > 0.$$

We assume that if the policyholder has an accident, a claim is made only if the damage done exceeds the subsequent increase in premiums that would be paid over the next two years.

(a) For each level of discount, calculate the probability that a policyholder will actually make a claim in a given year.

(b) Suppose there are 20,000 policyholders, all of whom initially start with no bonus. Determine the expected numbers in each of the discount levels once stability has been reached.

(c) Assuming all 20,000 policyholders have just paid full premiums, estimate the total premium income for the next year and compare it with the total expected premium once stability has been reached.

14. An NCD scheme for drivers in a town has three categories of discount levels: $E_0 = 0\%, E_1 = 35\%, E_2 = 70\%$. If a driver policyholder makes at least one claim during a year he/she moves down a discount level (or stays in E_0) and otherwise moves up one level (or stays at E_2). The probability that a policyholder makes one or more claims in a year varies with the discount level as follows: For those in E_0 it is p, for those in E_1 it is $0.7p$, while for those in E_2 it is $0.5p$.

(a) Write down the transition matrix for this NCD scheme in terms of p.

(b) Determine the steady state distribution for this NCD scheme in terms of p.

(c) Calculate the average premium payment per policyholder once a steady state has been reached in terms of p and M where M is the full premium (premium for someone starting in state E_0). Suppose that for the better drivers in Group A the value of $p = 0.1$ is appropriate, while that for those in Group B is $p = 0.2$. It is decided to use a full premium value of $M_A = 900$ for those in Group A. What should the value of M_B be for policyholders in Group B if it is desired that the average premium for Group B policyholders is to be 40% more than that for Group A? Comment on your answer.

15. A no claim discount scheme has three levels of discount: $0\%, 30\%$ and 50%. The rule for movement between discount levels is where one moves up a discount level (or stays at the 50% discount level) if no claims are made in a year, moves back one discount level following one claim, and returns to the 0% discount level if two or more claims are made in a year. The full premium for a year is $600. The loss distribution for damage is modeled by a lognormal random variable with mean $1200 and standard deviation $800. Damage resulting from an accident turns into a (first) claim for a year only if it exceeds the total extra premiums a person would have to pay over the next two years. The accident rate for any individual is assumed to be $\lambda = 0.25$.

(a) For each discount level, determine the smallest damage loss that will lead to a claim being made.

(b) For each discount level, determine the (first) claim incidence rate based on the calculations in part (a).

(c) This year there are 10,000 policyholders, half of whom paid the full premium and 25% of whom got a 30% discount. How many policyholders would one expect to get the full discount next year?

(d) What is the total expected premium income for next year?

16. An NCD scheme operates with four levels of discount: 0%, 20%, 30% and 40%. The rule for movement between discount levels is that one moves up one step (discount level) next year if no claims are made this year, and if one or more claims are made in a year then one drops to the next lower discount level in the following year (or stays at 0% discount). The full premium for an individual is $600 per year. When an accident occurs, the appropriate loss distribution X is lognormal with parameters $\mu = 5.8798927$ and $\sigma^2 = 0.2231435$.

   (a) For each discount level determine the size of damage below which it is not worthwhile making a claim, assuming a three-year time horizon. What are the mean and variance of a typical amount of damage?

   (b) For each level of discount, determine the probability that a claim will be made (assuming a three-year time horizon).

   (c) Assume the number of accidents an individual has in a year is a Poisson random variable with parameter 0.1. Find the matrix of transition probabilities for one-year movements in this NCD scheme.

17. Accidents for an individual in an NCD scheme occur at the (Poisson) rate of λ per year. The first accident in a year (if it occurs) will lead to a claim only if the damage incurred exceeds m, and this happens with probability $p = P(X > m)$. Assuming the soft transition rule is in effect, any accident subsequent to the first claim will be reported as a claim in any given year. If C_N is the random variable representing the number of claims for such an individual in a year, show that

$$E(C_N) = \lambda - \frac{1-p}{p}\left[1 - e^{-p\lambda}\right] = \lambda - \frac{1-p}{p}\,P(C_N \geq 1).$$

What would $E(C_N)$ be if $\lambda = 0.3$, the loss distribution for an accident is exponential with mean 1200 and the threshold for the first claim is $m = 800$?

7

Generalized Linear Models

7.1 Introduction to linear and generalized linear models

In 1972, Nelder and Wedderburn developed a theory of generalized linear models (GLMs) which unified much of the existing theory of linear modeling, and broadened its scope to a wide class of distributions. In particular, the theory encompasses the class of logit and probit models, which are useful models developed for handling binomial data, as well as Poisson regression and Poisson log-linear models for contingency table analysis. An important contribution to the theory of GLMs is their general iterated least squares algorithm for finding the maximum likelihood estimates of the parameters in these models. The introduction of generalized linear models has had a very significant impact on practical modeling in a wide variety of areas, particularly with the implementation of procedures for building models in statistical software such as GLIM, Genstat, S-Plus and R, SAS, Stata, Systat and SPSS. In actuarial science alone the theory has been used to model problems dealing with premium rating, mortality, multiple state models and claims reserving.

This chapter provides a brief introduction to generalized linear models. An excellent and thorough treatment of the topic is given in the 2nd edition of *Generalized Linear Models* by McCullagh and Nelder [41]. Dobson's 2nd edition of *An Introduction to Generalized Linear Models* [22] provides a very good but less technical introduction to the subject. For a good overview of the subject, the chapter by Firth [26] is well worth reading, and the text by Pawitan [49] gives interesting insight into modeling relationships more generally (but in particular by GLM). Haberman and Renshaw [27] give an interesting review of the applications of generalized linear models in actuarial science.

Modeling relationships between observations (responses) and variables is the essence of most statistical research and analysis. How is the incidence of lung cancer influenced by age, gender, smoking, diet and other variables? Are the numbers of automobile claims made by a driver related to age, education, gender, type of vehicle, engine size and daily usage? In what way is the size of an employer liability claim related to the personal characteristics of the employee (age, gender, salary) and the working environment (safety standards, hours of work, promotional prospects)? Constructing interpretable models for connecting (or linking) such responses to variables (which may be of a nominal, ordinal or interval nature) can often give added insight into the complexity of the relationship which may often be hidden in a huge amount of data. Strictly speaking, as many statisticians will admit (see, for example, Box [8]), all models are wrong – although some can be quite useful. *Parsimony* is an aspect of modeling which is highly desirable. Often a simple model with fewer explanatory variables is more useful (or parsimonious) than a complicated one which lacks intuition and is harder to interpret.

In the classical linear model, a random response Y is modeled via a relationship of the form

$$Y = \mathbf{x}^\mathbf{T}\boldsymbol{\beta} + \epsilon, \tag{7.1}$$

where $\boldsymbol{\beta}$ is a vector of unknown parameters, $\mathbf{x}$ is a vector of known explanatory variables and ϵ is an error term. Two fundamental aspects of such a model are the *linear* dependence of the response on the unknown parameters, and the *additive* error structure. Often one assumes that $E(\epsilon) = 0$ and $\text{Var}(\epsilon)$ is constant, in which case it follows that $\mu = E(Y) = \mathbf{x}^\mathbf{T}\boldsymbol{\beta}$. In the classical linear model the mean of the response is a linear function of the explanatory variables. In particular if y_i is the ith response corresponding to the vector of known explanatory variables (or covariates) $\mathbf{x}_i^\mathbf{T} = (x_{i1}, \ldots, x_{ip})$, then the model predicted value for y_i (using the *hat* notation for an estimate) is given by

$$\hat{y}_i = \hat{\beta}_1 x_{i1} + \cdots + \hat{\beta}_p x_{ip} \equiv \hat{\eta}_i.$$

Such a linear relationship between the mean μ_i of Y_i and the *linear predictor* η_i is both mathematically appealing and often readily interpretable, but there are many situations where such a linear connection or link may be either inappropriate or impossible.

If the unknown parameters $\boldsymbol{\beta}$ are allowed to vary freely, then the linear predictors η_i may in theory take real values outside the range of the response variable of interest. For example, if the response Y is a probability or a proportion, then the mean value of Y will be restricted to the unit interval. In other cases, the response Y may be positive (for example, when modeling counts) or restricted to a finite interval.

There are two somewhat natural methods to try, in order to overcome such a problem. One possibility is to transform the response Y with a function g and use linear modeling on $g(Y)$. With this approach, one is suggesting that the mean of a response Y_i is of the form:

$$E(g(Y_i)) = \sum_{j=1}^{p} \beta_j x_{ij}.$$

For example, the log function has been extensively used in various applications to transform data to an approximate normal (response) variable. In insurance the log transformation is often applied to claim size data, due to the inherent nature of claims to be both skewed and positive. Another approach to the problem is to use a transformation g to describe the relationship between the *mean* μ of Y, and linear combinations of the explanatory variables. In this case, we are linking the mean μ_i of an observation Y_i through a functional transformation g whereby

$$g(E(Y_i)) \;=\; g(\mu_i) \;=\; \sum_{j=1}^{p} \beta_j x_{ij}.$$

This approach is the essence of generalized linear modeling.

Logistic regression is a good example of a generalized linear model which is not a linear model in the classic sense. Suppose we are interested in the proportion π_x of insured male drivers of age x who will make an accident claim in the coming year. Letting Y_x be the sample proportion of insured drivers of age x who make a claim, then it is clear that the mean value $E(Y_x) = \pi_x$ is restricted in value to the interval $[0, 1]$. To model the proportion π_x as a linear function of the form $\pi_x = \beta_0 + \beta_1 x$ over the relevant age domain is usually too restrictive and inappropriate. That is, as x varies, $\pi_x = \beta_0 + \beta_1 x$ will often take values outside of the interval $[0, 1]$, and hence may not be interpretable as proportions (or probabilities). The logit of a probability or proportion π is the logarithm of the odds ratio $\pi/(1 - \pi)$, that is, $\mathrm{logit}(\pi) = \log[\pi/(1 - \pi)]$, and this is also referred to as the *logistic function*. Note that $\mathrm{logit}(\pi)$ is an increasing continuously differentiable function from $(0, 1)$ onto the real line R. In a linear logistic (logit, or logistic regression) model, one models logit (π) as a linear combination of appropriate explanatory variables. In our example, the link function g between the mean $\mu_x = \pi_x = E(Y_x)$ and the linear predictor $\eta_x = \beta_0 + \beta_1 x$, might be given by

$$g(\pi_x) \;=\; \mathrm{logit}(\pi_x) \;=\; \log \frac{\pi_x}{1 - \pi_x} \;=\; \beta_0 + \beta_1 x \;=\; \eta_x,$$

$$\text{or} \quad \pi_x \;=\; \frac{\exp(\beta_0 + \beta_1 x)}{1 + \exp(\beta_0 + \beta_1 x)} \;=\; \frac{\exp(\eta_x)}{1 + \exp(\eta_x)}.$$

Early uses of linear logistic models were in bioassay experiments and the analysis of survey data, but are now a common tool in many disciplines where proportional data is of interest. For example, we might consider a logit model if we are interested in modeling how the probability of an individual developing a critical illness (or having a car accident) is related to age, gender and various other health or risk factors. In mortality studies, the initial rate of mortality q_x at age x has been modeled by logistic regression (see [27]).

The Poisson distribution with mean μ is a basic and important distribution for count data, but has the well-known property that the mean and variance are the same. Therefore if the mean of a Poisson response variable depends linearly on a vector $\mathbf{x}$ of known explanatory variables, then its variance is nonconstant. Traditionally, one may attempt to model this situation via the classical approach to linear modeling by either transforming the response variable, or by using a weighted least squares approach where the weights are estimated (via iterated methods) from the data.

The generalized linear model approach would begin by noting that if Y is "Poisson" with mean μ, then although μ is positive it may not be reasonable to express it as a linear combination of predictor variables. One may, however, use a (multiplicative) log-linear model for the Poisson distribution, where we model

$$g(\mu) = \log \mu = \eta = \mathbf{x}^T \boldsymbol{\beta}, \quad \text{or equivalently,} \quad \mu = e^{\eta}.$$

Here the function linking the mean $E(Y) = \mu$ to the *linear predictor* η is the log function, which is a continuously differentiable function from $(0, \infty)$ onto the real line R. This example of a generalized linear model is also known as *Poisson (linear) regression*. Poisson regression can be used when we want to study how the number of claims made by an insured individual depends on various explanatory variables or predictors, or how the number of accidents at an intersection depends on weather, traffic intensity, hour of the day or day of the week. In the study of mortality, the force of mortality μ_x might be modeled via Poisson regression.

The method of probit analysis is one of the first examples of a generalized linear model that is not a linear model in the classical sense. It has its origins in work of Bliss [4] in 1935 in bioassay, where mortality is modeled on dose level. Suppose we are interested in the proportion Y_x of subjects that survive a drug or toxin given at a known dose level x (which is often measured in logarithmic units), and let $\pi_x = E(Y_x)$ denote the mean value of Y_x. In the probit model, one uses the inverse Φ^{-1} of the cumulative distribution function of the standard normal distribution as a link between the mean $\pi_x = \mu_x = E(Y_x)$ and the linear predictor $\eta_x = \beta_0 + \beta_1 x$, whereby

$$g(\pi_x) = \Phi^{-1}(\pi_x) = \eta_x, \quad \text{or} \quad \pi_x = \Phi(\beta_0 + \beta_1 x) = \Phi(\eta_x).$$

Note that no matter what the values of β_0 and β_1 are, $\Phi(\beta_0 + \beta_1 x)$ will still be a number between 0 and 1 – that is a probability. The function $g = \Phi^{-1}$ is a continuously differentiable $1 - 1$ mapping of $(0, 1)$ onto the real line R.

This brief introduction indicates how the theory of generalized linear models extends the classical linear model and also includes the logit, Poisson and probit models as special cases. The following points summarize some of the main ways in which this theory generalizes the classic approach to linear modeling:

- The relationship between the mean μ_i of a random observation Y_i and other explanatory variables $x_{i1}, \dots, x_{ip}$ may be more general than that in the linear model. Instead of assuming that $\mu_i = \beta_1 x_{i1} + \cdots + \beta_p x_{ip}$ (that is, the mean is a linear combination of the predictor variables), one assumes there is a function g (called the *link function*) connecting the mean μ_i with the *linear predictor* $\eta_i = \beta_1 x_{i1} + \ldots + \beta_p x_{ip}$ through $g(\mu_i) = \eta_i$.

- The variance of an observation may depend on the mean. In fact, we will see that $Var(Y_i) = (\phi/A_i) V(\mu_i)$, where A_i is a known prior weight and ϕ is a (often a *nuisance*) scale parameter independent of i. $V(\mu)$ is called the *variance function*, and it shows how the variance of an observation depends on its mean.

- The distribution of the response variable Y can be any member of the so-called *exponential family* of distributions. This family includes in particular the normal, Poisson, binomial and gamma distributions.

Our development of the generalized linear model begins with a review in Section 7.2 of normal linear models. Generalized linear models extend the class of linear models to a class of distributions known as exponential families, and these, together with the structural aspects of a GLM such as link functions and linear predictors are discussed in Section 7.3. The use of deviance as a tool in modeling, as well as residual analysis and goodness-of-fit, are discussed in Section 7.4. The reader should be aware that there are many important extensions of generalized linear models which we do not treat in this text, including generalized linear mixed models (GLMM). Such generalized linear models include *random* (in addition to what are called *fixed*) effects in the linear predictor.

7.2   Multiple linear regression and the normal model

In the classical normal linear model, one assumes that $\mathbf{Y} = (Y_1, \ldots, Y_n)$ are independent normal random variables, where $Y_i \sim N(\mu_i, \sigma^2)$ and

$$E(Y_i) = \mu_i = \mathbf{x}_i^T \boldsymbol{\beta}.$$

One traditionally writes this model in the form

$$\mathbf{Y} = \mathbf{X}\boldsymbol{\beta} + \boldsymbol{\epsilon},$$

where

$$\mathbf{Y} = \begin{bmatrix} Y_1 \\ \vdots \\ Y_n \end{bmatrix}, \quad \mathbf{X} = \begin{bmatrix} \mathbf{x}_1^T \\ \vdots \\ \mathbf{x}_n^T \end{bmatrix}, \quad \boldsymbol{\beta} = \begin{bmatrix} \beta_1 \\ \vdots \\ \beta_p \end{bmatrix}, \quad \boldsymbol{\epsilon} = \begin{bmatrix} \epsilon_1 \\ \vdots \\ \epsilon_n \end{bmatrix}$$

and the error terms ϵ_i are independent $N(0, \sigma^2)$ random variables. An observation Y_i is therefore split into a *systematic* component denoted by $\mathbf{x}_i^T \boldsymbol{\beta} = $

$\beta_1 x_{i1} + \cdots + \beta_p x_{ip}$, and a *random* component ϵ_i. The systematic component ($\eta_i = \mathbf{x_i^T}\boldsymbol{\beta}$) is a linear predictor for $E(Y_i)$. $\mathbf{X}$ is usually referred to as the $n \times p$ design matrix, where n is the number of random observations and p is the number of explanatory variables.

In this setting, the p explanatory (regressor, predictor or independent) variables may be categorical, continuous or a combination of both, and we refer to this class of models as the *general linear model*. If there is a constant in the linear model, then the first column of the design matrix $\mathbf{X}$ consists of all $1's$. The most basic such model is the simple linear regression model where

$$Y_i = \beta_1 + \beta_2 x_i + \epsilon_i.$$

The term *analysis of variance* or *ANOVA* is often used to describe a model where the explanatory variables are all categorical. Usually, one wants to compare groups defined by specifying different levels of these categorical variables or factors. For example, we might want to compare the log of a claim size (assuming claims are lognormally distributed) with respect to the gender and home address of a motorist. When there is a mixture of continuous and categorical variables in the normal linear model, one often uses the term *analysis of covariance* or *ANCOVA* to describe it. In such a model, one is interested in comparing groups defined by different levels of categorical variables, but also one wants to incorporate the possibility that the response within a group may need adjustment for certain *covariates* or continuous variables. In Example 7.1, an ANCOVA model for degree mark is investigated.

In a normal linear model, the mean $E(Y_i) = \mu_i$ is modeled as a linear function of the explanatory variables with coefficients $\boldsymbol{\beta}$. The (variance) parameter σ^2 is often referred to as a nuisance parameter, but of course may be useful in making inferences about μ_i. For a sample of observations $\mathbf{y}$, the likelihood function takes the form

$$L(\boldsymbol{\beta}, \sigma^2; \mathbf{y}) = \left(\frac{1}{2\pi\sigma^2}\right)^{n/2} \exp\left\{-\frac{1}{2\sigma^2} \sum_{i=1}^{n}(y_i - \mathbf{x}_i^T\boldsymbol{\beta})^2\right\}.$$

It is well known that the maximum likelihood estimates of $\boldsymbol{\beta}$ and σ^2 are given by

$$\hat{\boldsymbol{\beta}} = (\mathbf{X}^T\mathbf{X})^{-1}\mathbf{X}^T\mathbf{y} \quad \text{and}$$

$$\hat{\sigma}^2 = \frac{1}{n} \sum_{i=1}^{n}(y_i - \mathbf{x}_i^T\hat{\boldsymbol{\beta}})^2.$$

Usually, in the normal linear model, estimates of the parameters $\boldsymbol{\beta}$ are obtained by the least squares method, that is, by minimizing

$$\sum_{i=1}^{n}(y_i - \mathbf{x}_i^T\boldsymbol{\beta})^2,$$

with respect to $\boldsymbol{\beta}$. This gives results identical to those using the method of maximum likelihood, and these estimators are unbiased. In estimating σ^2, one normally uses a divisor of $(n-p)$ (the number of degrees of freedom) instead of n. The resulting estimator is an unbiased estimator of σ^2, and it is usually called the *mean residual sum of squares*, or the mean error sum of squares. The celebrated Gauss–Markov theorem states that the least squares (or maximum likelihood) estimates are the (BLUE) Best Linear (in the observations y_i) Unbiased Estimates of $\boldsymbol{\beta}$.

Often there are many possible explanatory variables that may be included in a model, and so it is important to identify a subset which explains a good amount of the variation in the response variable, yet leads to a parsimonious model. In multiple linear regression, one can make use of step-wise regression methods to identify suitable variables. In essence, one uses an analysis of variance table for a potential model, deciding on the basis of mean sums of squares and the χ^2 probability distribution whether or not a certain candidate variable (or set of variables) should be added to an existing model. In generalized linear models, however, one makes use of the concept of *deviance* to decide on which variables should be included in a model. The deviance function in GLM is one of the most important applications of the theory of likelihood through the use of the likelihood ratio statistic. The deviance may also be used to measure the lack-of-fit of a model.

Residual analysis is of course a crucial tool in modeling. If $\hat{\mu}_i$ is the fitted value for a given model corresponding to $\mathbf{x}_i^T$, then the ith residual is given by

$$\hat{\epsilon}_i \;=\; y_i - \mathbf{x}_i^T \hat{\boldsymbol{\beta}} \;=\; y_i - \hat{\mu}_i.$$

Unusual observations generally have large residuals, but it is often informative to standardize residuals in order to identify trends or patterns which might be of interest or concern. The *usual i^{th} standardized residual* is given by

$$r_i = \frac{\hat{\epsilon}_i}{\hat{\sigma}\sqrt{(1 - h_{ii})}},$$

where h_{ii} is the ith diagonal element in the so-called hat matrix $\mathbf{X}(\mathbf{X}^T\mathbf{X})^{-1}\mathbf{X}^T$. Another commonly used standardized residual is the *studentized residual* (see, for example, [23]).

In Example 7.1, a normal linear model for final degree mark for university students in an actuarial program is analyzed.

Example 7.1 Final university degree mark

A study was undertaken to develop a model to predict the final degree mark of university students in an actuarial program in Ireland on the basis of information available on their entry into the program (Ferguson [25]). Final degree performance can be measured in different ways, but for the purposes of this study attention centers on the overall degree mark Y (which is expressed as a % based on a weighted average of many course results obtained in the program). Table 7.1 shows how overall degree mark (%) is related to the *class* of degree obtained.

TABLE 7.1
University degree classification by overall degree mark (%).

Class of degree	Pass	Third	Lower second	Upper second	First
	P	III	II.2	II.1	I
Degree mark (%)	$[40, 50)$	$[50, 55)$	$[55, 62)$	$[62, 70)$	≥ 70

Entrance to the program is very competitive, and most students would expect to obtain a II.1 degree or higher. Marks over 70% are considered excellent (First class). Marks over 80% are infrequent and those over 90% rare. Extensive data were available on the 197 students who had already completed the program over a six-year period, including information on the following:

- *gender* (male or female)

- *age* (age in years at entry, ranging from $17 - 20$)

- *points* (entrance exam results on a scale of $0 - 640$)

- *home* (home address categorized as: Dublin, Ireland excluding Dublin (Non Dublin), or not Ireland (Other))

Figure 7.1 gives a basic plot of degree % by points at entry for the various categories of *home*. Most students have done reasonably well, with few Dublin students getting a % under 60.

On the basis that exam marks are often treated as normal variates, together with the fact that the response variable Y for each student is a weighted average of many exam scores, it was decided to try and fit a normal linear model for Y with respect to the four possible explanatory variables above. The variables *age* and *points* are continuous (and might be called covariates), while the variables *gender* and *home* are factors or categorical variables. Although the variable *points* can in theory take values anywhere in the range $[0, 640]$, because of the high quality of students the actual range was $[505, 640]$. The factor *gender* has two levels (male and female), while the factor *home* has three. After an analysis of several normal linear models, it was decided to use a model of the form:

$$E(Y) = \alpha_{home} + \beta_1 x_{points} + \beta_2 x_{points}^2 + \beta_3 x_{age}, \qquad (7.2)$$

where α_{home} takes different values for the various levels of the variable *home*, x_{points} is the number of points on entry, and x_{age} is age at entry. The covariate

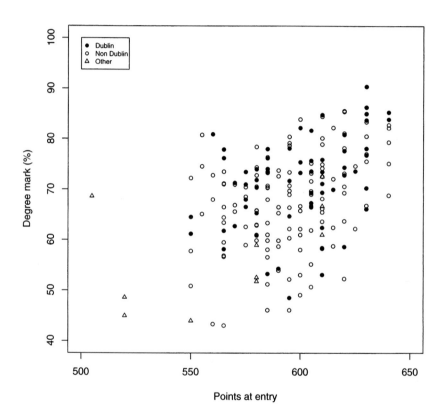

FIGURE 7.1
Final degree results and points at entry.

x_{points} enters the model as both a linear and quadratic predictor. Given knowledge of the variables x_{points} and x_{age}, there was no evidence to support using *gender* as an explanatory variable (*gender* was not significant). This result was not surprising, yet for various cultural reasons in the not too distant past in Ireland, females might not have been expected to perform on an equal basis in areas with a high mathematical content. No interaction terms were deemed appropriate for this model.

Using least squares (or equivalently, in this case the method of maximum likelihood), the following estimates were obtained for the relevant parameters:

$$\hat{\alpha}_D = 805.717737, \ \hat{\alpha}_{ND} = 802.66950, \ \hat{\alpha}_O = 794.99950,$$

$$\hat{\beta}_1 = -2.484114, \ \hat{\beta}_2 = 0.00222703, \ \hat{\beta}_3 = -2.5532705.$$

These estimates were obtained using the linear models function *lm* in the statistical package R with the data on each of the 197 students. Similar (identical) results should be obtained from other good statistical packages. For this model, the linear predictor η takes the form

$$E(Y) \ = \ \eta \ = \ \alpha_{home} + \beta_1\, x_{points} + \beta_2\, x_{points}^2 + \beta_3\, x_{age}$$

and we can use η to predict the overall degree mark for students with particular characteristics. In other words, the connection or *link* between the linear predictor η and the expected response $E(Y)$ is the identity link – that is, $E(Y) = \eta$ in this normal situation. With our model for degree performance, we have a linear predictor for an 17-year-old student from Dublin entering on 590 points of

$$\hat{\eta} = 805.717737 - 2.484114\,(590) + 0.00222703\,(590)^2 - 2.5532705\,(17)$$
$$= 71.91.$$

In particular, for such a student $E(Y) = 71.91\%$, which, if obtained, would result in a First-class (I) degree. A plot of the residuals by fitted values (given in Figure 7.2) supports a reasonable fit for the model.   ꟾ

7.3   The structure of generalized linear models

We now elaborate on many of the structural aspects of generalized linear models. We define and give examples of exponential families of distributions, discuss link functions and linear predictors. We also discuss factors and covariates, as well as interactions between possible predictor variables.

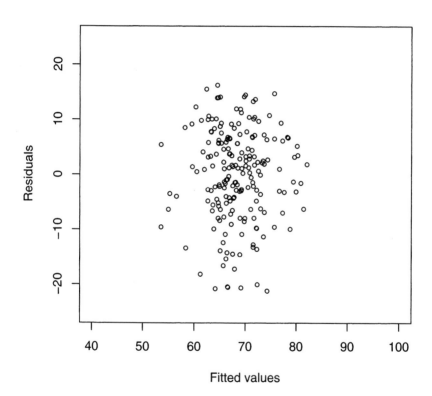

FIGURE 7.2
Residuals for degree mark with model (7.2).

7.3.1   Exponential families

The concept of an exponential family is fundamental to the theory of generalized linear models, but unfortunately the notation used varies considerably. To a large extent the notation used here follows that used by Venables and Ripley [57] and in the statistical packages R and S.

Many of the most useful one-parameter distributions used in modeling belong to what is called the *general exponential family*. The random variable Y has a distribution belonging to a one-parameter exponential family if it has a density (or probability) function f_Y which can be expressed in the form

$$f_Y(y; \theta, \phi) = \exp\left(A\left[y\theta - \gamma(\theta)\right]/\phi + \tau(y, \phi/A) \right), \tag{7.3}$$

or equivalently, the log of its density (log-likelihood function) $l(\theta)$ can be expressed as

$$l(\theta) = \log L(\theta, \phi; y) = \log f_Y(y; \theta, \phi) = A\left[y\theta - \gamma(\theta)\right]/\phi + \tau(y, \phi/A). \tag{7.4}$$

The parameter θ is referred to as the *natural* (or *canonical*) parameter, and (as we shall see) is related to the mean through $E(Y) \equiv \mu = \gamma'(\theta)$. ϕ is a scale or dispersion parameter which may be known, and is related to the variance of Y through $Var(Y) = \phi\gamma''(\theta)/A$. This allows the variance of Y to vary freely from the mean ($\mu = \gamma'(\theta)$) with the dispersion parameter ϕ, and this is an important property of exponential families in generalized linear modeling. γ and τ are known functions, and A is a known prior weight.

The score function U is the partial derivative of the log-likelihood function l of Y with respect to θ, and is denoted by $U(\theta) = (\partial/\partial\theta)\, l$. As a function of Y, it has mean 0 since

$$
\begin{aligned}
E_Y\left(\frac{\partial}{\partial\theta} l\right) &= \int f_Y \frac{\partial}{\partial\theta} \log f_Y \, dy \\
&= \int f_Y \frac{1}{f_Y} \frac{\partial f_Y}{\partial\theta} \, dy \\
&= \int \frac{\partial f_Y}{\partial\theta} \, dy \\
&= \frac{\partial}{\partial\theta} \int f_Y \, dy = \frac{\partial}{\partial\theta} 1 = 0.
\end{aligned}
\tag{7.5}
$$

Using expression (7.4) one may write

$$E_Y\left(\frac{\partial}{\partial\theta} l\right) = E_Y\left(\frac{A}{\phi}\left[Y - \gamma'(\theta)\right]\right),$$

which together with (7.5) implies that for distributions in an exponential family we have $E(Y) = \mu = \gamma'(\theta)$.

Moreover,

$$E_Y\left[\frac{\partial^2}{\partial\theta^2}l + \left(\frac{\partial}{\partial\theta}l\right)^2\right] = \int f_Y\left[\frac{\partial}{\partial\theta}\left(\frac{1}{f_Y}\frac{\partial f_Y}{\partial\theta}\right) + \left(\frac{1}{f_Y}\frac{\partial f_Y}{\partial\theta}\right)^2\right]dy$$

$$= \int f_Y\frac{1}{f_Y^2}\left[-\left(\frac{\partial f_Y}{\partial\theta}\right)^2 + \frac{\partial^2 f_Y}{\partial\theta^2}f_Y + \left(\frac{\partial f_Y}{\partial\theta}\right)^2\right]dy$$

$$= \int \frac{\partial^2 f_Y}{\partial\theta^2}dy$$

$$= \frac{\partial^2}{\partial\theta^2}\int f_Y\,dy = \frac{\partial^2}{\partial\theta^2}1 = 0.$$

The Fisher information $I(\theta)$ in Y for θ is defined by $I(\theta) = E(\frac{\partial}{\partial\theta}l)^2$, and therefore by the above we have two different expressions for this information given by

$$I(\theta) = E\left(\frac{\partial}{\partial\theta}l\right)^2 = -E\left(\frac{\partial^2}{\partial\theta^2}l\right).$$

Furthermore, it also follows that

$$E\left(\frac{\partial^2}{\partial\theta^2}l\right) = -\frac{\gamma''(\theta)}{\phi/A} \quad\text{and}\quad E\left(\frac{\partial}{\partial\theta}l\right)^2 = E\left(\frac{Y - \gamma'(\theta)}{\phi/A}\right)^2 = \frac{Var(Y)}{(\phi/A)^2}$$

and consequently,

$$Var\,(Y) = \frac{\phi}{A}\,\gamma''(\theta) \equiv \frac{\phi}{A}\,V(\mu).$$

We use $V(\mu) = \gamma''(\theta)$ to denote the *variance function* associated with the family. Note that when $\phi/A = 1$, the function $\gamma(\theta)$ has the property that $E(Y) = \gamma'(\theta)$ and $Var(Y) = \gamma''(\theta) = V(\mu)$. Consequently, $\gamma(\theta)$ is sometimes referred to as the cumulant function of Y.

We will now describe briefly four frequently used exponential families and their parameters.

Example 7.2 Normal family
If $Y \sim N(\mu, \sigma^2)$, then it has density function

$$f_Y(y) = \frac{1}{\sqrt{2\pi\sigma^2}}\,\exp\left(-\frac{(y-\mu)^2}{2\sigma^2}\right) \qquad\text{for all real } y.$$

By completing the square in the exponent, the log-likelihood function l may be represented in the form

$$l(\mu, \sigma^2 : y) = \frac{y\mu - \mu^2/2}{\sigma^2} - \frac{1}{2}\left(\frac{y^2}{\sigma^2} + \log 2\pi\sigma^2\right).$$

Using the notation

$$\theta = \mu,\ \phi = \sigma^2,\ \gamma(\theta) = \theta^2/2,\ A = 1 \ \text{ and } \ \tau(y,\phi) = -\frac{1}{2}\left(\frac{y^2}{\phi} + \log 2\pi\phi\right),$$

we see that Y belongs to an exponential family. Moreover,

$$E(Y) = \gamma'(\theta) = \theta \quad \text{and} \quad \text{Var}(Y) = \phi\gamma''(\theta) = \sigma^2 \cdot 1 = \sigma^2 V(\theta).$$

In particular, the natural parameter for the normal distribution $N(\mu,\sigma^2)$ is μ, and the scale or dispersion parameter is σ^2. ⬚

Example 7.3 Poisson family
Let Y have Poisson distribution with parameter μ, that is,

$$f_Y(y) = \mu^y \frac{e^{-\mu}}{y!} = \exp\left(y\log\mu - \mu - \log y!\right)$$

for $y = 0, 1, \ldots,$. Using

$$\theta = \log\mu,\ A = \phi = 1,\ \gamma(\theta) = e^\theta \ \text{and} \ \tau(y,\phi) = -\log y!,$$

it follows that the Poisson family of distributions is an exponential family with natural parameter $\theta = \log\mu$ and dispersion parameter $\phi = 1$. The suggestion is that $\log\mu$ is the natural parameter for the Poisson distribution, and not μ itself! In the Poisson GLM model, it is usually $\log\mu$ which is linked to linear combinations of explanatory random variables rather than μ. Note that

$$E(Y) = \gamma'(\theta) = e^\theta = \mu \quad \text{and} \quad \text{Var}(Y) = \phi\gamma''(\theta) = 1\,e^\theta = \mu = V(\mu).$$

In particular, the variance of a Poisson random variable is its mean and the dispersion parameter $\phi = 1$. This is in sharp contrast to the normal distribution, where the variance σ^2 is not in any way related to the mean μ. The normal distribution has dispersion parameter $\phi = \sigma^2$ and $V(\mu) = 1$. ⬚

Example 7.4 Binomial family
Let $S \sim B(n,p)$ be the number of successes in n Bernoulli trials. We know that the observed proportion of successes $Y = S/n$ is an unbiased estimator of the probability of success $p = \mu = E(Y)$. The probability function for Y may be expressed as

$$
\begin{aligned}
f_Y(y) &= P[Y = y] = P[S = ny] = f_S(ny) \\
&= \binom{n}{ny} p^{ny}(1-p)^{n-ny} \\
&= \exp\left\{ny\log\frac{p}{1-p} + n\log(1-p) + \log\binom{n}{ny}\right\} \\
&= \exp\left\{n\left[y\log(p/(1-p)) + \log(1-p)\right] + \log\binom{n}{ny}\right\}.
\end{aligned}
$$

Therefore using the notation

$$\theta = \text{logit } p, \ \gamma(\theta) = \log(1 + e^\theta), \ \phi = 1, \ A = n, \text{ and } \tau(y, \phi/n) = \log\binom{n}{ny},$$

it follows that the random variable Y (representing the proportion of successes in n Bernoulli trials with success probability $p = \mu$) belongs to an exponential family. The natural parameter for the binomial distribution is therefore $\theta = \text{logit } p = \log(p/1 - p)$, and not p. Note here that

$$E(Y) = \gamma'(\theta) = \frac{e^\theta}{1 + e^\theta} = \mu = p$$

and

$$\text{Var}(Y) = \frac{\phi}{n}\gamma''(\theta) = \frac{1}{n}\frac{e^\theta}{(1 + e^\theta)^2} = \frac{1}{n}\mu(1 - \mu) = \frac{1}{n}p(1 - p) = \frac{V(p)}{n}.$$

$\Box$

Example 7.5 Gamma family

The random variable Y has the gamma distribution ($Y \sim \Gamma(\alpha, \lambda)$) with parameters α and λ if it has density function given by

$$f_Y(y) = \frac{\lambda^\alpha y^{\alpha-1} e^{-\lambda y}}{\Gamma(\alpha)} \qquad \text{for } y > 0.$$

In this case, the mean of Y is given by $E(Y) = \mu = \alpha/\lambda$, while the variance is $Var(Y) = \alpha/\lambda^2$. There are other equivalent ways of reparameterizing the gamma distribution, and we have already seen that using the parameters α and $\mu = \alpha/\lambda$ instead of α and λ can be beneficial in maximum likelihood estimation from the gamma family. Using these parameters α and μ, we may re-express the density of Y as

$$f_Y(y) = \frac{(\frac{\alpha}{\mu})^\alpha y^{\alpha-1} e^{-\frac{\alpha}{\mu}y}}{\Gamma(\alpha)}$$

$$= \exp\left\{-\frac{\alpha}{\mu}y + \alpha \log \alpha - \alpha \log \mu + (\alpha - 1)\log y - \log \Gamma(\alpha)\right\}$$

$$= \exp\left\{\frac{y(-1/\mu) - \log \mu}{1/\alpha} + \alpha \log \alpha + (\alpha - 1)\log y - \log \Gamma(\alpha)\right\}.$$

Therefore using $\theta = -1/\mu$, $\gamma(\theta) = \log(-1/\theta) = -\log(-\theta)$, $\phi = 1/\alpha$, $A = 1$ and $\tau(y, \phi) = \alpha \log \alpha + (\alpha - 1)\log y - \log \Gamma(\alpha)$, it follows that the gamma family is an exponential family. We may write

$$E(Y) = \gamma'(\theta) = -1/\theta = \mu$$

and

$$Var(Y) = \phi \gamma''(\theta) = \frac{1}{\alpha} \frac{1}{\theta^2} = \frac{\mu^2}{\alpha} = \frac{V(\mu)}{\alpha},$$

where $V(\mu) = \mu^2$ is the variance function (recall that $Var(Y) = (\phi/A) V(\mu)$). In the special case when $\alpha = 1$, one obtains the exponential distribution where $Var(Y) = V(\mu)$. Note that the scale parameter $\phi = 1/\alpha$ for the Gamma family is actually the square of the coefficient of variation (standard deviation divided by the mean) of Y. □

In practice, in the use of GLM, we will usually assume that our data is a sample from an exponential family. We will then use the method of maximum likelihood to make inferences about relevant parameters. For n observations $\mathbf{y} = (y_1, \ldots, y_n)$ from an exponential family, the log-likelihood would have the form

$$l(\theta, \phi; \mathbf{y}) = \sum_{i=1}^{n} [A_i [y_i \theta_i - \gamma(\theta_i)]/\phi + \tau(y_i, \phi/A_i)], \tag{7.6}$$

where we assume that

$$E(Y_i) \equiv \mu_i = \gamma'(\theta_i) \text{ and } Var(Y_i) = (\phi/A_i)\gamma''(\theta_i) = (\phi/A_i) V(\mu_i).$$

For example, if we have a sample of n binomial observations where $y_i = s_i/n_i$ is the proportion of successes in n_i trials (with probability of success p_i), then the log-likelihood takes the form

$$l(\theta, \phi; \mathbf{y}) = \sum_{i=1}^{n} n_i[y_i\theta_i - \log(1 + e^{\theta_i})] + \log \binom{n_i}{n_iy_i}$$

$$= \sum_{i=1}^{n} n_i[y_i\text{logit}(p_i) + \log(1 - p_i)] + \log \binom{n_i}{n_iy_i}, \text{ where}$$

$$E(Y_i) = e^{\theta_i}/(1 + e^{\theta_i}) = p_i \quad \text{and}$$
$$Var(Y_i) = (1/n_i) e^{\theta_i}/(1 + e^{\theta_i})^2 = p_i(1 - p_i)/n_i.$$

The reader is advised to carefully note the distinction in this case between n (the number of binomial samples) and n_i (the number of Bernoulli trials in the ith sample).

7.3.2   Link functions and linear predictors

In GLM, we define a relationship between the mean response $E(Y) = \mu$ and a linear combination of selected explanatory variables $\mathbf{x}^T$ through a function g whereby

$$g(E(Y)) = \mathbf{x}^T \boldsymbol{\beta} = \eta.$$

Such a function g is called a *link* function, and although there are often several possibilities to use as a link in any given situation, for any given exponential family there is a *canonical* link function. In the expression (7.3) for the density (or probability) function of a random variable Y which belongs to a one-parameter exponential family, we referred to the parameter θ as the natural or canonical parameter. This parameter θ will be related to the mean μ of Y through some function g_N whereby $g_N(\mu) = \theta$. If we use this function g_N as our link to the linear predictor in the model, then our link is called the *canonical* or *natural* link.

Table 7.2 gives the canonical link and variance functions, as well as the scale parameters for the standard exponential families we have already considered.

TABLE 7.2

Canonical link and variance functions.

Family	Canonical link	Name	Variance $V(\mu)$	Scale ϕ
Normal	μ	identity	1	σ^2
Poisson	$\log \mu$	log	μ	1
Binomial	$\log[\mu/(1-\mu)]$	logit	$\mu(1-\mu)$	1
Gamma	$-1/\mu$	inverse	μ^2	$1/\alpha$

It is not obligatory to use the above canonical link functions when using a particular exponential family. It produces sensible results most of the time; however, each case should be judged on its own merits. For the binomial family the probit function $(g(\mu) = \Phi^{-1}(\mu))$ or the complementary log log function $(g(\mu) = \log\{-\log(1-\mu)\})$ are often used. For the Poisson family, instead of the (canonical) log link, one sometimes makes use of the identity $(g(\mu) = \mu)$ or square root $(g(\mu) = \sqrt{\mu})$ links. For the gamma family, it is common to use the link $1/\mu$ instead of $-1/\mu$ (although, of course, they both represent the same model), and on some occasions one uses the identity link or the log link. In general, one requires that the link function g be a one-to-one function which is differentiable. This is a technical requirement which is necessary to ensure that the numerical procedure used to calculate the maximum likelihood estimates works properly.

Having decided on a distribution for the data and a link function g, one decides which of the available explanatory variables (factors and covariates) have a significant effect on the response and hence should be included in a good model (this is often done by comparing many possible models). Interactions, either between factors or covariates, or between a factor and a covariate may also appear in the predictor. The linear predictor is $\eta = \mathbf{x}^T\boldsymbol{\beta}$, where the parameters $\boldsymbol{\beta}$ need to be estimated (for the GLM, this is done via maximum likelihood). Then, for any given set of explanatory variables, we can determine the corresponding linear predictor η and predict the value of the response variable to be $g^{-1}(\eta) = E(Y)$. Before going any further, we define exactly what we mean by factors, covariates and interactions.

7.3.3 Factors and covariates

Factors are *categorical* variables, and the values a factor takes are called its levels. For example, gender is a factor with two levels (male/female), smoking status is a factor with two levels (Y/N), and education achievement is a factor with say four levels (primary, secondary, tertiary and postgraduate). In a one-factor model we need to estimate one parameter for each level the factor takes. For each additional factor B (say with b levels) we add to the model, we need to estimate an extra $b-1$ parameters. We collectively refer to the parameters representing a factor, not involved in an interaction, as the *main effect* of the factor. *Covariates* are numerical variables. For example, age, policy duration, number of accidents in last three years, exam mark and salary are covariates. If a linear relationship exists between the linear predictor and the covariate, only one parameter for the variate needs to be estimated.

Example 7.6
An analyst, working in motor insurance, has developed a GLM to estimate the pure premium she should charge to a policyholder of a given age, gender and vehicle rating group (VRG). The company uses ten rating groups. It is known that the total claims an individual with a fixed explanatory vector (i.e., fixed age, gender type and VRG) makes is well modeled by a gamma random variable. Suppose the analyst finds no interactions between these factors and a linear effect of age. Then the final model will have a linear predictor of the form:

$$\eta = \alpha_i + \gamma_j + \beta \, x_{age},$$

where α_i is the parameter for the ith level of gender ($i = 1$ for male, 2 for female), γ_j is the parameter for the jth level of VRG ($1 \leq j \leq 10$) and β is the parameter for age (x_{age}).

Note here the model is overparameterized. We have defined 13 parameters but actually we only need 12 to completely specify the model. We will not worry about this here, however. In fact, for simplicity of interpretation, we will often write our models in this form.   ∎

7.3.4 Interactions

An *interaction* between two factors B and C exists when the effect on the linear predictor of changing the level of factor B varies depending on the level of factor C. Suppose factor B has b levels, and C has c levels. Then the model involving factors B and C but *not their interaction* has $b + c - 1$ parameters. However the model involving B and C and their interaction $B.C$ has bc parameters. Therefore adding an interaction term to the model results in estimating an extra $bc - (b + c - 1) = (b - 1)(c - 1)$ parameters!

Continuing with our example on motor insurance premium rating, suppose the analyst included in her previous model an interaction between gender and

VRG. This means that the difference (in the linear predictor) between males and females is nonconstant over VRG. The linear predictor for such a model can be written as:

$$\eta = \alpha_i + \gamma_j + (\alpha\gamma)_{ij} + \beta\, x_{age},$$

where the parameter $(\alpha\gamma)_{ij}$ represents the interaction effect at the ith level of gender and the jth level of VRG. Again notice that the model is overparameterized. The way it is written above suggests we need to estimate 20 interaction parameters. In fact, only $(2-1)(10-1) = 9$ can be estimated independently of the main effects.

We can also have an interaction between a covariate and a factor. In this case the effect of the variate changes at different levels of the factor. Suppose there is an interaction between a covariate x and a factor B (with b levels) in a model. If x has a linear relationship with η for a given level of B, then we will need to estimate b parameters for x (one for each level of B). For example, extending the previous model for motor insurance premiums, suppose the effect of age was different for each VRG. Then an interaction between age and VRG would exist and we could write

$$\eta = \alpha_i + \gamma_j + (\alpha\gamma)_{ij} + \beta\, x_{age} + \beta_j\, x_{age},$$

where β denotes the (average) effect of age and β_j is the difference between this (average) effect and that for VRG j $(1 \leq j \leq 10)$ of a one-year increase in age on the linear predictor η. All other parameters in the model are as before. Note once again that in writing the model in this (informative) way, it is strictly speaking overparameterized.

For a variate such as age, polynomial coefficients can appear in the linear predictor. The following describes a model containing the factors gender and VRG, and where there is a quadratic relationship between the linear predictor and age:

$$\eta = \alpha_i + \gamma_j + \beta_1\, x_{age} + \beta_2\, x^2_{age}.$$

Another way of looking at it however is to consider x^2_{age} as another variable, and in this way it is clear that η is "linear in the explanatory variables" (one of our requirements is that η is a linear combination of the parameters $\boldsymbol{\beta}$). It is also eligible to use a function of age as a variable. We may feel that there is a linear relationship between η and log(age). Again assuming the factors gender and VRG are in the model, a possible predictor would be:

$$\eta = \alpha_i + \gamma_j + \beta\, \log(x_{age}).$$

Table 7.3 gives some shorthand notation that is useful when fitting these models using a computer program such as SAS.

For the normal family, the canonical link function is the identity function, and in using this link one is modeling the mean itself as a linear combination of predictor variables. In Example 7.1 on modeling university degree mark, we

TABLE 7.3

Notation for linear predictors.

GLM model linear predictors	
Model predictor	Shorthand notation
$\eta = \alpha_i$	gender
$\eta = \alpha_i + \gamma_j$	gender + VRG
$\eta = \alpha_i + \gamma_j + \beta \, x_{age}$	gender + VRG + age
$\eta = \alpha_i + \gamma_j + (\alpha\gamma)_{ij} + \beta \, x_{age}$	gender + VRG+ gender.VRG + age
$\eta = \alpha_i + \gamma_j + (\alpha\gamma)_{ij}$ $\quad + \beta \, x_{age} + \beta_j \, x_{age}$	gender + VRG+ gender.VRG $\quad$ + age + age.VRG
$\eta = \alpha_i + \gamma_j + (\alpha\gamma)_{ij} + \beta_j \, x_{age}$	gender + VRG+ gender.VRG + age.VRG
$\eta = \alpha_i + \gamma_j + \beta_1 x_{age} + \beta_2 x_{age}^2$	gender + VRG + age + age^2
$\eta = \alpha_i + \gamma_j + \beta \log(age)$	gender + VRG+ log(age)

assumed a normal distribution for the data (degree marks), a linear predictor of the form

$$\eta \;=\; \alpha_{home} + \beta_1 \, x_{points} + \beta_2 \, x_{points}^2 + \beta_3 \, x_{age},$$

and an identity link between the mean μ and the predictor η (that is, $g(\mu) = \mu = \eta$). That is not to say that we could not have proceeded differently. For example, we could have tried to make use of information which was available on the year of entry (x_{year}) of each of the students into the program and hence (assuming significance of this covariate) used a linear predictor of the form

$$\eta^* \;=\; \alpha_{home} + \beta_1 \, x_{points} + \beta_2 \, x_{points}^2 + \beta_3 \, x_{age} + \beta_4 \, x_{year}.$$

Another variation would be to use a link function different to the identity link. We might have considered a log link where one models the log of the mean degree mark as a linear function of the predictors (that is, $\log(\mu) = g(\mu) = \eta$).

In practice, once one has decided on the particular exponential family to use as a distribution for the data, one usually makes use of the canonical link function and then decides on which variables to include in the linear predictor. Had our interest been, for example, on the *proportion* of students with given characteristics (gender, age, entry marks, home, etc.) who obtain a First-class honors degree (a degree mark in excess of 70%), then our approach would have been quite different. In this case, our data would be binomial, and we would probably make use of the either the logit (canonical) or probit link functions.

Example 7.7

A study is undertaken in order to devise a model to predict pass rates for young adults taking a test for a driving license. The data in Table 7.4 was collected from a local testing center on the basis of results over a 30-day period. Our observations for any given gender and age group are of the form Y/n where Y is the number passing the exam and n is the number taking the test.

TABLE 7.4

Driver testing results for males and females aged $[18 : 30]$.

Age	males		females	
	number n	pass	number n	pass
18	25	5	20	11
19	37	14	22	13
20	38	18	29	15
21	42	18	37	19
22	37	22	31	15
23	36	18	30	19
24	45	30	36	20
25	33	19	42	29
26	27	18	40	23
27	31	23	34	22
28	33	23	28	19
29	27	23	38	26
30	26	24	42	27

A graph of the pass rates by age for males and females is given in Figure 7.3. Pass rates appear to increase somewhat with age for males, and to a lesser extent for females. The overall pass rates (over all ages) are similar ($58.35\% = 255/437$ for males and $60.14\% = 258/429$ for females), with more variability for males. There are, of course, many factors that influence whether or not a person will pass such a test, and here we consider both gender and age as possible explanatory variables. In general terms, the linear predictor η we shall use in order to "link up" with the mean $\mu = E(Y/n)$ might have the form

$$\eta = \alpha_{gender} + \beta_{gender}\, x_{age},$$

where x_{age} is the age of individuals in the group of interest. The parameters α_{gender} and β_{gender} must be estimated, and in theory may take different values for males and females. Note, of course, that $\mu = E(Y/n)$ will necessarily be between 0 and 1, although if α_{gender} and β_{gender} are allowed to vary freely, then η can take any real value. For this binomial situation we consider the logistic function g for linking μ and η where $g(\mu) = \text{logit}(\mu) = \eta$.

An interesting aspect of this data (which could be ascertained by considering different models) is that gender on its own (or in addition to the variable age) is not a significant predictor, but that it *interacts* in a significant manner with

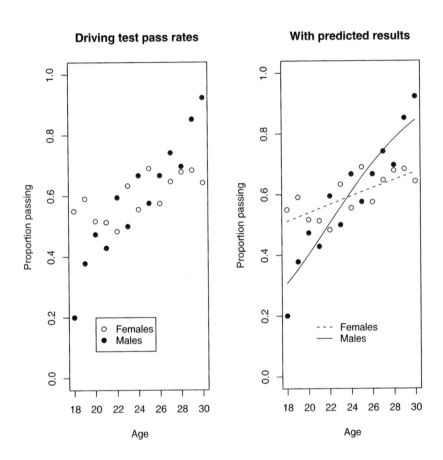

FIGURE 7.3

Driving test pass rates.

age as a predictor. Using the method of maximum likelihood for this model, the following estimates were obtained for the relevant parameters:

$$\hat{\alpha}_F = -0.968449, \quad \hat{\alpha}_M = -4.584923, \quad \hat{\beta}_F = 0.056405, \quad \hat{\beta}_M = 0.209524.$$

The above estimates were obtained using the linear models function lm in the package R with the data from Table 7.4, and similar (identical) results should be obtained from any other good statistical package. Note that estimates of β_{gender} for both males and females are positive, indicating that expected pass rates increase with age for both genders. We can summarize the above results by giving the linear predictors for females and males which are, respectively:

$$\hat{\eta}_F = -0.968449 + (0.056405)\, x_{age} \text{ and}$$
$$\hat{\eta}_M = -4.584923 + (0.209524)\, x_{age}.$$

For example, the linear predictor for a 22-year-old female would be -0.968449 $+ (0.056405)\, 22 = 0.272461$. However, we would normally be more interested in the model predicted *pass rate* for a 22-year-old female, and this is where the link function comes in. In our model, logit $E(Y/n) = $ logit $\mu = \eta$, or $\mu = e^{\eta}/(1 + e^{\eta})$. That is to say the inverse of the link function

$$g(\mu) = \text{ logit } \mu = \log\left[\frac{\mu}{1-\mu}\right] = \eta$$

is $g^{-1}(\eta) = e^{\eta}/(1+e^{\eta})$. Hence the model predicted pass rate for a 22-year-old female would be

$$\hat{\mu} = \frac{e^{0.272461}}{1 + e^{0.272461}} = 0.567697,$$

that is, about 56.8%. Plots of the model predicted pass rates for males and females are plotted in Figure 7.3. Use of the logit function allows one to make quick comparisons for the odds of passing. For example, using this model, the predicted odds of passing for a male would be

$$\frac{\hat{\mu}}{1 - \hat{\mu}} = e^{\eta_M} = e^{-4.584923 + (0.209524)\, x_{age}}.$$

In particular, for each additional year of age for a male, the odds of passing increase by a factor of $e^{(0.209524)} = 1.233091$ (approximately 1.23).

In the model used to predict driver pass rates, there is an interaction between the covariate *age* and the factor *gender*. In other words, the effect of age is different for the two genders, and this is illustrated effectively in Figure 7.3. A simpler model, but which in this case would not be appropriate, would be one where the linear predictor were of the form

$$\eta = \alpha_{gender} + \beta\, x_{age}.$$

That is, the effect of age is the same for both females and males, and plots of the linear predictors against age for females and males would give two parallel lines. (Note, however, that the corresponding predicted value curves for *pass rates* would not necessarily be parallel!)

When a variable (covariate or factor) is included in a model, it is called a *main effect*. When one includes an interaction term between two variables, one says there is an interaction effect and *usually includes both of the corresponding main effects in the model*. Clearly, it is possible to have interactions between two factors, two covariates, or a factor and a covariate in a model. Such interactions are order 2 interactions, and in models with many possible explanatory variables one may find significant interactions of order 3 or higher. Higher order interactions can often be challenging to interpret.   ☐

7.3.5   Minimally sufficient statistics

If we have a sample of n independent observations $(y_1, \ldots, y_n)$ where each y_i has distribution from a given exponential family with parameters (θ_i, ϕ), then the log-likelihood for the sample is

$$l(\theta, \mathbf{y}) = \sum_1^n A_i \left[\frac{y_i \theta_i - \gamma(\theta_i)}{\phi} \right] + \tau(y_i, \phi/A_i).$$

If g is a function such that $g(\mu_i) = \theta_i = \sum_1^p x_{ij} \beta_j$, then the log-likelihood for the parameters $(\beta_1, \ldots, \beta_p)$ takes the form

$$l(\boldsymbol{\beta}, \mathbf{y}) = \left(\sum_{j=1}^p \beta_j \sum_{i=1}^n A_i \frac{y_i x_{ij}}{\phi} - \sum_{i=1}^n \left[A_i \frac{\gamma(\theta_i)}{\phi} - \tau(y_i, \phi/A_i) \right] \right).$$

It then follows (assuming $\phi_1, \ldots, \phi_p$ are known) from the factorization theorem for sufficient statistics that

$$\left\{ \sum_{i=1}^n A_i \frac{y_i x_{ij}}{\phi} : j = 1, \ldots, p \right\}$$

is a set of minimally sufficient statistics for the parameters $(\beta_1, \ldots, \beta_p)$. Basically, this says that these statistics contain all of the information necessary to estimate $\boldsymbol{\beta}$, and knowing the individual values of the y_i gives us no more additional information in this regard. This is another reason that the function g with the property that $g(\mu) = \theta = \eta$ for an exponential family is called the *canonical link* for that family. As we have already seen, $E(Y) = \mu = \gamma'(\theta)$, from which it follows that the canonical link function g for a family satisfies $g = (\gamma')^{-1}$. The canonical link function and the variance function are related by $V(\mu) = (g^{-1})'(\mu)$ where $'$ denotes differentiation with respect to θ.

7.4 Model selection and deviance

Usually before a final model is selected, one has a number of candidate variables which may influence the response. In the final model a subset of these is selected which hopefully gives one the best possible (or at least a good) description of the underlying process. Two criteria that are used to make the choice are *goodness-of-fit* and *parsimony* (the quality of having few parameters yet fitting the data well). We would like the model to provide a good description of the data which was used to fit the model. Goodness-of-fit (or perhaps more accurately *lack-of-fit*) is often measured by deviance. One would also like the final model to be as simple as possible. Models with fewer parameters are easier to use. In addition, estimating parameters for variables or for factors that actually have no effect on the response is damaging to the model since it increases the variance of the parameter estimates, and hence decreases the accuracy of resulting inferences. Unfortunately, herein lies a conflict of interest! The more parameters one adds to the model, the better it will fit the data but the less parsimonious the resulting model will be. We need a way to resolve these two conflicting aims.

Although deviance may be used to give a measure of fit for a model, its main use is to compare possible models which are nested (that is, where the set of explanatory variables of one model is a subset of those of another). One may view the analysis of deviance as essentially a generalization of the classical method of analysis of variance.

7.4.1 Deviance and the saturated model

The *saturated model* provides a perfect fit for the given data. Such a model is also often referred to as a *maximal* or *full model*. The number of parameters (denoted by n_S) for the saturated model is in many cases the same as the maximum number of *covariate classes* when the data is categorical, or the number of *means* when the data is continuous. In some cases, it might actually be the number of observed data points n. In terms of goodness-of-fit, one cannot do any better than the saturated model, but in terms of parsimony it is essentially useless.

In order to compare the appropriateness of a model M with $p < n_S$ regression parameters for a set of data $\mathbf{y}$, we compare the maximum value of its likelihood function with that of the saturated model. The log-likelihood function $l(\theta; \mathbf{y})$ in a generalized linear model depends on the linear coefficients in the predictor $\eta = \beta_1 x_1 + \cdots + \beta_p x_p$ by way of the link function $g(\mu) = \eta$. One finds maximum likelihood estimates of the coefficients in the linear predictor, which by the invariance property of the method of maximum likelihood yield ML estimates of the μ_i.

Let L_S and L_M be, respectively, the maximum values of the likelihood functions for the data in hand for the saturated model S and the model under consideration M. Using $l_S = \log L_S$ and $l_M = \log L_M$, we consider the statistic (known as the log of the likelihood ratio statistic) given by $\log \lambda_{S,M} = \log(L_S/L_M) = l_S - l_M$. Now $L_S \geq L_M$, and hence $l_S - l_M \geq 0$ since (assuming that the same exponential family distribution and link function is used for both models) the saturated model provides a complete description of the data. If the model M fits the data reasonably well, then $L_S \doteq L_M$ and $\log \lambda_{S,M}$ should be reasonably small. Large values of $\log \lambda_{S,M}$ suggest a poor fit.

Let us assume for the moment that we are working with a model where the dispersion parameter $\phi = 1$ (for example, when the response is Poisson or binomial). The maximum likelihood estimate for $E(Y_i) = \mu_i = \gamma'(\theta_i)$ in the saturated model is clearly y_i itself, or, equivalently, the estimate for θ_i in the saturated model is $\theta(y_i) \equiv (\gamma')^{-1}(y_i)$. Let $\hat{\theta}_i$ be the maximum likelihood estimate of θ_i under the model M. Then the *deviance* of the model M based on the data $\mathbf{y}$ (or twice the log-likelihood ratio statistic for testing the model M relative to the saturated model S) takes the form (see Venables and Ripley [57])

$$D_M = 2 (l_S - l_M)$$

$$= 2 \sum_{i=1}^{n_S} A_i \left[\{y_i \theta(y_i) - \gamma(\theta(y_i))\} - \left\{y_i \hat{\theta}_i - \gamma(\hat{\theta}_i)\right\} \right], \qquad (7.7)$$

where

$$D_M(y_i) = 2 A_i \left[(y_i \theta(y_i) - \gamma(\theta(y_i))) - \{y_i \hat{\theta}_i - \gamma(\hat{\theta}_i)\}\right]$$

is the individual contribution of the observation y_i to the overall deviance D_M. As we shall see later, d_i is sometimes used for $D_M(y_i)$.

Example 7.8 Deviance for the Poisson GLM
In the Poisson GLM with n independent Poisson observations,

$$E(Y_i) = \mu_i = \gamma'(\theta_i) = e^{\theta_i} \text{ for } i = 1, \ldots, n.$$

Using the natural parameter $\theta = \log(\mu)$, the log-likelihood function is

$$l(\theta, \mathbf{y}) = \sum_{1}^{n} [y_i \theta_i - e^{\theta_i} - \log y_i!] = \sum_{1}^{n} [y_i \log \mu_i - \mu_i - \log y_i!]. \qquad (7.8)$$

In the saturated model, the maximum likelihood estimate for θ_i is

$$\theta(y_i) = (\gamma')^{-1}(y_i) = (\exp')^{-1}(y_i) = \log(y_i).$$

Therefore the maximum value of the log-likelihood function for the saturated model is

$$l_S = \sum_{1}^{n} [y_i \log y_i - y_i - \log y_i!].$$

Letting $\hat{\mu}_i = e^{\hat{\theta}_i}$ (equivalently, $\log(\hat{\mu}_i) = \hat{\theta}_i$) be the maximum likelihood estimates for μ_i for the model of interest M, then

$$l_M = \sum_1^n [y_i \log \hat{\mu}_i - \hat{\mu}_i - \log y_i!].$$

The deviance for model M takes the form (we assume the scale parameter $\phi = 1$)

$$D_M = 2\,(l_S - l_M)$$

$$= 2 \sum_{i=1}^n \left[\{y_i\theta(y_i) - \gamma(\theta(y_i))\} - \left\{y_i\hat{\theta}_i - \gamma(\hat{\theta}_i)\right\} \right]$$

$$= 2 \sum_{i=1}^n [\{y_i \log y_i - y_i\} - \{y_i \log \hat{\mu}_i - \hat{\mu}_i\}]$$

$$= 2 \sum_{i=1}^n [y_i\,(\log y_i - \log \hat{\mu}_i) - (y_i - \hat{\mu}_i)]. \tag{7.9}$$

Assume the model M has a constant term ($\log(\mu_i) = \beta_1 + \sum_{j=2}^p \beta_j x_{ij}$). The score statistic U_1 for β_1 is the partial derivative of l with respect to β_1, and hence using (7.8), $U_1 = (\partial/\partial\beta_1)\,l = \sum_1^n (y_i - \mu_i)$. For maximum likelihood estimates, the score statistics (partials of the log-likelihood function l with respect to the parameters β_i) must be zero. In particular, $U_1 = 0$, and hence $\sum_1^n y_i = \sum_1^n \hat{\mu}_i$. The deviance D_M therefore reduces to

$$D_M = 2 \sum_{i=1}^n y_i(\log y_i - \log \hat{\mu}_i) \qquad \left(\text{since } \sum_{i=1}^n (y_i - \hat{\mu}_i) = 0\right)$$

$$= 2 \sum_{i=1}^n y_i \log\,(y_i/\hat{\mu}_i) \;=\; 2 \sum_{i=1}^n o_i\,\log(o_i/e_i),$$

where $o_i = y_i$ and $e_i = \hat{\mu}_i = \exp(\hat{\beta}_1 + \sum_{j=2}^p \hat{\beta}_j x_{ij})$ are, respectively, the observed and estimated (predicted or expected) values. ⬜

The deviance D_M may be thought of as a measure of distance between the saturated model (basically, the data in hand) and a proposed model M. The simplest possible model is the so-called constant model where $E(Y_i) = \mu$ is constant for all i, and is sometimes referred to as the *null* model M_0 (with one parameter – the unknown mean). The deviance of the null model is usually called the *null deviance*.

When the scale parameter ϕ differs from one, the quantity D_M (where the ML parameter estimates take account of ϕ) is still usually referred to as the deviance, even though it no longer is twice the likelihood ratio statistic (see Venables and Ripley [57], or McCullagh and Nelder [41] for further details).

When $\phi \neq 1$, one often considers the *scaled deviance* given by D_M/ϕ. For the normal model M where $Y_i \sim N(\mu, \sigma^2)$ and

$$E(Y_i) = \mu_i = \mathbf{x_i^T}\boldsymbol{\beta} = \beta_1 x_{i1} + \cdots + \beta_p x_{ip},$$

the deviance takes the form $D_M = \sum_1^n (y_i - \hat{\mu}_i)^2$ (see Problem 2) and moreover,

$$\frac{D_M}{\sigma^2} = \frac{\sum_1^n (y_i - \hat{\mu}_i)^2}{\sigma^2} \sim \chi^2_{n-p}.$$

Hence in the normal case where the variance is externally known, the (scaled) deviance can be used to test if the model M is acceptable. When the scale parameter ϕ is unknown, it is common practice to use an estimate.

The scale parameter (or scale factor) may be used to model overdispersion in the model. Overdispersion is a relatively common phenomena resulting from observing more variation in one's data than might be expected under a given model. In particular, it is often observed when dealing with count data (and using the binomial or Poisson exponential families where, in theory, the scale parameter is one, but variation in the data suggests higher variation than expected). Possible causes for overdispersion include the presence of outliers and a lack of explanatory variables in the model.

If the data is not normal and ϕ is known, then D_M/ϕ may be approximately χ^2_{n-p} when M is a p parameter model (see Firth [26] for a discussion on the binomial model). Note that the (scaled) null deviance is approximately χ^2 distributed with $n_S - 1$ degrees of freedom when the *null* model is valid! If the distribution is normal, then D_{M_0}/σ^2 has the (exact) χ^2_{n-1} distribution.

7.4.2   Comparing models with deviance

Suppose we know that the correct model for our data should definitely contain the p variables: $x_1, x_2, \ldots, x_p$, but that we would like to know whether (some of) the q variables $x_{p+1}, \ldots, x_{p+q}$ should also appear in the correct model. Let β_i be the parameter corresponding to the variable x_i in the model. We let M_1 denote the model where $\beta_i = 0$ for $i > p$, and M_2 the (larger) model where $\beta_i \neq 0$ for at least one $i > p$. The likelihood ratio test of $H_0 : M_1$ versus $H_A : M_2$ is based on the test statistic

$$2 \log \frac{L_{M_2}}{L_{M_1}} = 2(l_{M_2} - l_{M_1}) \stackrel{.}{\sim} \chi^2_q, \tag{7.10}$$

where L_{M_2} is the maximum value of the likelihood under the model with $p+q$ parameters with corresponding log-likelihood l_{M_2} (similarly for L_{M_1} and l_{M_1}). Essentially, if the extra variables in the larger model truly have no effect, then twice the difference between the (maximized) log-likelihoods of the two models will have, approximately, a χ^2 distribution with q degrees of freedom!

As previously mentioned, the deviance is a measure of goodness-of-fit of the model. Intuitively speaking, this makes sense, for if the model of interest (with only p variables) does not fit the data well, the model will most likely predict fitted values quite different from the observed values. In other words, the data will not be consistent with the model and hence the likelihood (or probability) of the observed data (and also the log-likelihood l_{M_1}) will be low under the model being considered. The saturated log-likelihood l_S will then be substantially larger than l_{M_1}, and the deviance will be large. Using similar reasoning, it is sensible that the deviance should be small when the model fits the data well. Therefore in testing the model M_1 versus M_2, we have from Equation (7.10) that when M_1 is valid

$$2\left(l_{M_2}-l_{M_1}\right) \;=\; 2\left(l_S-l_{M_1}\right)-2\left(l_S-l_{M_2}\right) \;=\; D_{M_1}-D_{M_2} \equiv \Delta D \stackrel{.}{\sim} \chi_q^2. \quad (7.11)$$

Hence under the null hypothesis that $\beta_i = 0$ for $i \geq p+1$, we have that the difference in deviance ΔD between the smaller and larger model is (approximately) a χ^2 random variable with q degrees of freedom. If the null hypothesis is false (with M_2 being the correct model), we might expect the difference in deviance to be larger than that under the null hypothesis. At the α percent significance level, it makes sense then to conclude that some of the extra variables should be in the correct model if $\Delta D = D_{M_1} - D_{M_2} > \chi_{1-\alpha,q}^2$. (i.e., for a 5% test we would include the extra terms in our model if the difference in deviance ΔD exceeds the 95% quantile of the χ^2 distribution with q degrees of freedom). It should be noted that models are usually compared by testing one extra term at a time!

Example 7.9 Poisson regression model for absenteeism
A study is to be made on absenteeism by employees in a factory, and in particular to investigate if the rate of absenteeism differs by gender. Samples of m male and $n - m$ female employees are to be selected, and then their records will be used to find how many days the respective individuals were absent for work during the past year. We shall assume that the observations $Y_1, \ldots, Y_m, \ldots, Y_n$ can be viewed as independent Poisson random variables with $E(Y_i) = \mu_i$ where

$$\log \mu_i = \begin{cases} \beta_0 & i = 1, \ldots, m \\ \beta_0 + \beta_1 & i = m+1, \ldots, n. \end{cases} \quad (7.12)$$

We will use M_1 to denote the Poisson GLM model given by (7.12), and M_0 to denote the null model (where $\log \mu_i = \beta_0$ for all $i = 1, \ldots, n$). From Equation (7.8), it follows that the log-likelihood function is of the form

$$l(\beta; \mathbf{y}) = -me^{\beta_0} - (n - m)e^{\beta_0+\beta_1} + \beta_0 \sum_{i=1}^{n} y_i + \beta_1 \sum_{i=m+1}^{n} y_i - \log\left(\prod_{i=1}^{n} y_i!\right).$$

Solving the (score) equations

$$\frac{\partial l}{\partial \beta_0} = 0 = -me^{\beta_0} - (n - m)\,e^{\beta_0+\beta_1} + \sum_{1}^{n} y_i$$

$$\frac{\partial l}{\partial \beta_1} = 0 = -(n-m)\, e^{\beta_0 + \beta_1} + \sum_{m+1}^{n} y_i,$$

one obtains the maximum likelihood estimators $\hat{\beta}_0 = \log(\bar{y}_M)$ and $\hat{\beta}_1 = \log(\bar{y}_F / \bar{y}_M)$, where $\bar{y}_M = \sum_1^m y_i / m$ and $\bar{y}_F = \sum_{m+1}^{n} y_i / (n-m)$ are, respectively, the mean number of absences for males and females.

For model M_1, the fitted values are therefore $\hat{\mu}_i = \bar{y}_M$ for $i = 1, \dots, m$, and $\hat{\mu}_i = \bar{y}_F$ for $i = m+1, \dots, n$. Hence using (7.9), the deviance D_{M_1} is

$$2 \left(\sum_1^m y_i (\log y_i - \log \bar{y}_M) - (y_i - \bar{y}_M) + \sum_{m+1}^{n} y_i (\log y_i - \log \bar{y}_F) - (y_i - \bar{y}_F) \right)$$

$$= 2 \left(\sum_1^m y_i \log (y_i / \bar{y}_M) + \sum_{m+1}^{n} y_i \log (y_i / \bar{y}_F) \right).$$

For the null model M_0, the maximum likelihood estimate of β_0 is $\log(\bar{y}) = \log(\sum_1^n y_i / n)$, and hence the deviance is

$$D_{M_0} = 2 \left(\sum_1^n y_i \log y_i - \sum_1^n y_i \log \bar{y} \right) = 2 \left(\sum_1^n y_i \log(y_i / \bar{y}) \right).$$

Suppose the following data is collected on the number of days absent from work during the past year from a sample of 24 employees:

$$\mathbf{y} = (12, 9, 7, 5, 7, 4, 6, 9, 3, 9, 4, 10, 13, 11, 8, 7, 4, 5, 6, 8, 11, 7, 9, 6).$$

The first 12 components in $\mathbf{y}$ represent male employees and the remainder are for females. Is gender a significant factor in modeling days absent for workers in this company, on the basis of this data? In other words, is M_1 a reasonable model here?

Table 7.5 gives the individual contributions

$$d_i = 2 \left[y_i (\log y_i - \log \hat{\mu}_i) - (y_i - \hat{\mu}_i) \right]$$

to the deviance for the models M_0 and M_1 using the expression (7.9) for a Poisson GLM. For this data, $\bar{y}_M = 7.08$, $\bar{y}_F = 7.92$ and $\bar{y} = 7.5$.

Table 7.6 gives the deviances for the null model M_0 and the model M_1 including gender as a factor. One sees that both of the models fit reasonably well (remember that the mean of a χ^2 distribution is its degrees of freedom). The drop in deviance ΔD from the null model M_0 to M_1 is only 0.556, which is not significant relative to a χ^2 distribution with 1 degree of freedom. This suggests that there is no need to include β_1 in the model for work absence, or in other words, that gender is not a significant factor.

The following is summary output from R for this Poisson GLM model labeled M1 with data WorkAbsence. "glm" is the procedure in R used to fit

TABLE 7.5

Deviance contributions in models M_0 and M_1 for work absence.

			Deviance contribution					Deviance contribution	
i	A	G	M_0	M_1	i	A	G	M_0	M_1
1	12	M	2.28009	2.81856	13	13	F	3.30120	2.72879
2	9	M	0.28179	0.47731	14	11	F	1.42583	1.06968
3	7	M	0.03410	0.00098	15	8	F	0.03262	0.00087
4	5	M	0.94535	0.68360	16	7	F	0.03410	0.11049
5	7	M	0.03410	0.00098	17	4	F	1.97113	2.37193
6	4	M	1.97113	1.59506	18	5	F	0.94535	1.23801
7	6	M	0.32228	0.17484	19	6	F	0.32228	0.50680
8	9	M	0.28179	0.47731	20	8	F	0.03262	0.00087
9	3	M	3.50226	3.01187	21	11	F	1.42583	1.06968
10	9	M	0.28179	0.47731	22	7	F	0.03410	0.11049
11	4	M	1.97113	1.59506	23	9	F	0.28179	0.14191
12	10	M	0.75364	1.06348	24	6	F	0.32228	0.50680

TABLE 7.6

Models for work absence with and without gender.

Model	Deviance	df	ΔD	Δdf	$\chi^2_{\Delta df, 0.95}$
M_0	22.789	23			
M_1	22.233	22	0.556	1	3.84

generalized linear models, specified by giving a symbolic description of the linear predictor and a description of the error distribution. Initially, summary statistics for the deviance residuals are given. The estimate for the gender coefficient β_1 is 0.1112, which is not significant. The *residual deviance* is what we have labeled D_{M_1}. The AIC (which may be useful in comparing models that are not hierarchical) is 117.69, and the number of scoring iterations used for convergence for the parameter estimates is 3.

The AIC is what is known as Akaike's information criterion, and is sometimes used to compare models (particularly, competing nonnested models). For a model M it is defined to be $-2l_M + 2p$, where l_M is the maximized log-likelihood and p is the number of parameters in the model M. It takes account of both fit (through l_M) and parsimony (through p), and hence low values of the AIC are to be preferred.

```
> summary(M1)
Call: glm(formula = absence ~ gender, family = poisson,
data=WorkAbsence) Deviance Residuals:
     Min        1Q      Median        3Q         Max
-1.73548   -0.74063   -0.03138    0.69087    1.67885
 Coefficients: Estimate Std. Error   z value    Pr(>|z|)
 (Intercept)    1.9577      0.1083    18.072    <2e-16 ***
  gender1       0.1112      0.1492     0.746     0.456
```

```
Signif. codes:  0 '***' 0.001 '**' 0.01 '*' 0.05 '.' 0.1 ' ' 1
(Dispersion parameter for poisson family taken to be 1)
    Null deviance: 22.789  on 23  degrees of freedom
Residual deviance: 22.233  on 22  degrees of freedom AIC: 117.69
Number of Fisher Scoring iterations: 3
```

⬜

7.4.3 Residual analysis for generalized linear models

Having selected a model, perhaps using an analysis of deviance as outlined in the last section, we need to check if the model assumptions are justified and if the model provides a good fit to the data.

A basic assumption for a GLM is that the data points $y_1, y_2, \ldots, y_n$ are independent random observations from the same exponential family. We should also check to see that the particular GLM we have proposed is an appropriate one, that we have used a reasonable (natural?) link function, and that we have included in the model all the available significant factors/variables, keeping in mind the parsimony principle. Often we can detect some flaws in a model by the use of residual plots.

Residuals are based upon differences between observed data points and fitted values predicted by the model. The ith Pearson residual for a given model is given by

$$r_{P_i} = (y_i - \hat{\mu}_i)/\sqrt{Var(\hat{\mu}_i)},$$

where $\hat{\mu}_i = g^{-1}(\hat{\eta}_i)$ and $Var(\hat{\mu}_i) = (\phi/A_i)\, V(\hat{\mu}_i)$ (here $Var(\hat{\mu}_i)$ is the estimated variance of Y given mean $\hat{\mu}_i$). The ith deviance residual on the other hand is given by

$$r_{d_i} = \text{sgn}\,(y_i - \hat{\mu}_i)\,\sqrt{d_i},$$

where d_i is the contribution of the ith data point to the (scaled) deviance for the model, and $\text{sgn}(y_i - \hat{\mu}_i) = 1$ if $y_i - \hat{\mu}_i > 0$ and -1 otherwise.

The Pearson residuals are often skewed for non-normal data, and this makes the interpretation of such residual plots more difficult. The deviance residuals are more likely to be approximately normally distributed, and hence they are often preferred (for insurance and actuarial applications). One can show that for normally distributed data, the Pearson and deviance residuals are the same.

Under the assumptions for a GLM, we expect the residual plots with a good model to show no pattern. After model selection, one should plot the residuals against both factors and variates in the model. Any trend picked up may indicate that an effect that should have been included in the model has been missed. Data points with exceptionally large residuals may be outliers. A histogram of the residuals can be used to test the distributional form assumed in the GLM.

Example 7.10

The data in Table 7.7 (from Pawitan [49]) gives the number of traffic accidents at eight different locations in a city, both before and after modifications were made in traffic controls. Note that before the changes in traffic controls were made, the total number of accidents was recorded either over eight or nine years depending on location, but after changes it was over only two or three years. For this type of data, it is natural to consider a Poisson response model, but we should take account of the fact that the observations were made over different lengths of time (years).

One question of interest is surely whether or not the traffic control measures have improved things in terms of reducing the rate of accidents? Here years is an exposure variable. Before the changes, accidents were occurring in the various locations at a rate of $124/68 = 1.824$ per year, while after the changes the rate was reduced to $15/18 = 0.833$ per year. Is this observed reduction in rate of accidents $(0.833/1.824 = 0.457)$ actually significant?

TABLE 7.7
Traffic accidents in eight city locations before and after traffic control improvements.

Location	Before changes ($j = 0$)		After changes ($j = 1$)	
	Years	Accidents	Years	Accidents
1	9	13	2	0
2	9	6	2	2
3	8	30	3	4
4	8	30	2	0
5	9	10	2	0
6	8	15	2	6
7	9	7	2	1
8	8	13	3	2

We consider Poisson GLM models where the number of accidents in location i at time j (where $j = 0$ and $j = 1$ indicate before and after changes, respectively) is Poisson with parameter $\lambda_{ij} = n_{ij}\,\mu_{ij}$. Here n_{ij} is the *known* exposure (in years) associated with a given location i and time j.

Let M_1 represent the model where time (before and after) is a factor represented by

$$\log \lambda_{ij} = \log n_{ij} + \log \mu_{ij} = \log n_{ij} + \beta_0 + \beta_1 j.$$

$\log n_{ij}$ is an example of what is called an *offset* parameter, since it should probably be included in the model but does not need to be estimated!

The output from a GLM procedure in R for M_1 (where $\log n_{ij}$ is an offset parameter) is given below, and indicates that time is significant. Note that the estimate of the time parameter β_1 (labeled *After*) is $\hat{\beta}_1 = -0.7831$, indicating that the accident rate *after* relative to *before* is reduced by a factor of $\exp(-0.7831) = 0.457$, which is the same as that observed directly from the data! A residual analysis would show that there are some unusual observations – in particular, those in locations $2, 3, 4$ before the changes. The residual deviance is 62.331 on 14 degrees of freedom, which is rather large.

```
Deviance Residuals:
     Min        1Q    Median        3Q       Max
 -2.9578   -1.8258   -0.4910    0.4056    3.5264
 Coefficients:Estimate Std. Error z value Pr(>|z|)
 (Intercept)    0.6008     0.0898   6.690 2.23e-11 ***
   After    -0.78310.2727    -2.872  0.00408       **
Signif. codes:   0 '***' 0.001 '**' 0.01 '*' 0.05 '.' 0.1 ' ' 1
(Dispersion parameter for poisson family taken to be 1)
     Null deviance: 72.378  on 15  degrees of freedom
Residual deviance: 62.331  on 14  degrees of freedom
AIC: 115.98
```

There are many possible explanations for this rather poor fit. One is that there is too much variation here with respect to a Poisson distribution (overdispersion), and another is that there are not enough variables in the model to explain the variation. With the information available here, one could also consider a model M_2 incorporating the factor *location*, where

$$\log(\mu_{ij}) = \beta_0 + \beta_1 j + \tau_i,$$

and τ_i is a location parameter. A table of deviances for these candidate models (including the null model M_0 where $\log(\mu_{ij}) = \beta_0$) is given in Table 7.8. This suggests that location should be added as a factor to the model, but also that even for M_2 there is some evidence of overdispersion (as the deviance D_{M_2} is still larger than $\chi^2_{7,0.95}$).

TABLE 7.8
Models for traffic accidents before and after changes.

Model	Deviance	df	ΔD	Δdf	$\chi^2_{\Delta df, 0.95}$
M_0	72.378	15			
M_1	62.331	14	10.047	1	3.84
M_2	18.368	7	43.963	7	14.07

A residual plot of the deviance residuals for model M_2 is given in Figure 7.4, and this indicates in particular two outliers. A deviance residual of -2.401 occurs in location 4 after the changes (where the yearly rate of accidents fell from 3.75 to 0). The other outlier occurs in location 6 after the changes (with a deviance residual of 2.262), where an increase in the yearly rate of accidents from 1.88 to 3 occurred. ☐

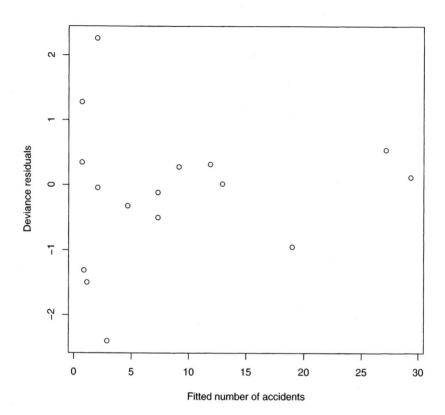

FIGURE 7.4
Deviance residuals and fitted values for accidents in model M_2.

Example 7.11

Table 7.9 gives data on automobile accidents from the the state of Florida in 1988 (see p. 201 in [1]). Subjects involved in accidents were classified according to whether or not they were wearing a seat belt, whether or not they were ejected from the vehicle, and whether or not they suffered a fatal injury. For example, of the 411,594 individuals involved in an accident who were wearing a seat belt, only 483 (or 0.12%) died as a result. One would be interested in modeling how fatality in an accident is related to wearing a seat belt and/or being ejected from the vehicle. We consider here a logistic regression model, that is, a generalized linear model using the binomial family with the logit link. Here $\theta = p$ is the probability of a fatality and $\log(p/(1-p)) = \text{logit}(p)$

is the log of the odds ratio of this probability.

TABLE 7.9
Automobile accident mortality data in Florida (1988).

Seat belt used	Ejected	Fatal injury	Nonfatal injury	Total	%
Yes	Yes	14	1,105	1,119	1.25
Yes	No	483	411,111	411,594	0.12
No	Yes	497	4,624	5,121	9.71
No	No	1,008	157,342	158,350	0.64

Below is a source file for the package R to build a GLM model (lmFL) for the fatality rate using the factors belt and eject. Initially, the *data frame* Florida is constructed. The package R accepts data for the binomial GLM model in two forms. If the response is a vector, it is assumed to be a binary vector of $0's$ and $1's$ (and, in this case, $n_i = 1$ for all i, that is, the vector is the result of a sequence of Bernoulli trials). The other possibility is a two-column matrix where the first column is the number of successes and the second holds the number of failures (that is, the data represents the numbers of successes and failures in a set of binomial observations). Here it is clearly easier to use the second method, and the matrix Florida$Ymat is constructed (and added to the data frame Florida by using Florida$) by binding together (using the function *cbind*) the two vectors *fatalinj* (numbers of fatalities) and *inj* (number of nonfatal injuries).

R Source File for GLM model of Florida Automobile accidents.

```
Florida<-data.frame(belt<-factor(c("Y","Y","N","N")),
    eject<-factor(c("Y","N","Y","N")),
    fatalinj<-c(14,483,497,1008),
    inj<-c(1105,411111,4624,157342))
Florida$Ymat<-cbind(Florida$fatalinj,Florida$inj)
lmFL<-glm(Florida$Ymat ~ belt+eject,
    family=binomial,data=Florida)
```

There are other models which one should consider, and in particular one should look at the saturated model (that is, where there is an interaction term *belt*eject*). One would find that the interaction term is not significant. The binomial GLM model $Y \sim$ seat + eject fits very well, as can be seen from a summary of the model given below. This is a logistic model with two factors and no interaction.

```
Call: glm(formula = Florida$Ymat ~ belt + eject,
     family =binomial, data = Florida)
Deviance Residuals: [1]   -1.6132    0.3142    0.3256   -0.2165
Coefficients: Estimate   Std. Error   z value    Pr(>|z|)
  (Intercept)  -5.04362     0.03120    -161.65    <2e-16 ***
   beltY       -1.71732     0.05402     -31.79    <2e-16 ***
   ejectY       2.79779     0.05526      50.63    <2e-16 ***
---
Signif. codes:
   0 '***' 0.001 '**' 0.01 '*' 0.05 '.' 0.1 ' ' 1
(Dispersion parameter for binomial family taken to be 1)
     Null deviance: 3567.723  on 3  degrees of freedom
Residual deviance:    2.854  on 1  degrees of freedom
     AIC: 38.039
Number of Fisher Scoring iterations: 3
```

With no interaction term one is stating that the odds ratio relating fatality to seat belts is the same both for individuals who are ejected and for those who are not. Similarly, the odds ratio relating fatality and ejection from vehicle is the same whether one is wearing a seat belt or not. The coefficient for the factor belt is -1.71732, suggesting that the logit decreases by -1.71732 on wearing a seat belt, or equivalently, that the odds ratio is reduced by a factor of $e^{-1.71732} = 0.17955$. In other words, the odds (ratio) of a fatality is $1/0.17955 = 5.56956 \doteq 5.57$ times higher for a person not wearing a seat belt than for one who is (no matter whether they were ejected or not). Similarly, the logit increases by 2.79779 when one is ejected from a vehicle, or equivalently that the odds of a fatality are $16.40834 \doteq 16.41$ times higher when ejected from the vehicle than when not.

Note that the residual deviance is 2.854 with 1 degree of freedom, and that the deviance residuals are modest, indicative of a good fit. A further analysis of residuals will reveal a high Cook's distance (indicative of an influential observation) for the situation when a seat belt is not used and one is not ejected.

Predicted values of the probabilities of fatalities using this model are given in Table 7.10, as well as the observed proportions. Remember that if logit $(p) = \eta$, then $p = e^\eta/(1 + e^\eta)$. Hence, for example, for a person not wearing a seatbelt and who is not ejected, the predicted probability of a fatality is $e^{-5.04362}/(1 + e^{-5.04362}) = 0.00641$.

□

TABLE 7.10
Actual (and model predicted) *percentages* of
fatalities in Florida automobile accidents.

Seat belt	Ejected	Not ejected
Yes	1.25 (1.86)	0.12 (0.12)
No	9.71 (9.57)	0.64 (0.64)

7.5 Problems

1. Consider the model for final degree performance treated in Example 7.1.

   (a) What is the predicted performance for a 19-year-old from outside Ireland who enters university with 600 points?

   (b) For what range of age would a Dublin student entering on 610 points expect to obtain a First-class degree?

   (c) Given the applicable range for *age* [17,20] and *points* [505,640] for this model, what is the maximum predicted performance? What would the minimum predicted performance be?

2. Show that for a (normal GLM) model M where

$$E(Y_i) = \mu_i = \mathbf{x}_i^T \beta, \quad Y_i \sim N(\mu_i, \sigma^2), \quad i = 1, \ldots, n,$$

   the deviance D_M takes the form

$$D_M = \sum_1^n (y_i - \hat{\mu}_i)^2.$$

3. Show that the deviance contribution for an observation y_i in a Poisson GLM (with the canonical link) given by

$$D_M(y_i) = d_i = 2 \left[y_i \left(\log y_i - \log \hat{\mu}_i \right) - \left(y_i - \hat{\mu}_i \right) \right]$$

   is always nonnegative.

4. Let $S_i \sim B(n_i, p_i)$ and $Y_i = S_i/n_i$ for $i = 1, \ldots, m$. Show that the deviance in a binomial GLM model M with the logit link function for the data $\mathbf{Y} = \mathbf{y}$ is of the form

$$D_M = 2 \sum_{i=1}^m \left\{ n_i \, y_i \log \frac{y_i}{\hat{\mu}_i} + n_i \left(1 - y_i \right) \log \frac{1 - y_i}{1 - \hat{\mu}_i} \right\},$$

   where $\hat{\mu}_i$ is the maximum likelihood estimate of $E(Y_i) = E(S_i/n_i) = p_i = \mu_i$.

5. A binomial generalized linear model was used by a bank to predict the probability of success of new personal investment packages. The model took the form

$$\log\left(\frac{p}{1-p}\right) = \alpha_i + \gamma\,x_\$ + \beta_i\,x_\$$$

where α_i and β_i are the parameters representing whether the product is principally property, equity or bond based. γ is the parameter for the variate $x_\$$ (in \$million), representing the marketing budget for the product. Maximum likelihood estimates of these parameters are given by

	$i = 1$ (property)	$i = 2$ (equity)	$i = 3$ (bond)
$\widehat{\alpha}_i$	-0.562	0.126	-2.243
$\widehat{\beta}_i$	0.031	0.007	0.082

$$\widehat{\gamma} = 0.003$$

What is the predicted probability of a new equity product with a \$2 million budget being a success? At which budget level will a property and a bond-based product have an equal probability of success? This model is said to be overparameterized. Briefly suggest a suitable reparameterization which would reduce the number of parameters required.

6. In a study undertaken to model the reliability of car engines by the motor industry, the time to failure of these engines was modeled by the density function

$$f(t) = \frac{4}{\mu^2}\,t\,e^{-2t/\mu}.$$

(a) Show that this density function is a member of the exponential family of distributions. Determine the natural parameter and the canonical link function, and find the variance function, $V(\mu)$.

(b) A GLM with an inverse link function was fitted with a linear predictor $\eta = \alpha_i + \beta\,x_{size}$, where α_i is the parameter for the ith car manufacturer and β is the parameter for the engine size, x_{size}. The ML estimators of these parameters were calculated

$$\widehat{\alpha}_1 = -0.0712, \quad \widehat{\alpha}_2 = -0.0565, \quad \widehat{\alpha}_3 = -0.0678 \text{ and } \widehat{\beta} = -0.0207,$$

where there existed only three car manufacturers, $i = 1, 2, 3$. Using this model, find the estimated mean survival time for a car produced by manufacturer 3 with a 2.5 liter engine? Express the ratio of mean survival times for cars produced by manufacturers 1 and 2 as a function of the engine size, x_{size}.

7. In a study undertaken to model automobile accident rates for young adults, a Poisson (regression) GLM model is fitted with linear predictor of the form

$$\xi = \alpha_G + \beta_G\,x,$$

where α_G and β_G are parameters for the factor gender and the variate age x (in years). ML estimates of these parameters are given by $\hat{\alpha}_F = -0.72$, $\hat{\alpha}_M = -0.54$, $\hat{\beta}_F = -0.06$, $\hat{\beta}_M = -0.05$ where F and M represent females and males, respectively.

(a) What is the predicted automobile accident rate for a 25-year-old male?

(b) Express the ratio of accident rates for females over males as a function of age. What is the maximum value of this ratio in the $[20, 35]$ age range?

8. In a study undertaken to model the pass rates for young adults taking a driving test, a binomial GLM with logistic link function was used with linear predictor $\eta = \alpha_G + \beta_G x$ where α_G and β_G are parameters for the factor gender and x is the variate age (in years). ML estimates of these parameters gave $\hat{\alpha}_F = -0.968449$, $\hat{\alpha}_M = -4.584923$, $\hat{\beta}_F = 0.056405$ and $\hat{\beta}_M = 0.209524$, where F = female and M = male.

(a) What is the estimated pass rate for a 20-year-old female using this model?

(b) What are the predicted odds (ratio of probability of success to probability of failure) for a 20-year-old male?

(c) Briefly comment on the effect of age in pass rates.

9. Statistics on the number of deaths (in thousands) due to traffic accidents and the corresponding number of registered automobile vehicles (in hundred thousands) over a 24-year period are given in Table 7.11.

TABLE 7.11
National annual automobile deaths and registered vehicles.

Year	Deaths (y)	Vehicles (x_1)	Year	Deaths (y)	Vehicles (x_1)
1	37.3	62.9	13	52.9	99.1
2	39.4	65.5	14	55.1	103.2
3	38.6	67.8	15	55.8	107.7
4	37.1	68.8	16	54.2	111.4
5	38.1	72.2	17	54.8	116.7
6	38.3	75.2	18	56.6	122.6
7	39.5	76.7	19	55.3	129.9
8	40.3	79.8	20	47.7	134.8
9	42.4	83.8	21	46.1	138.2
10	48.1	87.5	22	46.9	144.2
11	49.1	92.2	23	49.2	149.1
12	53.1	96.2	24	51.3	153.7

In year 20, a new law was introduced limiting the speed limit for driving. It is decided to fit a normal linear model of the form $E(Y_i) = \beta_0 + \beta_1 x_i + \beta_2 x_i^2$, where Y_i and x_i are, respectively, the number of deaths and registered automobiles for year i. When using such a normal linear model, we know that the maximum likelihood (and least squares) estimators of $\boldsymbol{\beta}^T = (\hat{\beta}_0, \hat{\beta}_1, \hat{\beta}_2)$ are given by

$$\boldsymbol{\beta} = (\mathbf{X}^T\mathbf{X})^{-1}\mathbf{X}^T\mathbf{y}$$

$$= (\mathbf{X}^T\mathbf{X})^{-1}(1127.2, \sum_1^{24} x_i y_i, \sum_1^{24} x_i^2 y_i).$$

Find the estimates $\boldsymbol{\beta}$ directly by finding $(\mathbf{X}^T\mathbf{X})^{-1}$ and $\mathbf{X}^T\mathbf{y}$. Determine the residual sum of squares for this model, and obtain an estimate of σ^2. Make a plot of the deaths versus number of registered vehicles, and include a plot of the predicted curve for deaths. Would this model be useful for predicting future deaths?

10. Using the model developed for predicting pass rates for male and female drivers in Example (7.7), what are the linear predictor and the predicted pass rate for a 20-year-old male? At what age does the predicted pass rate for males exceed that of females?

11. Construct a binomial GLM model for the driver testing data in Table 7.4, using a probit link function instead of the logit link function. How do the predictors from such a model (including an age by gender interaction term) compare with those given in Example 7.7?

12. The representation of a random variable in an exponential family given by Equation (7.3) is not unique. Show that if the density of Y can be represented in the form (7.3) with natural parameter θ, then it can also be expressed in a similar way with natural parameter $\theta^* = \theta/2$ by using $\gamma^*(\theta^*) = \gamma(2\theta^*)/2$ and $A^* = 2A$.

13. Let Y have the negative binomial distribution with parameters k and and p, that is, Y is the number of failures observed in order to attain (a predetermined) k successes in a sequence of Bernoulli trials with probability of success p. Such a distribution is often considered when modeling count data where the variance is larger than the mean (an overdispersed Poisson). Show that the negative binomial family is an exponential family where the natural parameter is $\theta = \log(1-p)$.

14. A new surgical technique for removal of a certain type of malignant tumor in women was performed on 50 women, and survival beyond one year was the variable of interest. Information was available on each of the women with respect to age in years, smoking status and number of children. For the ith individual, $y_i = 1$ if the patient survived one

year, and otherwise $y_i = 0$. It is natural to consider modeling Y_i as a Bernoulli random variable, where the probability of success μ_i depends on age, smoking status and number of children. It was decided to use a linear predictor of the form

$$\eta = \beta_0 + \beta_1 \, x_{age} + \beta_2 \, x_{children},$$

where β_0 takes one value (β_S) for smokers and another (β_{NS}, which is not necessarily different) for nonsmokers. Using a binomial GLM model with the canonical link, maximum likelihood estimates of the parameters above were

$$\hat{\beta}_S = 0.92, \quad \hat{\beta}_{NS} = 1.01, \quad \hat{\beta}_1 = -0.005, \quad \text{and} \quad \hat{\beta}_2 = -0.107.$$

(a) What would you predict as the survival rate for a 45-year-old non-smoking woman who had two children? What would the corresponding probability be for a woman with the same characteristics except that she is a smoker?

(b) The odds of surviving (to not surviving) is of the form e^η. For a woman with a given smoking status and age, by what factor does the odds of surviving change with each additional child?

(c) According to this model, smoking decreases the odds of survival by how much?

15. Employee accident numbers in two large manufacturing companies (A & B) were compiled over a two-year period with the intention of comparing rates between the two companies and looking for trends over time. The unit of time used was two months, with the results over two years appearing in Table 7.12. The Poisson distribution is a natural one to consider for counting accident data of this type. Although no information is provided here on *exposure* of the companies to accidents (such as number of employees), it would normally be of interest. For example, if Company B had twice as many employees as Company A (and otherwise working conditions were similar), then we might naturally expect the accident rate in Company B to be approximately twice that of Company A. Exposure information of this type can be taken into consideration in GLM modeling through the use of offset parameters.

TABLE 7.12
Industrial accidents over time in companies A and B.

Time Period (Unit = 2 Months)											
1	2	3	4	5	6	7	8	9	10	11	12
Company A	8	6	10	18	11	19	13	19	17	21	16	21
Company B	12	19	14	15	23	27	19	29	37	27	35	26

A plot of the data reveals a slight upward trend in accidents over time for both companies. Consider and analyze a multiplicative model where the accident rate is proportional to time for each company, or in other words, that $\log \mu = \beta_0 + \beta_1 \log t + x_C$ where x_C is an indicator for company.

16. Table 7.13 contains information on type of ship, year of construction, period (of operation), service (in months) and number of damage incidents (inc) caused by waves to forward section of cargo vessels. The data for this table was provided by J. Crilley and L.N. Hemingway of Lloyd's Register of Shipping. This is a classic data set, which both appears as a data set in the software packages S-plus (ships) and R, and also is analyzed by McCullagh and Nelder ([41], pp. 204 − 208).

It is of interest to know how the risk of damage is related to the three variables of type of ship, year and period. Note that although there are 40 entries in the table, 5 of them (these are denoted with an *) have no exposure time (months of service = 0) and hence cannot have any accidents. There is also 1 entry (denoted with **) which is accidently 0 due to a missing exposure time. Therefore we have in fact 34 observations on which to base a model. The response is the number of damage incidents, and hence it is natural to consider a Poisson GLM model where the number of incidents is Poisson with parameter $\lambda_{ijk} = n_{ijk} \mu_{ijk}$. Here n_{ijk} is the known exposure (aggregate months of service) associated with a given type i, year j and period k, and hence is what is termed an offset parameter. The variables type of ship, year and period being factors, analyze the data and in particular consider the following model:

$$\log \text{ E(number of incidents)}_{ijk} = \log \lambda_{ijk}$$
$$= \log n_{ijk} + \text{ effect (type)}$$
$$+ \text{ effect (year) } + \text{ effect (period) .}$$

TABLE 7.13

Damage incidents to ships.

Type	Year	Period	Service	Inc	Type	Year	Period	Service	Inc
A	60-64	60-74	127	0	C	70-74	60-74	783	6
A	60-64	75-79	63	0	C	70-74	75-79	1948	2
A	65-69	60-74	1095	3	C	75-79	60-74	0	*0
A	65-69	75-79	1095	4	C	75-79	75-79	274	1
A	70-74	60-74	1512	6	D	60-64	60-74	251	0
A	70-74	75-79	3353	18	D	60-64	75-79	105	0
A	75-79	60-74	0	*0	D	65-69	60-74	288	0
A	75-79	75-79	2244	11	D	65-69	75-79	192	0
B	60-64	60-74	44882	39	D	70-74	60-74	349	2
B	60-64	75-79	17176	29	D	70-74	75-79	1208	11
B	65-69	60-74	28609	58	D	75-79	60-74	0	*0
B	65-69	75-79	20370	53	D	75-79	75-79	2051	4
B	70-74	60-74	7064	12	E	60-64	60-74	45	0
B	70-74	75-79	13099	44	E	60-64	75-79	0	**0
B	75-79	60-74	0	*0	E	65-69	60-74	789	7
B	75-79	75-79	7117	18	E	65-69	75-79	437	7
C	60-64	60-74	1179	1	E	65-69	75-79	437	7
C	60-64	75-79	552	1	E	70-74	75-79	2161	12
C	65-69	60-74	781	0	E	75-79	60-74	0	*0
C	65-69	75-79	676	1	E	75-79	75-79	542	1

8

Decision and Game Theory

8.1 Introduction

All around us, in all aspects of life, decisions continually need to be made. The decision makers may be governments, boards of management, political parties, concert promoters, team captains or even individuals like ourselves. In many situations, a decision has to be made about *which* of the *possible actions* will be taken in the face of considerable uncertainty. Some decisions can be made quickly without much forethought and analysis. In other cases, the consequences of the different actions taken may be quite variable, and it may be worthwhile analyzing the various possibilities with considerable care.

Many decisions have to be made where the decision maker has only partial control of the resulting outcome. This may be due to other players (or decision makers) influencing the result, who have other objectives in mind and could be considered as the *competition*. There are other situations, however, where there may be no obvious competitors, but where there is uncertainty due to an unknown force that has a significant impact on the outcome of any decision made. We shall generally refer to this unknown force as *nature*. For example, the success of a company expansion may depend on the health of the economy in the coming year, and we may refer to the future state of the economy as the *state of nature*. A decision to go ahead with an outdoor concert may well depend on the weather forecast, and again we may refer to the upcoming weather as the state of nature.

When the decision makers are conscious that their actions affect the choices, consequent payoffs and possible behavior of the others, we have what we shall call a *game*. A game is said to be *cooperative* if the players can make binding agreements about possible actions and strategies, otherwise it is a *noncooperative* game. The following are some examples of decisions being made on a daily basis. Which of these would you consider to be games?

1. A political party is contesting a forthcoming election and must decide on their policy regarding a sensitive constitutional issue regarding permanent residency and citizenship of foreign nationals.

2. A manufacturer of personal computers is invited by a university to submit computer price packages for exclusive selling rights to the student

body.

3. A management executive body is preparing a salary package (with share options) for a new CEO.

4. A traveler to a rural area in Africa is advised to buy malaria tablets for the journey. The tablets are, however, relatively expensive and have some side effects.

5. A young couple in their twenties has just celebrated the birth of their first child. Both parents are very healthy with good jobs. They are considering the possibility of purchasing life cover for themselves and/or the child at this early stage.

6. As a graduation present, a student received a new car from her uncle. She already has established three years' worth of *no claims* driving. She is not sure, however, if she should purchase comprehensive insurance (in addition to the compulsory third party, fire and theft) on the car, which would add an additional 300 euro to the price of the insurance.

Of these six decision making scenarios, the first three may be viewed as games. The political party will clearly make its decision in light of what it believes the other parties might do, with the ultimate objective of attracting the most followers. The computer manufacturer wants to secure the contract and make a profit, in spite of knowing that it might have to lower prices to compete with other manufacturers. The management executive body knows that the package offered will influence the decision to accept the post and will probably have some impact on the resulting decisions the new CEO will make for the company. The later three decisions are made under uncertainty, but with no clearly intelligent competitor or decision maker influencing the ultimate action. The traveler to Africa knows that the decision to buy tablets hinges on the unknown exposure to mosquitoes (state of nature) she will encounter, while the young couple can only imagine the (noncompetitive) future in terms of their family health and welfare. The new car owner has to make a decision on the basis of the likelihood of her (possibly random) involvement in a future accident with the car.

The action or strategy which a decision maker ultimately takes will, of course, depend on the *criterion* adopted for *making a decision*. In any given situation there may be several possible criteria to consider. In Section 8.2, we will concentrate on the *minimax criterion*. We will consider zero-sum two-person games, but also give examples of variable-sum games where cooperation would be useful to both parties (as in the classical Prisoner's Dilemma problem) but in fact often does not exist. In Section 8.3 we consider the general area of decision making under risk, using both the minimax and Bayes criteria in situations where one tries to cope with the unknown state of nature. We do not assume that nature is either competitive or intelligent. However, the degree of uncertainty in nature might be reduced by experimentation or the collection of additional sample information. Of course, this often must be done at a cost, hence one may want to determine what it is worth paying to get it! In Section 8.4, the concept of utility is briefly treated as an alternative value system to a strictly monetary one.

The foundations of decision and game theory can be a useful tool for the insurance analyst. Some interesting references for applications in insurance are Borch [6] (game theory and automobile insurance), Lemaire [34] (cost allocations), Lemaire [36] (cooperative game theory and insurance), and Pollack [50] (game theory and reinsurance).

8.2 Game theory

Many of the ideas in game theory, including much of the commonly used terminology and the idea that conflict can be mathematically modeled, originate with the classic (1944) text *The Theory of Games and Economic Behavior* by von Neumann and Morgenstern [58]. The basic components in a game include the players, the actions and/or strategies, and the payoffs. By an *action*, we shall mean a choice or option for one of the players. We shall also sometimes refer to an action as a *pure strategy*, whereby with probability 1 that action is taken! Often, however, a strategy for a player can be something more general; for example, it may involve the player making a decision about which action to take on the basis of some information that comes to hand, or on the basis of observing some random device (a mixed or random strategy). Each person has a set of possible actions or options open to them, and subsequent to the selection of an action by each player, there is a resulting payoff for each player. When the payoffs for each possible set of actions (or strategies) by the n players sum to 0, then the game is called a *zero-sum* n-person game, and otherwise it is called a *variable-sum* game. There may be any number n of players in a game, although in this treatment we shall restrict consideration to two-player (two-person) games. In game theory we generally assume that the other players (or opponents) are intelligent, and in many cases working against us (noncooperative competitors). They will often be trying to outmaneuvre us, and in some cases would not be beyond *spying* on us.

It should be noted that *game theory* itself does not necessarily prescribe the *ideal strategy* for each player. If, however, the players have *specified objectives or criteria in mind (for example, to minimize their maximum expected loss)*, then the theory outlines the strategy that the players should take. On the other hand, if these objectives do not exist or are not important to the players, then the theory will be useless in describing their ideal behavior. Our study of game theory may seem somewhat idealistic or simplistic, but understanding the essential aspects of the theory can be very useful for decision making in the *social sciences, economics, political science, business, industrial relations and military strategy* – to name but a few. Our treatment begins with a discussion of zero-sum two-person games using the minimax and Bayes criteria for deciding on a strategy. The classical minimax theorem states that in such a game, there are optimal (usually mixed or random) strategies for both players, yielding what is called the *value ν* of the game. The Prisoner's Dilemma provides a classic example of a variable-sum game where competition hurts both players and cooperation may be hard to come by.

8.2.1   Zero-sum two-person games

Assume that the two players A and B in a zero-sum game have actions $\mathcal{A} = \{a_1, \ldots, a_r\}$ and $\mathcal{B} = \{b_1, \ldots, b_c\}$ open to them, respectively. The results or payoffs of the game may be represented by an $r \times c$ matrix $\mathcal{P}$ of losses for player B, where a loss for B is a gain for A and vice versa (hence the term *zero-sum*). In the matrix $\mathcal{P}$ the payoff $L_{i,j}$ *represents the loss to player B (or equivalently the gain to player A)* when A chooses a_i and B chooses b_j. How the payoff matrix for a game is viewed or presented (for example, as losses to B, gains to A, or otherwise) is often a matter of preference. When each player has two possible actions, the payoff matrix (of losses for B) has the form given in Table 8.1.

TABLE 8.1
Payoff matrix $\mathcal{P}$ for a
zero-sum two-person
game.

		Player B	
		b_1	b_2
Player	a_1	$L_{1,1}$	$L_{1,2}$
A	a_2	$L_{2,1}$	$L_{2,2}$

$$\mathcal{P}$$

Usually, we assume that players A and B must make their choice of action without any knowledge of what the other player is doing, and that once a player has made her (his) choice, it cannot be changed. In general, we will be asking: What *actions* or *strategies* will A and B select? We can answer this assuming A and B reason in certain (rational) ways. Once we know this, we can determine what is called *the value ν of the game – that is, the expected payoff given that A and B are using their "optimal" strategies*.

In some situations, we may eliminate certain actions or strategies for a player as they are *dominated* by others. One action for a player is said to dominate another action, if the choice of the first action always (no matter what the opponent does) leads to a payoff which is at least as good as that when using the second action, and in some circumstances better.

Example 8.1

The matrix $\mathcal{P}_1$ below (Table 8.2) is a payoff matrix for a zero-sum two-person game involving Alison (A) and Brian (B), both of whom are intelligent people. This is a loss matrix for B, hence if B selects b_3 and A selects a_2, then B loses 3. On the other hand, if B chooses b_2 and A chooses a_1 then B loses -1, or gains 1.

TABLE 8.2
Loss matrix $\mathcal{P}_1$ for Brian.

		Brian			
		b_1	b_2	b_3	b_4
	a_1	3	-1	1	4
Alison	a_2	1	3	3	2
	a_3	5	1	2	7

B (Brian) is interested in minimizing his loss to A (Alison). No matter what choice A makes, the loss B will incur by choosing b_2 is less than (or equal to) the loss incurred by choosing b_3. We say that action b_2 *dominates* action b_3 for player B, or, equivalently, that action b_3 is *dominated* by action b_2. Similarly, b_1 is always better for B than b_4 is, and b_4 is dominated by b_1. B will never choose b_3 or b_4 (they are *inadmissable*), thus for all practical purposes the game reduces to the 3×2 (loss) payoff matrix $\mathcal{P}_{11}$ given in Table 8.3 (both A and B being intelligent realize this).

TABLE 8.3
Loss matrix $\mathcal{P}_{11}$.

		Brian	
		b_1	b_2
	a_1	3	-1
Alison	a_2	1	3
	a_3	5	1

Of course, A is interested in maximizing her gain from B. She further realizes that in choosing strategy a_3 she is guaranteed a higher payoff (or gain) than if she uses strategy a_1, no matter what B chooses to do. Again, using the terminology above, strategy a_3 dominates strategy a_1 for her. Therefore the game further reduces to the 2×2 payoff matrix $\mathcal{P}_{12}$ given in Table 8.4.   □

TABLE 8.4
Loss matrix $\mathcal{P}_{12}$.

		Brian	
		b_1	b_2
Alison	a_2	1	3
	a_3	5	1

Disregarding dominated strategies in Example 8.1 has reduced the complexity of the problem, but still has not revealed the optimal strategies for A and B. To solve the game (find "optimal" strategies for A and B, and the corresponding payoff) we must make an assumption about the *decision making criteria* of the players. There are various criteria which an intelligent person might consider in picking a strategy. A very basic but somewhat pessimistic (and risk averse) criteria to use is the *minimax* criterion. When using the minimax criterion, a player determines the worst that can happen for each possible strategy (in fact, the player "expects" the worst), and then selects that strategy which minimizes this quantity. If a payoff matrix possesses what is called a *saddle point*, then the minimax strategy will be a *pure* strategy for both players. Otherwise, *random* or *mixed* strategies may be employed. Other criteria that are sometimes used are the *minimax regret criterion* (see [14]) and the very optimistic *maximin criterion*. The minimax regret criterion uses a minimax approach to a regrets matrix (as opposed to a payoff matrix). When using the maximin criterion, one determines for each possible strategy the best that can happen (in fact, "expecting" the best to happen), and then selects that strategy which maximizes this quantity. Our treatment of game theory, however, will concentrate on the minimax criterion.

8.2.2   Minimax and saddle point strategies

If players A and B are using the *minimax* criterion in a game (with payoff matrix of losses for B), then B will choose his strategy by *minimizing his maximum expected loss* while A will try to maximize her minimum expected gain. Under *minimax*, we know that players adopt a pessimistic view, considering (and expecting) only the worst possible outcome for each strategy. The strategy giving the *least worst* outcome is then selected.

Example 8.2

Players A and B compete in a zero-sum game with payoff matrix $\mathcal{P}_2$ given in Table 8.5. If B is using the minimax criterion, then he decides between b_1, b_2 and b_3 by comparing the worst possible loss he would incur when choosing each of these actions. For example, if B uses action b_1, then he could (and expects to) lose as much as 6, while similarly, for actions b_2 and b_3 he could lose as much as 8 and 4, respectively. The least of these possible worst losses is therefore 4, and hence this implies that b_3 is the action or *pure strategy* which *minimizes the maximum loss incurred by B.*

Similarly, we can show that a_2 is the action which maximizes the minimum gain for A. Therefore when A and B use the minimax criterion to decide on a pure strategy, A will go for a_2 and B for b_3. B has an expected maximum loss of 4 (expected *loss ceiling*), while A has an expected minimum gain (expected *floor gain*) of 4. The *value of the game is therefore* 4. If player B were offered 5 to play the game, he would probably do so (since he would have an expected gain of at least $5 - 4 = 1$). $\quad\square$

TABLE 8.5

Loss matrix $\mathcal{P}_2$ for player B.

		Player B		
		b_1	b_2	b_3
	a_1	-2	8	3
Player A	a_2	6	7	4
	a_3	2	0	-4

In Example 8.2 the payoff matrix possesses what is called a *saddle point*, and in such a case one need not go through such an argument to find the minimax solution to the game. A saddle point for a loss payoff matrix, if it exists, will be the *lowest payoff in its row, and the highest in its column.* In the payoff matrix $\mathcal{P}_2$, note that the payoff $L_{2,3}$ is such a point. The minimax theorem in game theory (which we will discuss later) implies that if the payoff matrix for a game has a saddle point, then the minimax strategies for both players A and B are the pure strategies determined by it. Moreover, such a *saddle-point strategy* both minimizes B's expected (maximum) loss and maximizes A's expected (minimum) gain, and furthermore the payoff at the saddle point is the *value ν of the game.* In Example 8.2, since $L_{2,3} = 4$ is a saddle point, B would (use the pure strategy and) always select b_3 while A would always select a_2. Moreover, the value of the game is 4.

A saddle point in a payoff matrix is a *point of equilibrium* for a game in the sense that if both A and B are employing saddle-point based strategies, then neither A nor B can benefit from knowing that their opponent is using the minimax criterion. For example, suppose that in the game with payoff matrix $\mathcal{P}_2$, B (who wants to use a minimax strategy) has been told that A is going to use a minimax strategy. Then this information will not change B's decision to use the action (or pure strategy) b_3. In fact, any move away from this saddle-point strategy for B will worsen the situation, assuming A stays put at her saddle-point strategy a_2.

Another interesting aspect of a saddle-point strategy is that it is a *spy-proof* strategy if both players are capable of spying. A player in a game may know that their opponents are intelligent, but other than that they might not know much more about how they are thinking. In some important games (for example, in setting next year's premium for a motor policy in a competitive market) a player might be worried about the possibility that an opponent is spying. A minimax strategy is said to be *spy-proof* in the sense that none of the players can benefit from knowing what the others are doing. For instance, in Example 8.2, suppose A is considering using a_1 instead of a_3. Knowing this (by spying), B will consider moving from b_3 to b_1. Now if A knows this (again by spying), then she will consider moving on to a_2. However, then in turn B (knowing this) will move back to b_3 and both are now at the minimax strategies.

Not all payoff matrices for zero-sum two-person games have a saddle point, as the following example shows.

Example 8.3

Table 8.6 gives the payoff matrix $\mathcal{P}_3$ in terms of losses for player B when playing against A. Y would be a saddle point if and only if it were both larger than 7 and less than -5, which is impossible. Similarly, X could only be a saddle point if both $X \geq 4$ and $X \leq 1$, which again is impossible. On the other hand $L_{3,3} = 4$ would be a saddle point if $X \leq 4$. □

TABLE 8.6
Payoff matrix $\mathcal{P}_3$.

		Player B			
		b_1	b_2	b_3	b_4
	a_1	Y	-5	3	12
Player A	a_2	7	6	X	1
	a_3	5	9	4	7

Consider the above game when $Y = 1$ and $X = 0$, with the resulting loss matrix $\mathcal{P}_{31}$ for player B as given in Table 8.7. Then $L_{3,3} = 4$ is a saddle point, the value of the game is 4, and the minimax pure strategies are a_3 for A and b_3 for B. Note that if B knows that A is using the minimax strategy, he cannot improve things by moving from b_3 to any other pure strategy (in fact, it only gets worse in the minimax sense). Also, since a saddle point exists here, the minimax strategy is *spy-proof* in the sense that if both players can spy on the other, then both will still stick to the minimax strategy. For example, if A knows that B is thinking of using strategy b_1 instead of strategy b_3, she will move to a_2. But, knowing this, B will then move to back to b_3, and then in turn A will move back to a_3, yielding once again the minimax solution!

TABLE 8.7
Payoff matrix $\mathcal{P}_{31}$.

		\multicolumn{4}{c}{Player B}			
		b_1	b_2	b_3	b_4
	a_1	1	-5	3	12
Player A	a_2	7	6	0	1
	a_3	5	9	4	7

8.2.3 Randomized strategies

What if no saddle point exists in the payoff matrix of a zero-sum two-person game? If the payoff table for a game does not have a saddle point, then the minimax strategy is not spy-proof. Hence if one player knows what the other intends to do, then this information may be used to her advantage (for example to use a nonminimax strategy). With this in mind, a player may wish to *randomize* their selection of strategies. In this case, a player will not know in advance what the exact maximum loss may be; however, one may calculate the *expected maximum loss* in order to decide on an appropriate *randomized strategy*.

Example 1 (continued): Consider again the payoff matrix $\mathcal{P}_{12}$ (Table 8.8) for Alison and Brian (after the dominated strategies have been dropped), who are conservative in nature and want to apply the minimax criteria for decision making.

TABLE 8.8
Payoff matrix $\mathcal{P}_{31}$.

Player		\multicolumn{2}{c}{Brian}	
		b_1	b_2
	a_2	1	3
Alison	a_3	5	1

No saddle point exists in the resulting payoff matrix, and b_2 is the pure strategy which minimizes Brian's maximum loss resulting in an expected (maximum) loss of 3. Alison is indifferent to strategies a_2 and a_3, as both have an expected (minimal) gain of 1. These are, however, *not* the optimal strategies for either player when using the minimax criterion. In fact, both players may improve the situation if they consider a strategy which uses a randomized method (mixed strategy) to choose between possible pure actions or strategies. The use of randomization in the selection of your actions has the appealing aspect that it counteracts to a certain extent the knowledge that your opponent is intelligent and may try to reconstruct how you think. (For example, if you are playing a tennis match and your opponent knows that your powerful serve is your greatest strength, you still might want to randomly deliver a slow serve to keep him alert.)

Consider the randomized strategy for Brian (see Table 8.9), in which he randomly chooses between b_1 (with probability q) and b_2 (with probability $1 - q$). With such a strategy, the maximum loss to Brian is no longer deterministic, but in fact is a random quantity or variable. If Alison chooses a_2, Brian can *expect* to lose (on the average) $L_{a_2}(q) = 1q + 3(1-q) = 3 - 2q$, while if Alison chooses a_3, B can expect to lose $L_{a_3}(q) = 5q + 1(1 - q) = 1 + 4q$.

TABLE 8.9
Mixed strategy for Brian.

		$q \times b_1 \ + (1 - q) \times b_2$
Alison's	a_2	$L_{a_2}(q) = 3 - 2q$
Action	a_3	$L_{a_3}(q) = 1 + 4q$
		Expected loss for Brian

Using a minimax criterion, Brian wants to use a strategy which minimizes the maximum expected loss, that is, to select q to minimize

$$max[L_{a_2}(q), L_{a_3}(q)] = max[3 - 2q, 1 + 4q].$$

Since $L_{a_2}(q)$ is a decreasing function of q on the interval $[0, 1]$ while $L_{a_3}(q)$ is increasing, the function $max[3 - 2q, 1 + 4q]$ is minimized where the two functions of q meet or intercept. Solving, one obtains $q = 1/3$, and Brian has reduced his expected (maximum) loss from 3 to 7/3 by employing a randomized strategy (randomly selecting b_1 with probability $1/3$ and b_2 with probability $2/3$). Similarly, a randomized strategy for Alison will increase her expected (minimum) gain from 1 to 7/3. What precisely is this optimal randomized strategy for A (see Problem 2)? *We say that 7/3 is the value of the game.*

In general, when no saddle point exists, players should employ randomized strategies with probabilities chosen to adhere to the minimax theorem.

In general, if a payoff matrix $\mathcal{P}$ represents the losses for B in a zero-sum game with player A, then using (pure) minimax strategies, B has an expected maximum loss of $\nu_{B(pure)} = min_j[max_i(L_{i,j})]$, while A has an expected minimal gain of $\nu_{A(pure)} = max_i[min_j(L_{i,j})]$. It is easy to show that $\nu_{A(pure)} \leq \nu_{B(pure)}$, and that, in fact, they are equal if and only if the matrix $\mathcal{P}$ has a saddle point. When they are not equal at least one of the players (normally both) can improve their situation by using randomized strategies. Suppose player B has possible actions $\{b_1, b_2, \ldots, b_c\}$ open to him and $\mathbf{q} = (q_1, q_2, \ldots, q_c)$ is a probability vector. B may use the mixed strategy determined by the vector $\mathbf{q}$ whereby he selects to take action b_j with probability q_j for $j = 1, \ldots, c$. If similarly, player A randomly selects the possible actions $\mathcal{A} = \{a_1, \ldots, a_r\}$ with respective probabilities $\mathbf{p} = (p_1, \ldots, p_r)$, then the expected payoff (loss for B and gain for A) is given by $\sum p_i L_{i,j} q_j$.

Let Q and P represent the sets of possible random strategies for B and A, respectively. Player B would worry about what particular strategy $\mathbf{p}$ player A will use, and hence would try to find a strategy $\mathbf{q}$ to minimize $[max_{\mathbf{p} \in P} \sum p_i L_{i,j} q_j]$. This would result in an expected (maximum) loss of $\nu_B = min_{\mathbf{q} \in Q}[max_{\mathbf{p} \in P} \sum p_i L_{i,j} q_j]$. Similarly, the best mixed strategy for player A is that $\mathbf{p} \in P$ which maximizes $[min_{\mathbf{q} \in Q} \sum p_i L_{i,j} q_j]$, yielding an expected minimum gain of $\nu_A = max_{\mathbf{p} \in P}[min_{\mathbf{q} \in Q} \sum p_i L_{i,j} q_j]$ for player A.

The *Minimax theorem* is the most important result in game theory, which says that ν_A *and* ν_B *are equal* (see [46]). Hence there exist mixed strategies $\mathbf{p}^*$ and $\mathbf{q}^*$ for players A and B, respectively, which are optimal for both, and where the expected (maximum) loss for B equals the expected (minimum) gain for A. This common value ν is called the *value of the game*, where

$$max_{\mathbf{p} \in P}\left[min_{\mathbf{q} \in Q} \sum p_i L_{i,j} q_j\right] = \nu_A$$
$$= \nu$$
$$= \nu_B = min_{\mathbf{q} \in Q}\left[max_{\mathbf{p} \in P} \sum p_i L_{i,j} q_j\right].$$

Example 8.4

The payoff matrix $\mathcal{P}_4$ in Table 8.10 represents a zero-sum game between two intelligent players A and B. Here a positive payoff represents a gain to player A (and consequently, a loss gain to player B).

Assuming a minimax criterion for both players, b_3 is an inadmissable strategy for B, and there is no saddle point. Player A is considering a mixed strategy where she randomly uses a_1 with probability p and otherwise uses a_2. Her expected gain if B uses b_1 is $10p + 3(1-p) = 7p + 3$, while if B uses b_2 it is $-4p + 7(1-p) = 7 - 11p$. As player A is conservative, she considers the minimum of these two quantities, and wants to find the value of p maximizing it. The maximum over p of $min[7p + 3, 7 - 11p]$ occurs at $p = 2/9$ where the two lines intersect. Therefore the minimax (mixed) strategy for A is to select

TABLE 8.10

Payoff matrix $\mathcal{P}_4$.

		Player B		
		b_1	b_2	b_3
Player	a_1	10	-4	12
A	a_2	3	7	5

a_1 with probability $p = 2/9$ (and a_2 with probability 7/9). Similarly, B would select b_1 with probability $q = 11/18$ and otherwise select b_2. The value of the game is $\nu = 41/9$.   ⬚

Example 8.5

Table 8.11 gives the payoff matrix $\mathcal{P}_5$ for a zero-sum game between two intelligent players A and B. The entries represent losses to player B, and hence gains to player A. Clearly b_3 is inadmissable for B as it is dominated by b_1. Player A has therefore a pure minimax strategy of a_2 with expected minimum gain of 1, while B has a pure minimax strategy of b_1 with expected maximum loss of 3. The value of the game ν therefore satisfies $1 \leq \nu \leq 3$.

TABLE 8.11

Payoff matrix $\mathcal{P}_5$.

		Player B		
		b_1	b_2	b_3
	a_1	3	-2	5
Player A	a_2	1	4	2
	a_3	-2	5	0

Randomly selecting b_1 with probability q (and b_2 with probability $1 - q$), B wants to find q which minimizes function $max_q[5q - 2, 4 - 3q, 5 - 7q]$ (which is plotted in **bold** in Figure 8.1). This minimum occurs at $q = 3/4$, leading to a game value of $\nu = 5(3/4) - 2 = 7/4$. On the other hand, player A wants to select a_1, a_2 and a_3 with probabilities p_1, p_2 and $p_3 = 1 - p_1 - p_2$, in order to maximize $min_{(p_1, p_2)}[5p_1 + 3p_2 - 2, -7p_1 - p_2 + 5]$. This must occur when these two expressions are equal, or where $12p_1 + 4p_2 = 7$. The function $5p_1 + 3p_2 - 2$ is maximized on the line segment joining $(7/12, 0)$ and $(3/8, 5/8)$ at $(p_1, p_2) = (3/8, 5/8)$, giving (also) a game value of 7/4 when $p_1 = 3/8$ and $p_2 = 5/8$. The optimal mixed strategy for A is therefore to select a_1 with probability 3/8 and a_2 with probability 5/8 (and to never select a_3).   ⬚

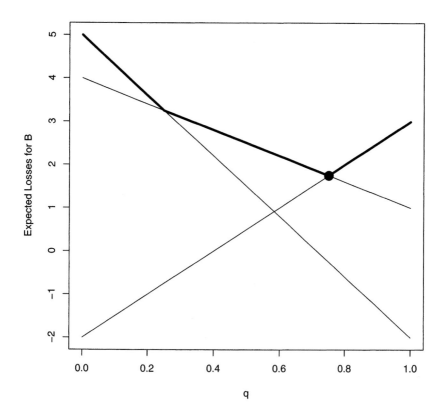

FIGURE 8.1

Optimizing q for player B (Example 8.5) by minimizing max[5q-2, 4-3q, 5-7q].

8.2.4   The Prisoner's Dilemma and Nash equilibrium in variable-sum games

In a two-person zero-sum game the loss to one player is a gain to the other, and hence one need only know one of these quantities to determine the other. In a variable-sum game this need not be the case, and the payoffs (which we will assume are gains) for A and B, respectively, will be written in the form $L_{i,j} = (\alpha, \beta)$. When $\alpha + \beta = 0$ for all i and j, then we have a zero-sum game. The following classic example of a two-person variable-sum game is interesting and somewhat paradoxical.

Example 8.6 **Prisoner's dilemma**

Two prisoners A and B have a dilemma. Having been accused jointly of a serious crime and arrested, they are about to be questioned separately about their alleged participation in the crime. In confessing to involvement, they know they face a sentence of as much as 15 years. If they both confess then each gets a sentence of 10 years, while if they both deny each sentence is 3 years. On the other hand, if one confesses and the other does not, then the confessor gets off free (with no sentence) while the other gets a sentence of 15 years. This is an example of a variable-sum two-person game, as the sum of the payoffs is not constant and varies with the actions taken by the prisoners.

Prisoner A thinks that if B is considering denying involvement (Deny), then he will certainly prefer a payoff of 0 to −3, and hence will confess to the crime. On the other hand, if he thinks that B is going to confess, then he would certainly prefer −10 to −15 and thus will also confess. This thinking is indicated by the downward arrows ↓ in Table 8.12. Similarly, B will decide to confess both when A decides to Deny or Confess (indicated by the right arrows →) in Table 8.12, due to the symmetry of the situation. Hence no matter what one player does, the other will prefer to confess. However, it is clear that the payoffs for both would be better if they both (cooperated and) denied the crime, and herein lies the dilemma!   ▯

TABLE 8.12
The Prisoner's dilemma!

		Prisoner B		
		Deny		Confess
	Deny	(-3, -3)	→	(-15, 0)
Prisoner A		↓		↓
	Confess	(0, -15)	→	(-10, -10)

Payoffs for Prisoners A and B: (gain for A, gain for B)

The pair of actions (Confess, Confess) in the Prisoner's dilemma is an example of what is called a point of Nash equilibrium ([43], [44]). At such a point, knowing that the other is intending to confess does not lead a prisoner to change his own decision to confess. A Nash equilibrium point in a game represents a steady state in the sense that any one player will not move away from this point knowing what the other is doing. Although such a point is not necessarily an optimal point for all players (as in the Prisoner's dilemma), when they exist they represent a type of solution for the game.

For a variable-sum two-person game where player A can take actions $\mathcal{A} = \{a_1, \ldots, a_r\}$ and B can take actions $\mathcal{B} = \{b_1, \ldots, b_c\}$, we let $L_{ij} = (\alpha_{ij}, \beta_{ij})$ represent the payoffs to A and B when they, respectively, take actions a_i and b_j. If player A prefers action a_k over a_i when B takes action b_j, we denote this preference by $(a_k, b_j) \geq_A (a_i, b_j)$. Similarly, we write $(a_i, b_k) \geq_B (a_i, b_j)$ if player B prefers action b_k to b_j when A is taking action a_i. If player A knows that player B is intending to use action b_j, then the best response (best possible action to take) for A is the (set of) action(s) $A(b_j) = \{a_l \epsilon A : (a_l, b_j) \geq_A (a_i, b_j) \; \forall a_i \epsilon A\}$. This set-valued function is called the *best response function* for player A (note that $A(b_j)$ would be the empty set ϕ when there is no best response for A), and, similarly, we can define the best response function for player B.

The pair of actions (a_*, b_*) for players A and B, respectively, represents a *Nash equilibrium point* for the game if

$$(a_*, b_*) \geq_A (a_i, b_*) \text{ for all } i = 1, \ldots, r \text{ and}$$
$$(a_*, b_*) \geq_B (a_*, b_j) \text{ for all } j = 1, \ldots, c,$$

or, equivalently, $a_* \epsilon A(b_*)$ and $b_* \epsilon B(a_*)$. In a variable-sum two-person game, there may be several Nash equilibrium points, but also there may be none.

In the Prisoner's dilemma, the best response functions for A and B satisfy A(Deny) = A(Confess) = B(Deny) = B(Confess) = {Confess}, and hence (Confess, Confess) is the unique Nash equilibrium point for the game. In a zero-sum two-person game with a saddle point, the saddle-point actions represent the unique Nash equilibrium point for the game.

The following Example 8.7 has two Nash equilibria, and historically has been known as the *Battle of the Sexes* (see Luce and Raiffa [39]) or *Bach or Stravinsky* (see Osborne and Rubenstein [45]). Pollack [50] argues that most reinsurance transactions may be viewed as (Prisoner's dilemma type) games between the cedant and reinsurance companies.

Example 8.7 Rugby or Football?

Two brothers Alan and Bart both are avid sports fans, but Alan's favorite is rugby while football is Bart's favorite. They do enjoy going to sporting events together, however, and must (separately) make a decision for this coming Saturday when simultaneously important games are being played locally in both rugby and football. The payoff matrix for the game is given in Table

8.13, and the brothers are interested in cooperating to the benefit of both. Given the payoff matrix (of gains to both) below, what should each brother do?

TABLE 8.13
Rugby or Football?

| | | Bart | |
		Rugby	Football
Alan	Rugby	(20, 15)	(10, 10)
	Football	(10, 10)	(15, 20)

Payoffs for Alan and Bart: (gain for Alan, gain for Bart)

The best response function for Alan is given by A(Rugby) = {Rugby} and A(Football) = {Football}. The best response function for B is identical. There are two optimal strategies in this game, namely (*Rugby, Rugby*) and (*Football, Football*), and both are Nash equilibria. ▯

8.3 Decision making and risk

In a game theoretic situation, we determine our strategy and make a decision on what to do assuming that our opponents are intelligent and generally speaking in competition with us. There are, however, many scenarios where knowledge of unknown factors could be informative in deciding on a decision strategy, but where it is perhaps inappropriate to assume that we have intelligent opponents working against us! Statistical inference (for example, the estimation and testing of means, variances and regression coefficients from some population) may sometimes be viewed as a game between *Nature* (which controls the features of the relevant population) and the *Statistician* (who tries to guess or estimate nature's choice θ on the basis of sample information). We shall refer to θ as the *state of nature*, which in some cases is a population parameter of interest. The statistician (or other decision maker) will then make a decision or take a course of action on the basis of the observed sample and some decision making criterion. For example, a decision to go ahead with an outdoor concert may depend on the weather (state of nature θ) on the day in question. A quality weather forecast (sample X) may be helpful in making this decision.

In such a decision making scenario, we define a *loss function* which details the penalty incurred by the statistician for each combination of the population characteristic θ and the statistician's choice of action a. Unlike the games considered in the previous section, we do not necessarily assume that *Nature* is our competitor, or is trying to outdo us. In many cases, we may have additional information (sample information X) which may be useful in inferring something about what nature is doing, and hence in deciding on a strategy to employ.

The statistician normally uses a *decision function* d in deciding upon his estimate of the unknown state of nature θ. Supplied with some sample information X, we let $d(x)$ be the statistician's choice of action which results from observing $X = x$. Mathematically, d is a function from the sample space to the set of possible actions or strategies $\mathcal{A}$ for the statistician. A number of possible decision functions usually exist, many of which are inadmissable or otherwise not appropriate. *Decision theory* tries to identify the best decision function in terms of the possible losses to the statistician, using some criterion like *minimax* or *Bayes*.

Example 8.8 Fair Die or an Ace-Six Flats?

A statistician must decide whether or not a six-sided die is Fair or has been weighted to be an Ace-Six Flats die. With a loaded *Ace-Six Flats die*, the probabilities of a 1 or 6 are both $1/4$, while the other possibilities $(2, 3, 4, 5)$ have equal probabilities of $1/8$. She is not allowed to examine the die, but the die is to be rolled once and she will be told that either the result is an Ace or a Six $(X = 1)$, or that it is some other number $(X = 0)$. She incurs a loss of 1 if she makes the wrong decision. Table 8.14 gives the payoff matrix for this decision making scenario.

TABLE 8.14
Ace-Six Flats?

	Statistician decides	
State of nature	a_1 = Ace-Six Flats	a_2 = Fair
θ_1 (Ace-Six Flats)	$L_{1,1} = L(\theta_1, a_1) = 0$	$L_{1,2} = L(\theta_1, a_2) = 1$
θ_2 (Fair die)	$L_{2,1} = L(\theta_2, a_1) = 1$	$L_{2,2} = L(\theta_2, a_2) = 0$

Losses for statistician

The sample information X indicates whether a toss of the die gives an Ace or Six $(X = 1)$ or not $(X = 0)$, hence there are only four decision functions for the statistician as given in Table 8.15. For example, d_1 is the decision function which decides that the die is Ace-Six Flats when $(X = 1)$, and otherwise decides the die is Fair. What is the most appropriate decision function d for the statistician to use?   □

TABLE 8.15
Decisions functions for Ace-Six Flats.

	Decision function			
Sample X	d_1	d_2	d_3	d_4
$X = 1$ (1 or 6 observed)	a_1	a_1	a_2	a_2
$X = 0$ (2, 3, 4 or 5 observed)	a_2	a_1	a_2	a_1

The *risk function* of a decision function (or strategy) is a useful tool in comparing decision functions. *The risk function for decision function d_j is a function of the state of nature θ, defined to be the expected loss incurred if we use d_j when the state of nature is θ.* We denote the *risk function* for d_j when the state of nature is θ_i by

$$R(d_j, \theta_i) = E_X[L(\theta_i, d_j(X))].$$

For example,

$$
\begin{aligned}
R(d_1, \theta_1) &= E[L(\theta_1, d_1(X))] \\
&= L(\theta_1, d_1(1)) \cdot P(X = 1 \mid \theta_1) + L(\theta_1, d_1(0)) \cdot P(X = 0 \mid \theta_1) \\
&= L(\theta_1, a_1) \cdot 1/2 + L(\theta_1, a_2) \cdot 1/2 \\
&= 0 \cdot 1/2 + 1 \cdot 1/2 = 1/2, \quad \text{and} \\
R(d_1, \theta_2) &= E[L(\theta_2, d_1(X))] \\
&= L(\theta_2, d_1(1)) \cdot P(X = 1 \mid \theta_2) + L(\theta_2, d_1(0)) \cdot P(X = 0 \mid \theta_2) \\
&= 1 \cdot 1/3 + 0 \cdot 2/3 = 1/3.
\end{aligned}
$$

Table 8.16 gives the values of the risk functions for the decision functions d_1, d_2, d_3 and d_4.

TABLE 8.16
Risk functions for Ace-Six Flats.

	Decision function			
State of nature	d_1	d_2	d_3	d_4
θ_1 (Ace-Six Flats)	1/2	0	1	1/2
θ_2 (Fair)	1/3	1	0	2/3

A look at the table of risk functions indicates that we can discard decision function d_4 (d_4 is *inadmissable*), since it is dominated by d_1. This is not too surprising since d_4 is an unusual decision procedure in that it says conclude the die is Ace-Six Flats if and only if one does not observe an Ace or a Six! We can therefore restrict consideration to Table 8.17, which gives the risk functions for (d_1, d_2, d_3). Note, however, that we still have not chosen a *best* decision function! We can decide between d_1, d_2 and d_3 on the basis of either of two criteria – minimax or Bayes.

TABLE 8.17
Reduced set of risk functions for
Ace-Six Flats.

State of nature	Statistician		
	d_1	d_2	d_3
θ_1 (Ace-Six Flats)	1/2	0	1
θ_2 (Fair Die)	1/3	1	0

8.3.1 The minimax criterion

The statistician should choose the strategy which *minimizes the maximum value of the risk function.* That is, we choose d_i, where $max_\theta R(d_i, \theta) = min_j[max_\theta R(d_j, \theta)]$, or in other words the maximum value of the *risk function* for d_i is less than that for any other strategy d_j. This criterion is clearly analogous to choosing the minimax strategy in game theory.

In the Ace-Six Flats example, the statistician should use decision function d_1 under the minimax criterion. In other words, the minimax decision would be to say the die is Ace-Six Flats if either an Ace or Six is observed with a throw, and otherwise to say the die is fair. In "games" against nature we generally do not employ randomized strategies, since we do not consider nature a malevolent opponent (an opponent who tries to maximize the statistician's loss). In this example, the die is either Ace-Six Flats or Fair (balanced). Nature does not think about its strategy to try to beat the statistician!

8.3.2 The Bayes criterion

The *Bayesian* approach to statistical inference is the basis for the Bayes criterion in decision theory. *We regard the unknown state of nature θ as a random variable* with some probability distribution (the *prior* distribution of θ). We then choose the decision function which minimizes the expected value of the risk function (where the expectation is taken over θ). That is, we choose d_i where

$$E_\theta(R(d_i, \theta)) = min_j E_\theta(R(d_j, \theta)).$$

In Example 8.8, a prior distribution on θ would be of the form:

$$P(\theta = \text{Ace-Six Flats}) = p \text{ and } P(\theta = \text{Fair}) = 1 - p.$$

Hence

$$E_\theta(R(d_1, \theta)) = (1/2) \cdot p + (1/3) \cdot (1 - p) = 1/3 + 1/6p$$
$$E_\theta(R(d_2, \theta)) = 0 \cdot p + 1 \cdot (1 - p) = 1 - p, \text{ and}$$
$$E_\theta(R(d_3, \theta)) = 1 \cdot p + 0 \cdot (1 - p) = p.$$

The Bayes decision function will depend on the prior distribution on θ, or in other words on the value of p. Clearly, the Bayes decision function is d_3 if $0 < p < 2/5$, d_1 if $2/5 < p < 4/7$, and d_2 otherwise. For example, if one felt that there is a 50% chance that the die is Ace-Six Flats, then the Bayes decision rule would be to say that the die is Ace-Six Flats if a 1 or 6 is observed, and Fair otherwise.

Example 8.9

A veterinarian is asked to inspect a farm animal for bovine tuberculosis (TB), and ultimately decide whether or not to declare it a *reactor* (in which case it must be destroyed) – otherwise, it is deemed to be healthy. The state of nature here is the true health of the animal, and we let θ_1 denote that the animal has TB and θ_2 that the animal does not. The veterinarian performs a test on the animal prior to making a decision, with a result of $X = 1$ indicating a positive test (suggesting the animal has TB) and $X = 0$ indicating a negative test. Unfortunately, as is the case with many tests, the test is good but not perfect. For example, it is assumed that $P(X = 1 \mid \theta_1) = 0.9$ and that $P(X = 0 \mid \theta_2) = 0.8$. Hence in particular there is a 20% chance of a false positive and a 10% chance of a false negative.

The situation may be described by the payoff (loss) matrix for the owner of the animal given in Table 8.18. The losses have been coded, and note for example that the biggest loss occurs if the animal is diseased but it is not detected. The smallest loss occurs in the situation where the animal does not have TB and the test support this.

TABLE 8.18

		Veterinarian's decision	
		R = Reactor	NR = Not Reactor
State of	θ_1 (TB)	$L(\theta_1, R) = 2$	$L(\theta_1, NR) = 6$
animal	θ_2 (No TB)	$L(\theta_2, R) = 4$	$L(\theta_2, NR) = 0$

As the test (sample) information X takes only two possible values ($X = 1$ if the test is positive and $X = 0$ if it is negative), there are only four possible decision functions as given in Table 8.19.

TABLE 8.19

	Decision functions for veterinarian			
Sample (Test) X	d_1	d_2	d_3	d_4
X = 1 (positive)	R	R	NR	NR
X = 0 (negative)	NR	R	NR	R

For example, d_1 is the decision function which decides that the animal is a reactor (R) if the test is positive and not a reactor (NR) otherwise. Now

$$R(d_1, \theta_1) = E[L(\theta_1, d_1(X))]$$
$$= L(\theta_1, d_1(1)) \cdot P(X = 1 \mid \theta_1) + L(\theta_1, d_1(0)) \cdot P(X = 0 \mid \theta_1)$$
$$= L(\theta_1, R) \cdot 0.9 + L(\theta_1, NR) \cdot 0.1$$
$$= 2 \cdot 0.9 + 6 \cdot 0.1 = 2.4.$$

Table 8.20 gives the values of the risk functions for the decision functions d_1, d_2, d_3 and d_4. Note in particular that d_4 is inadmissable (dominated by d_1). The minimax decision strategy is d_1.

Suppose that $p = P(\theta = \theta_1 = TB)$ is an appropriate indicator of the incidence of TB. Then

$$E_\theta[R(d_1, \theta)] = 1.6p + 0.8, \quad E_\theta[R(d_2, \theta)] = 4 - 2p, \quad \text{and} \quad E_\theta[R(d_3, \theta)] = 6p.$$

One may therefore show that if p is small ($p < 2/11$), then d_3 is the Bayes decision rule and hence one should (irrespective of the test) declare the animal to be free of TB. On the other hand, if $2/11 < p < 8/9$, then d_1 is the decision rule, while if $8/9 < p$ then the Bayes decision rule is d_2 (declare the animal to be a reactor irrespective of the test).

Hence the Bayesian strategy is clearly dependent on the value p which one assumes for the incidence of TB. Note also that the Bayesian strategy also clearly depends on the values in the (loss) payoff matrix, and different values in this matrix (representing more realistic losses?) could lead to different decisions. ▯

TABLE 8.20

		Risk functions for veterinarian			
		d_1	d_2	d_3	d_4
State of	θ_1 (TB)	2.4	2	6	5.6
animal	θ_2 (No TB)	0.8	4	0	3.2

Example 8.10

The IT section of a large company is to invest in a new computer system for the company's employees, and the decision will ultimately be one concerning the size of the system as follows: $a_1 \rightarrow$ large system, $a_2 \rightarrow$ medium system, or $a_3 \rightarrow$ small system. There is a question as to how the employees will take to the new system, and Table 8.21 gives a fair summary of the payoff (profit) to the company in dollars over the next year depending on the size of the system purchased. Amongst the three possible actions the IT section may take, a_3 (go with the small system) has the largest minimum gain (of 130,000), and hence is the minimax action (or pure strategy).

TABLE 8.21
IT Gains (profit) matrix.

		Size of system purchased		
Nature		large	medium	small
θ_1	(high acceptance)	410,000	310,000	210,000
θ_2	(low acceptance)	-30,000	50,000	130,000

Assume that prior information suggests that there is a 60% chance that the employees will find the new system acceptable. Then the expected profits ($\$000's$) for the actions a_1, a_2 and a_3 are 234, 206 and 178, respectively. Thus the Bayes action would be a_1 (that is, to go for the large system).

A local marketing company can be used to make a prediction concerning the acceptance of the system to employees. On the basis of an employee survey they will make a prediction X, where $X = 1$ indicates that the employees will generally be favorable to the system and $X = 0$ indicates that they will not. Table 8.22 (based on the past record of the marketing company) indicates the probability of a correct prediction as a function of the state of nature θ.

TABLE 8.22
Market prediction.

		Market prediction	
Nature		$P(X = 1)$	$P(X = 0)$
θ_1	(high acceptance)	0.8	0.2
θ_2	(low acceptance)	0.1	0.9

The nine possible decision functions which may be used (depending on the sample information X observed) are given in Table 8.23. The corresponding risk functions for each of these as well as the mean value of the risk function corresponding to the prior distribution $(P(\theta_1) = 0.6, P(\theta_2) = 0.4)$ are given in Table 8.24.

For example,

$$R(d_2, \theta_1) = L(\theta_1, a_1)P(X = 1 \mid \theta_1) + L(\theta_1, a_2,)P(X = 0 \mid \theta_1)$$
$$= 410(0.8) + 310(0.2) = 390$$

and

$$E_\theta[R(d_2, \theta)] = 390(0.6) + 42(0.4) = 250.8.$$

TABLE 8.23
IT decision functions.

Survey	IT decision functions								
	d_1	d_2	d_3	d_4	d_5	d_6	d_7	d_8	d_9
$X = 1$	a_1	a_1	a_1	a_2	a_2	a_2	a_3	a_3	a_3
$X = 0$	a_1	a_2	a_3	a_1	a_2	a_3	a_1	a_2	a_3

TABLE 8.24
IT risk function values ('000s).

Nature	$R(d_1, \theta)$	$R(d_2, \theta)$	$R(d_3, \theta)$	$R(d_4, \theta)$	$R(d_5, \theta)$
high θ_1	410	390	370	330	310
low θ_2	-30	42	114	-22	50
Bayes risk	234.0	250.8	267.6	189.2	206.0

Nature	$R(d_6, \theta)$	$R(d_7, \theta)$	$R(d_8, \theta)$	$R(d_9, \theta)$
high θ_1	290	250	230	210
low θ_2	122	-14	58	130
Bayes risk	222.8	144.4	161.2	178.0

One can see from Table 8.24 that the minimax strategy (the strategy which in this case maximizes the expected minimum payoff) is d_9, while the Bayes strategy (the one which maximizes the expected risk with respect to the prior on θ) is d_3. The minimax strategy is clearly conservative, going for the smallest system no matter what the marketing prediction is. The Bayes strategy goes for the large installation if the market company predicts a high acceptance rate amongst employees, and otherwise goes for the small system. □

8.4   Utility and expected monetary gain

Faced with making a decision in the face of uncertainty, one often selects the action or strategy which maximizes expected monetary gain. However, in many situations this does not yield the most satisfactory action.

Example 8.11

A decision maker is to decide between three courses of action $\{a_1, a_2, a_3\}$, and depending on the true state of nature θ, will gain an amount as indicated in payoff Table 8.25. The expected monetary gain for both a_1 and a_2 is 0, while that of a_3 is negative (in fact, action a_3 is clearly dominated by a_2). Using the criterion of maximizing expected utility, it follows that a_1 and a_2 are equally acceptable and both preferable to a_3. For most individuals however, a likely loss (with probability 0.5) of 500,000 would have very serious negative consequences, and therefore a_2 would definitely be preferable to a_1!   ▯

TABLE 8.25
Gains for actions in Example 8.11.

		Actions			
		a_1	a_2	a_3	$P(\theta)$
State of	θ_1	500,000	50	40	0.5
nature	θ_2	-500,000	-50	-50	0.5

Example 8.12

The payoffs for a decision problem with two possible options $\{a_1, a_2\}$ are given in Table 8.26. Both options have beneficial and high expected monetary gains. Using the criterion of maximizing expected monetary gains, one would decide on option a_2 over a_1, yet most people would in fact go for a_1. How does one explain this behavior?   ▯

TABLE 8.26
Gains for options in Example 8.12.

		Options		
		a_1	a_2	$P(\theta)$
State of	θ_1	3,000,000	5,000,000	0.5
nature	θ_2	1,000,000	0	0.5

Example 8.13

There are 5,245,786 = $\binom{42}{6}$ possible winning combinations in a version of the Irish National Lottery game LOTTO on any given night. Rules stipulate that one must make at least two selections on any given go, and this can be done for 2 euro. If the jackpot is 1,000,000 euro (and one ignores both smaller prizes and the fact that the jackpot might be shared when won), then Table 8.27 presents a simple analysis of the situation for someone wishing to play at the minimal level (paying 2 euro to do so). The expected monetary gain in playing is therefore $E(a_2) = -1.619$, which in particular is negative! So why do so many people play LOTTO?

TABLE 8.27
Gains for Irish National Lottery LOTTO.

		Options		
		Don't play	Play	
		a_1	a_2	$P(\theta)$
State of	θ_1 (win)	0	1,000,000 -2	2/5,245,786
nature	θ_2 (lose)	0	-2	1 - 2/5,245,786

▯

Example 8.14

A student has just purchased a new bike with a value of $400, and insurance cover for theft is available for a premium of $25. Chances that the bike will be stolen in the next year are estimated to be about 0.05. The decision has to be made whether or not to purchase insurance for the bike, and Table 8.28 gives a simple overview of the situation. Since $E(a_1) = -20 > -25 = E(a_2)$, the no insurance option maximizes expected monetary gain, yet many people buy bike and other types of insurance. Why is this the case?   ▯

TABLE 8.28
Gains table for bike insurance.

		Options		
		No insurance	Buy insurance	
		a_1	a_2	$P(\theta)$
State of	θ_1 (no theft)	0	-25	0.95
nature	θ_2 (theft)	-400	-25	0.05

Although these examples are simple, they do suggest that maximizing expected monetary gain may not always yield the best or most desirable action to take. The idea of a utility function is often used to provide an alternative criterion for deciding on an appropriate action in the face of uncertainty. The utility function for an individual (or group or company) will normally be related to the monetary value of a possible action, but it also may take into account other aspects of its value.

8.4.1   Rewards, prospects and utility

In any decision making situation, one is faced with a set of possible consequences – which may be termed rewards, payoffs, losses, etc. Let us label this set by $\mathcal{R}$. By a *prospect* we shall mean a probability distribution on the set of rewards $\mathcal{R}$, and we will use $\mathcal{M}$ to denote the set of all prospects. In particular, we may view actions or strategies as prospects. In Example 8.14, one could write $\mathcal{R} = \{-400, -25, 0\}$ and view the action of not buying insurance as the prospect that puts the probability distribution (0.05, 0, 0.95) on $\mathcal{R}$. Generally speaking, one assumes that in a decision making scenario the set of prospects $\mathcal{M}$ for the decision maker has the following properties:

- If $\mathcal{M}_1$ and $\mathcal{M}_2$ are two prospects, then either $\mathcal{M}_1$ is preferred to $\mathcal{M}_2$ (written $\mathcal{M}_1 > \mathcal{M}_2$), $\mathcal{M}_1 < \mathcal{M}_2$, or they are equally preferable ($\mathcal{M}_1 = \mathcal{M}_2$).

- $\mathcal{M}_1 \geq \mathcal{M}_2$ and $\mathcal{M}_2 \geq \mathcal{M}_3 \Rightarrow \mathcal{M}_1 \geq \mathcal{M}_3$.

- If $\mathcal{M}_1 > \mathcal{M}_2 > \mathcal{M}_3$, then there exist probabilities p and q such that $[\mathcal{M}_1, \mathcal{M}_3 : p, 1 - p] > \mathcal{M}_2$ and $\mathcal{M}_2 > [\mathcal{M}_1, \mathcal{M}_3 : q, 1 - q]$. By $[\mathcal{M}_1, \mathcal{M}_3 : p, 1 - p]$ we shall mean the prospect which is the $p : 1 - p$ mixture of $\mathcal{M}_1$ and $\mathcal{M}_2$.

- If $\mathcal{M}_1 > \mathcal{M}_2$ and $\mathcal{M}_3$ is another prospect, then $[\mathcal{M}_1, \mathcal{M}_3 : p, 1 - p] > [\mathcal{M}_2, \mathcal{M}_3 : p, 1 - p]$ when $0 < p < 1$.

It can be shown that when the set $\mathcal{M}$ of prospects satisfies the above properties, then there is a (real valued) utility function $u : \mathcal{M} \to R$ for the decision maker such that

- $\mathcal{M}_1 > \mathcal{M}_2 \Rightarrow u(\mathcal{M}_1) > u(\mathcal{M}_2)$

- If $\mathcal{M}^* = [\mathcal{M}_1, \mathcal{M}_2 : p, 1 - p]$ then $u(\mathcal{M}^*) = p\, u(\mathcal{M}_1) + (1 - p)\, u(\mathcal{M}_2)$

- If $\mathcal{M}^*$ is a prospect with mass or density function $p_{\mathcal{M}^*}$ on $\mathcal{R}$, then $u(\mathcal{M}^*) = \int u(r) p_{\mathcal{M}^*}(r) dr$.

Essentially, u assigns to each prospect (corresponding to a strategy or action) a real number in such a way that preferences for the individual are preserved. Hence the decision maker will select the strategy giving the highest utility. Note that the utility of a prospect can be viewed (and is often calculated) as an expected utility over the set of rewards, thus decisions about what action (or prospect) to take are made using the criterion of *maximizing expected utility*.

It should be clear that if one knows the utility function on the set of rewards $\mathcal{R}$, then one may determine it on any prospect with this knowledge. In Example 8.14, if $\mathcal{M}$ is the prospect on $\mathcal{R} = \{-400, -25, 0\}$ with probability distribution (p_1, p_2, p_3), then $u(\mathcal{M}) = p_1 u(-400) + p_2 u(-25) + p_3 u(0)$. If $a_1 = \mathcal{M}_1$ is the prospect or action of not buying insurance while $a_2 = \mathcal{M}_2$ is that of buying insurance, then to say that the decision maker prefers $\mathcal{M}_2$ to $\mathcal{M}_1$ means that for her utility function u,

$$
\begin{aligned}
u(\mathcal{M}_2) &= 0.95\, u(-25) + 0.05\, u(-25) \\
&= u(-25) \\
&> 0.95\, u(0) + 0.05\, u(-400) \\
&= u(\mathcal{M}_1).
\end{aligned}
$$

Determining an appropriate utility function for an individual can be done by assigning utility values to individual concrete rewards and then assessing the value on various prospects. The utility function for a decision maker is clearly not unique, for given a utility function u, the function $\alpha u + \beta$ for any constants $\alpha > 0$ and β will be as useful in decision making as u itself.

For most individuals the utility function is a concave function of monetary rewards, and this reflects the conservative nature of the majority of people. Such a person is called a *risk avoider* (and is risk averse), since he or she generally prefers the action with lower variability even when deciding between two prospects with the same expected monetary return. For example, suppose that $\mathcal{R} = \{10, 30, 40, 50\}$ and $\mathcal{M}_1$ and $\mathcal{M}_2$ are prospects on $\mathcal{R}$ with respective distributions $(1/3, 0, 0, 2/3)$ and $(0, 1/3, 2/3, 0)$. The expected monetary return on each is $110/3$, but the variation in the return on $\mathcal{M}_1$ is greater. For the risk avoider (with a strictly concave increasing utility function u), this would be reflected in the inequality

$$
\begin{aligned}
u(\mathcal{M}_1) &= (1/3)\, u(10) + (2/3)\, u(50) \\
&< (1/3)\, u(30) + (2/3)\, u(40) \\
&= u(\mathcal{M}_2).
\end{aligned}
$$

The risk avoider also has a decreasing marginal value for money (an additional 1000 is less enticing to the individual the wealthier she becomes). If the utility function for an individual is convex, then such a decision maker is called a *risk taker*. In some (rare) cases the utility function is linear in monetary value (and hence decision making reduces to selecting the strategy with the smallest expected monetary value).

8.4.2   Utility and insurance

The fact that most individuals are risk avoiders underlies the basis of insurance and the resulting premiums charged to purchase it. Suppose that an individual with wealth w is subject to a random loss X and is interested in insuring against this loss. What is the maximum amount P that such a person is willing to pay? The insurance company will find it necessary to charge an amount at least as large as $E(X)$ in order to make a profit, as well as to cover claims and overheads.

A classic inequality of Jensen says that if u is a concave function and Y is any random variable, then $E[u(Y)] \leq u[E(Y)]$ (and in fact the inequality is strict if u is strictly concave and Y is not constant). For a risk averse individual with utility function u, the amount P is the solution to $u(w - P) = E(u(w - X))$. Because of Jensen's inequality we have

$$u(w - P) = E(u(w - X)) \leq u(w - E(X)).$$

As u is an increasing function it follows that $w - P \leq w - E(X)$ and consequently that $P \geq E(X)$. This is an important conclusion, for it says that most individuals are willing to pay more in premiums than the expected value of the loss they want to be insured for!

A sufficient condition for a function u to be strictly concave is $u''(x) < 0$. An attractive risk averse (and strictly concave) utility function to work with (partially because of its mathematical tractability) is the exponential utility function of the form $u(x) = a - b e^{-cx}$ where a, b and c are constants ($b, c > 0$).

Example 8.15
An individual is considering insurance for a possible loss X which is exponentially distributed with mean 25. He will make his decision on the basis of expected utility where his utility function is given by $u(x) = 1 - e^{-0.01x}$. If w is his current wealth, then the maximum premium P he is willing to pay to be relieved of this loss is the solution to $E(u(w - P)) = E(u(w - X))$. Letting $M_X(t) = 0.04/(0.04 - t)$ be the moment generating function of X, then

$$E(u(w - X)) = 1 - E(e^{-0.01(w - X)})$$
$$= 1 - e^{-0.01\,w} E(e^{0.01X}) = 1 - e^{-0.01\,w} M_X(0.01).$$

Therefore it follows that

$$e^{0.01P} = M_X(0.01) = 0.04/(0.04 - 0.01) \text{ or } P = 100 \log(4/3) = 28.77.$$

Note both that $P > E(X) = 25$, and P is independent of the initial wealth w.   □

Example 8.16
Sarah is a decision maker with wealth of 50,000 who uses an exponential utility function of the form $u(x) = 1 - e^{-0.001x}$ for $x \in R$. She is liable for a loss X

which is an exponentially distributed random variable with mean 400. What is the maximum amount of money she is willing to pay to be relieved of 75% of this loss?

Let us denote this value (premium) by $P_{0.75}$. Finding this value on the basis of expected utility, we want to solve

$$E[u(50,000 - P_{0.75} - 0.25X)] = E[u(50,000 - X)].$$

Hence

$$1 - e^{-0.001(50,000)} e^{0.001P_{0.75}} E[u(e^{0.00025X})] = 1 - e^{-0.001(50,000)} E[u(e^{0.001X})]$$

$$\text{or} \quad e^{0.001P_{0.75}} M_X(0.00025) = M_X(0.001)$$

$$\Rightarrow \quad P_{0.75} = 1000 \left(\log \frac{M_X(0.001)}{M_X(0.00025)} \right) = 405.47.$$

$\Box$

Generally speaking, when a risk averse individual is subject to a loss, she is willing to pay more than the expected value of that loss to be insured for it. In some cases, the individual cannot afford (or does not want) to pay to be insured for the complete loss X, but is willing to spend an amount P to be insured for part of it. Standard types of insurance that cover only part of a loss X are proportional insurance and excess of loss (or stop-loss) insurance.

More generally, an insurance company might cover the amount $I(X)$ of any loss X incurred by the policyholder where $0 \le I(X) \le X$. In proportional insurance, there is a proportion α (where $0 \le \alpha \le 1$) such that $I(X) = (1 - \alpha)X$ for any loss X. Here the insurance company pays $(1 - \alpha)\%$ of any claim X and the insured pays the rest. In excess of loss (or stop-loss) insurance there exists a value d such that $I(X) = 0$ if $X < d$, and otherwise $I(X) = X - d$ which is the excess over d. In this case the loss payment for the insured does not exceed (or is *stopped* at) d. The insurance company in effect is treating d as a deductible!

One may naturally ask if there is an optimal type of insurance which one may purchase for a given amount of money (or premium) P? A classic result of Arrow [2] shows that under certain restrictions the best type of insurance that a risk averse individual can purchase is an excess of loss (or stop-loss) insurance policy.

THEOREM 8.1 Arrow's theorem on optimal insurance

A risk averse individual with utility function u (where $u'' < 0$) is subject to a random loss X, and is willing to spend an amount P to be relieved of part of this loss. Various types of insurance $I(X)$, all of which have the same expected cost $C_P = E[I(X)]$ for the insurer, are available for purchase at a price of P. Then the type of policy maximizing expected utility for the insured *individual is the excess of loss (or stop-loss policy) $I_{d^*}(X)$ where*

$$I_{d^*}(X) = \begin{cases} 0 & \text{if } X < d^* \\ X - d^* & \text{if } X \geq d^* \end{cases}$$

and d^* is the solution to $C_P = \int_d^{+\infty} (x - d) f_X(x) \, dx$.

Arrow's theorem is important and intuitively appealing, but it has its limitations for practical use. Firstly, it is not realistic to assume that there will be many different types of insurance on offer to a person, all of which have the same expected cost C_P and fixed premium P. Secondly, in Arrow's theorem the amount P to be spent on insurance is fixed, while in practice one is more likely to ask how much should I spend (subject to a given maximum) on insurance for a potential loss X? In the following, a sketch of a proof of Arrow's theorem is given.

PROOF Arrow's theorem on optimal insurance

Since the risk averse utility function u is strictly concave and $u'' < 0$, the first derivative u' of u is decreasing! Strictly concave functions may be viewed as functions that are in some sense *turning downwards* (even though they may be increasing). Furthermore, it is not difficult to show (and graphically illustrate) that u has the property that for any y and z,

$$u(y) - u(z) \leq (y - z) \, u'(z). \tag{8.1}$$

We use $I_{d^*}(X)$ to indicate the stop-loss insurance and $I(X)$ another on offer where $E[I(X)] = E[I_{d^*}(X)] = C_P$. We want to show that

$$E[u(w - X + I(X) - P)] - E[u(w - X + I_d^*(X) - P)] \leq 0. \tag{8.2}$$

Letting

$$y = w - x + I(x) - P \quad \text{and} \quad z = w - x + I_{d^*}(x) - P,$$

it follows from Equation (8.1) that

$$u(w - x + I(x) - P) - u(w - x + I_{d^*}(x) - P)$$
$$\leq \quad [I(x) - I_d^*(x)] \, u'(w - x + I_{d^*}(x) - P). \tag{8.3}$$

We next show that

$$[I(x) - I_{d^*}(x)] \, u'(w - x + I_{d^*}(x) - P) \leq [I(x) - I_{d^*}(x)] \, u'(w - d^* - P) \tag{8.4}$$

by considering the various possibilities for (the sign) of $[I(x) - I_{d^*}(x)]$.

- If $I(x) - I_{d^*}(x) = 0$, then clearly (8.4) is true since both sides are 0.

- If $I(x) - I_{d^*}(x) < 0$, then in particular, $I_{d^*}(x) > 0$ and $I_{d^*}(x) = x - d^*$. Therefore $-x + I_{d^*}(x) = -d^*$, and again both sides of (8.4) are equal.

- Assume now that $I(x) - I_{d^*}(x) > 0$. It is always true that $I_{d^*}(x) \geq x - d^*$, which implies that $-d^* \leq -x + I_{d^*}(x)$. Since u' is decreasing it follows that $u'(w - x + I_{d^*}(x) - P) \leq u'(w - d^* - P)$. Therefore using that $I(x) - I_{d^*}(x) > 0$, the inequality (8.4) is valid.

Now using both (8.3) and (8.4), it follows that

$$u(w - x + I(x) - P) - u(w - x + I_{d^*}(x) - P) \leq [I(x) - I_{d^*}(x)]\, u'(w - d^* - P).$$

Taking expectations with respect to X and remembering that (since our assumption is that all available insurance contracts have the same expected cost for the insurer) $E[I(X) - I_{d^*}(X)] = 0$, it follows that (8.2) holds, thus proving Arrow's theorem. ■

8.5 Problems

1. Which of the following decision making situations could be described as games?

   (a) A poor pensioner has received a surprise gift of \$50 which he should probably save for the forthcoming winter. However, he still dreams of that elusive holiday in Spain which he never got earlier on, and he is tempted to spend it on the national lottery game LOTTO. What would you recommend?

   (b) A young student walking home very late at night can take a short-cut through a (poorly lit) city park which would save at least 15 minutes in her homeward journey. It is raining and no one else seems to be around, so should she take the shorter route (with the additional risk of traversing this dark walkway), or should she decide to take the route she normally does?

   (c) Local farmers in a community meet to determine annual output of various crops and who should plant what.

   (d) The minister of justice in a country is considering cutting back on overtime hours for prison employees.

2. For the following payoff matrix $\mathcal{P}_{12}$ of Example 8.2 (continued) for Alison and Brian, the value of the game is $7/3$ and the optimal strategy for Brian is to randomly select b_1 with probability $q_1 = 1/3$ and b_2 with probability $q_2 = 2/3$. What is the optimal strategy for Alison?

		Brian	
		b_1	b_2
Alison	a_2	1	3
	a_3	5	1

$$\mathcal{P}_{12}$$

3. Deliveries of cigarettes are made at 10 AM to two warehouses (W_1 and W_2) each Monday morning, and a local thief knows there is an opportunity for a robbery in either place for a few minutes after a delivery. The cigarette company has employed one security agent, but clearly he can only monitor one delivery at a time, and the two warehouses are a good distance apart. The weekly delivery to W_1 is worth 100,000, while a smaller delivery worth 25,000 is made to W_2. If the thief arrives at a warehouse around 10 AM when the security agent is present, he can escape with no loss, but has insufficient time to get to the other warehouse. On the other hand, if he arrives at one warehouse and the agent is covering the other he has the ability to rob the delivery. Which warehouse is the thief more likely to target, and with what probability? On the average, what is the value of the cigarettes stolen? What is the optimal strategy for the security agent?

		Thief	
		W_1	W_2
Security	W_1	0	25
agent	W_2	100	0

$$\mathcal{P}_{Thief}$$

4. Edward is playing a game (against Francine) with the following loss matrix. Both Edward and Francine are using the minimax criterion for deciding on a *pure* strategy. Under what conditions on X and Y does the matrix have a saddle point? If $Y = 11$ and $X = 7$ what is the value of the game, and which are the optimal strategies for Edward and Francine?

		Edward		
		I	II	III
	1	4	5	Y
Francine	2	X	9	8
	3	6	2	10

5. Katherine and Daniel (intelligent beings) are playing a zero-sum game with the following payoff matrix. Here a positive payoff represents a loss to Katherine and a gain to Daniel. Assuming they both choose their

strategies using the minimax criterion, show that the value of the game is 5.

Player		Katherine		
		I	II	III
	i	0	2	1
Daniel	ii	4	1	3
	iii	5	7	5

6. The following payoff matrix represents a zero-sum game between two intelligent players, Bertie and Mary. An entry in this payoff matrix represents a *loss* to Bertie (or equivalently, a gain to Mary). Assuming players choose their strategy with the objective of minimizing their expected maximum loss, find Bertie's optimum strategy. If Bertie will receive 4 if he participates in the above game, and he uses the strategy found above, should he play? What is Mary's optimal strategy? Find the value of the game.

		Bertie		
		I	II	III
	1	2	3	4
Mary	2	3	4	5
	3	7	-1	8

7. Consider the following loss (payoff) matrix for Ann when playing against Bob in a zero-sum game. For what values of X and Y does the payoff matrix have a saddle point? If $X = 7$ and $Y = 1$, what is the value of the game when minimax strategies are used?

		Ann			
		I	II	III	IV
	1	4	5	7	Y
Bob	2	11	X	8	13
	3	6	2	10	7

Consider the above game now where $X = 9$ and $Y = 6$, that is, the loss matrix for Ann is given by

		Ann			
		I	II	III	IV
	1	4	5	7	6
Bob	2	11	9	8	13
	3	6	2	10	7

Determine the minimax randomized strategy for Ann and the value of the game.

8. Consider the following *loss (payoff) matrix* for player B when playing against opponent A in a zero-sum game.

		Player A		
		a_1	a_2	a_3
	b_1	- 2	1	6
Player	b_2	0	4	1
B	b_3	3	5	0
	b_4	6	5	2

Find the optimal minimax decision strategies for A and B, and determine the value of the game.

9. Richie is playing a game against Mort, and they are fierce competitors. The following matrix is a gains matrix for Mort in a zero-sum game with Richie, and both players are using the minimax criterion for deciding on a strategy. Find the optimal (randomized) strategies for Richie and Mort, and determine the value of the game. Do you think Richie would play the game if offered 5 to do so?

		Richie			
		I	II	III	IV
	1	1	3	6	2
Mort	2	-3	0	4	9
	3	5	7	7	1

10. The following payoff matrix is for a zero-sum game between two intelligent players, Andrew and Barbara. Here a positive payoff represents a loss to Andrew and a gain to Barbara. Assuming both players choose their strategies using the minimax criterion, find the strategy Barbara will use and the value of the game.

		Andrew		
		I	II	III
	1	8	-5	9
Barbara	2	2	6	7

11. The payoff matrix for friends Albert and Bill (both avid sports fans) is that below, with respect to selecting what to do in the case of conflicting rugby and football matches. Albert is a little keener on sports in general than Bill. What are the Nash equilibria for this game, and what is the optimal strategy?

		Bill	
		Rugby	Football
Albert	Rugby	(25, 20)	(10, 12)
	Football	(12, 10)	(20, 15)

Payoffs for Albert and Bill: (Gain A, Gain B)

12. The management of a new sports club is considering how many members should be allowed to join. It must consider variable costs over which it has little or no control, yet which will influence the profit (euro) made per member in the club on an annual basis. Assume that there are three levels of these variable costs that may arise in the near future, which may be classified as: θ_1 (High), θ_2 (Normal or most likely) or θ_3 (Low). One must decide on the following levels of membership for the club: $a_1 = 1500$, $a_2 = 1400$ or $a_3 = 1200$. Table 8.29 gives the profit per member which is expected depending on the level of variable costs and the number of members.

Assuming a prior distribution on the variable cost levels of the form $(P(\theta_1) = 0.15, P(\theta_2) = 0.55, P(\theta_3) = 0.30)$, what is the Bayes criterion action for the number of members to allow? What is the minimax action on the membership for the club?

TABLE 8.29
Sports club membership.

		Variable costs		
		θ_1	θ_2	θ_3
Membership	a_1	79	88	105
level	a_2	82	99	110
	a_3	87	101	114

13. The probability of success p in a binomial experiment is known to be either 0 or 1/3. A statistician must decide on the value of p. To aid the decision, she can observe the outcome (denoted by X) of 1 Bernoulli trial with probability of success p. The statistician has four possible decision functions d_1, d_2, d_3 and d_4, where d_3 and d_4 are defined by:

$$d_3(0) = 0, \quad d_4(0) = 1/3, \quad d_3(1) = 0, \quad d_4(1) = 1/3.$$

The statistician will suffer a loss of 1 if she makes an incorrect decision, and otherwise no loss is suffered. Define d_1 and d_2, the other possible decision functions. Table 8.30 gives the value of the risk function $R[d_i, p]$ (for d_3 and d_4) for each possible true parameter value p. Before

TABLE 8.30

Binomial experiment.

Value of p	d_1	d_2	d_3	d_4
		Risk functions		
0			0	1
1/3			1	0

observing X, the statistician felt that 0 and 1/3 were equally likely to be the value of p. Complete the rest of this table. Given this information, what decision function should the statistician use according to the Bayes criterion?

14. Tod (a dog) is a 12-year-old labrador who likes his creature comforts. Every night Tod has the choice of two chairs to sleep on: an old green one in the office, or a brand new exclusive "parker knoll" ("PK") armchair in the living room. However, Tod gets punished if grandpa catches him sleeping on either chair in the morning. Only if grandpa chooses the wrong room to enter first, does Tod have the time to climb down from the chair before grandpa sees him. Estimates (in units) of Tod's pleasure or utility (assume Tod's pleasure = Grandpa's displeasure) are given in the following table:

Pleasure	Utility
Sleeping on PK	12
Caught on PK	-10
Sleeping on green	3
Caught on green	-4

What strategy should Tod and grandpa use according to the minimax criterion? Tod has utility 0 if he sleeps in his basket. Should he sleep in his basket?

15. A statistician is to observe the results (number of successes) in two Bernoulli trials. He knows that the success probability p is either 1/4 or 1/2, and he is trying to decide between these two values. Let X be his observation. Although there are many possible decision rules, he proposes to use one of the following four possibilities:

Sample X	d_1	d_2	d_3	d_4
		Decision functions		
$X = 0$	$p = 1/4$	$p = 1/4$	$p = 1/4$	$p = 1/2$
$X = 1$	$p = 1/2$	$p = 1/4$	$p = 1/4$	$p = 1/2$
$X = 2$	$p = 1/2$	$p = 1/2$	$p = 1/4$	$p = 1/2$

If he incorrectly concludes that $p = 1/4$, he suffers a loss of 1. If he incorrectly concludes that $p = 1/2$, he suffers a loss of 2. Find the risk function for each of these decision functions, and the decision function the statistician would use if he follows the minimax criterion. The statistician has a "prior" feeling that p is equally likely to be 0.25 or 0.50. What decision function would he choose when using the Bayes criterion?

16. A reinsurer decides to use a continuous uniform distribution on the interval $(0, \theta)$ to model claim size X. She wishes to estimate θ on the basis of a single observation of X and using a decision function of the form $d(X) = kX$. If the loss incurred is proportional to the absolute value of the error, find the value of k which minimizes the (expected) risk.

17. A game between a statistician and nature is set up, where two independent observations, X and Y from a uniform distribution on $[0, 1]$ are sampled. Consider X and Y in such a way that they split a unit line segment into three pieces. The statistician must decide whether a triangle can be formed from the pieces resulting. He suffers a loss of 2 if he incorrectly concludes a triangle will not be formed, and a loss of 1 if he incorrectly concludes a triangle will be formed. No loss is incurred if the statistician makes the correct decision.

Assume the statistician is forbidden to observe X or Y before making his decision. If he follows a Bayesian strategy, what will he decide?

18. The IT section of a large company is to invest in a new computer system for the company's employees, and the decision will ultimately be one concerning the size of the system as follows: $a_1 \rightarrow$ large system, $a_2 \longrightarrow$ medium system or $a_3 \longrightarrow$ small system. There is a question as to how the employees will take to the new system, and Table 8.31 gives a fair summary of the payoff (profit) to the company in thousands of euro over the next year depending on the size of the system purchased.

TABLE 8.31
IT company payoff.

Nature	Action		
	a_1	a_2	a_3
High acceptance θ_2	550	430	310
Low acceptance θ_1	−80	30	160

(a) What is the minimax action (a_1, a_2 or a_3) for the size of the computer system to be purchased?

(b) Prior information suggests that there is a 60% chance that the employees will find the new system highly acceptable. If this is the case, what is the Bayes decision strategy?

(c) A local marketing company can be used to make a prediction concerning the acceptance of the system to employees. On the basis of an employee survey, they will make a prediction X where $X = 1$ indicates that the employees will generally be highly favorable to the system and $X = 0$ indicates that they will not. Table 8.32 (which is based on the marketing company's past record) indicates the probability of a correct prediction depending on the state of nature θ, and Table 8.33 gives the possible decision functions. What is the Bayes decision rule for the company?

TABLE 8.32
Marketing prediction probabilities.

Nature	$P(X = 1)$	$P(X = 0)$
High acceptance θ_2	0.8	0.2
Low acceptance θ_1	0.3	0.7

TABLE 8.33
Decision functions for IT company.

	Possible decision functions								
Survey result	d_1	d_2	d_3	d_4	d_5	d_6	d_7	d_8	d_9
$X = 1$	a_1	a_1	a_1	a_2	a_2	a_2	a_3	a_3	a_3
$X = 0$	a_1	a_2	a_3	a_1	a_2	a_3	a_1	a_2	a_3

19. A statistician is observing values from a $B(3, p)$ distribution, and she knows that p is equal either to $1/3, 1/2$ or $2/3$. One observation $X \sim B(3, p)$ is to be made, and the two decision functions given in Table 8.34 are under consideration. Assume that the loss function when estimating p by p^* is $|p^* - p|$. Determine the risk function for each of these decision functions. Using both the minimax and Bayes criteria (assume each of the three possible values for p are initally equally likely) for decision making, find which of the two decision functions d_1 and d_2 is to be preferred.

20. Gavin is subject to a random loss X which is exponential with mean \$400, and he has a utility function of the form $u_I(x) = 10 - e^{-0.002x}$. The Ajax insurance company uses a utility function of the form $u_C(x) =$

TABLE 8.34

Decision functions for p.

	d_1	d_2
X = 0	p=1/3	p =1/3
1	p=1/3	p =1/3
2	p=1/2	p =2/3
3	p=2/3	p =2/3

$1 - e^{-0.001x}$, and it is willing to consider insuring Gavin for half of his loss X. Suppose that the premium P which is ultimately paid by Gavin is the mean of the maximum he would be willing to pay to be relieved of half of this loss and the minimum the Ajax company would be willing to accept to cover half of his loss. What is this premium P?

21. Elaine with initial wealth a_0 and utility function $u(x) = 1 - e^{-0.005x}$ is considering insurance for two independent random losses X_1 and X_2 which she will incur in the coming year. X_1 is exponentially distributed with parameter $\lambda_1 = 0.01$ and X_2 is a gamma random variable with parameters $n = 2$ and $\lambda_2 = 0.02$ ($X_2 \sim \Gamma(2, 0.02)$). An insurance company is willing to insure Elaine for one or both of these losses. The decisions are to be made on the basis of expected utility. Determine the values G_1, G_2 and G_{1+2} which represent the maximum Elaine would be willing to spend to be insured for losses X_1, X_2 and X_1+X_2, respectively.

References

[1] Agresti, A., *Categorical Data Analysis*, John Wiley and Sons, New York, 1990.

[2] Arrow, K.J., Uncertainty and the welfare of medical care, *Amer. Econ. Rev.*, 55, 941, 1963.

[3] Beard, R.E., Pentikäinen, T., and Pesonen, M., *Risk Theory*, 3rd ed., Chapman and Hall, London, 1984.

[4] Bliss, C.I., The calculation of the dosage-mortality curve, *Ann. Appl. Biol.*, 22, 134, 1935.

[5] Bonsdorff, H., On the convergence rate of bonus malus systems, *ASTIN Bull.*, 22, 217, 1992.

[6] Borch, K., Application of game theory to some problems in automobile insurance, *ASTIN Bull.*, 2, 208, 1962.

[7] Bornhuetter, R.L., and Ferguson, R.E., The actuary and IBNR, *Proc. Cas. Act. Soc.*, 59, 181, 1972.

[8] Box, G., Robustness in the strategy of scientific model building, in *Robustness in Statistics*, Launer. R.L., and Wilkerson, G.N., Eds., Academic Press, New York, 1979, 201.

[9] Bowers, N.L., et al., *Risk Theory*, Society of Actuaries, Schaumburg, IL, 1986.

[10] Brockman, M.J., and Wright, T.S., Statistical motor rating: making effective use of your data, *Journ. Inst. Act.*, 119, 457, 1992.

[11] Bühlmann, H., Experience rating and crediblity, *ASTIN Bull.*, 4, 119, 1967.

[12] Bühlmann, H., and Straub, E., Glaubgewürdigkeit für Schadensätze, Bull. Swiss Assoc. Act., 70, 111, 1970, translated as *Crediblity for Loss Ratios* (by C.E. Brooks), ARCH, 1972.

[13] Chernoff, H., and Lehmann, E.L., The use of maximum likelihood estimates in χ^2 tests for goodness-of-fit, *Ann. Math. Stat.*, 25, 579, 1954.

[14] Chernoff, H., and Moses, L., *Elementary Decison Theory*, John Wiley and Sons, New York, 1959.

[15] Colgan, G., An examination of no claim discount systems in motor insurance using S-Plus, (unpublished) M.A. thesis, University College Dublin, Dublin, 1998.

[16] Conover, W.J., *Practical Nonparametric Statistics*, John Wiley and Sons, New York, 189, 1980.

[17] Coutts, S.M., Motor insurance rating: an actuarial approach, *Journ. Inst. Act.*, 111, 87, 1984.

[18] Cramér, H., On the mathematical theory of risk. *Festskrift Skand.*, Stockholm, Sweden, 1855, 1930.

[19] Daykin, C.D., Pentikäinen, T., and Pesonen, M., *Practical Risk Theory for Actuaries*. Monographs on Statistics and Applied Probability, Applied Probability, Chapman and Hall, London, 53, 1996.

[20] Dickson, D., *Insurance Risk and Ruin*, Cambridge University Press, Cambridge, 2005.

[21] Dickson, D., and Waters, H., *Ruin Theory*, Notes for the Institute and Faculty of Actuaries, 1992.

[22] Dobson, A., *An Introduction to Generalized Linear Models*, 2nd ed., Chapman & Hall: Texts in Statistical Science, 2001.

[23] Everitt, B., *The Cambridge Dictionary of Statistics*, 2nd ed., Cambridge University Press, Cambridge, 2005.

[24] Feller, W., *An Introduction to Probability Theory and Its Applications*, Vol. 1, Wiley, New York, 1957.

[25] Ferguson, J., The BAFS degree – A study of the years 1991 – 1997, (unpublished) M.A. thesis, University College Dublin, Dublin, 2001.

[26] Firth, D., Generalized linear models, in *Statistical Theory and Modelling*, Hinkley, D.V., Reid, N., and Snell, E.J. Eds., Chapman and Hall, London, 1991, chap. 3.

[27] Haberman, S., and Renshaw, A., Generalized linear models and actuarial science, *The Statistician*, 45, 4, 407, 1996.

[28] Herzog, T.N., *Introduction to Credibility Theory*, ACTEX Publications, Winsted, CT, 1996.

[29] Hossack, I.B., Pollard, J.H., and Zehnwirth, B., *Introductory Statistics with Applications in General Insurance*, 2nd ed., Cambridge University Press, Cambridge, 1999.

[30] Jewell, W.S., Credible means are exact Bayesian for simple exponential families, *ASTIN Bull.*, 8, 77, 1974.

[31] Kendall, M.G., and Stuart, A., *The Advanced Theory of Statistics, Inference and Relationship* Volume 2, 4th ed., Oxford University Press, New York, 30, 1979.

[32] Klugman, S.A., Panjer, H.H., and Willmot, G.E., *Loss Models – From Data to Decisions*, John Wiley and Sons, New York, 1998.

[33] Larson, H.J., *Introduction to Probability Theory and Statistical Inference*, 3rd edition, John Wiley and Sons, New York, 1982.

[34] Lemaire, J., An application of game theory: cost allocation, *ASTIN Bull.*, 14, 61, 1984.

[35] Lemaire, J., *Automobile Insurance: Actuarial Models*, Kluwer, Boston, 1985.

[36] Lemaire, J., Cooperative game theory and its insurance applications, *ASTIN Bull.*, 21, 17, 1991.

[37] Lemaire, J., *Bonus-Malus systems in Automobile Insurance*, Kluwer, Boston, 1995.

[38] Longley-Cook, L.H., An introduction to credibility theory, *Proc. Cas. Act. Soc.*, 49, 194, 1962.

[39] Luce, R.D., and Raiffa, H., *Games and Decisions*, John Wiley and Sons, New York, 1957.

[40] Lundberg, O., Über die Theorie der Ruckversicherung, in *Transactions of the First International Congress of Actuaries*, 2, 877, 1909.

[41] McCullagh, P., and Nelder, J.A., *Generalized Linear Models*, 2nd ed., Monographs on Statistics and Applied Probability 37, Chapman and Hall, London, 1989.

[42] Moore, D.S., Tests of chi-squared type, in *Goodness-of-Fit Techniques*, D'Agostino, R.B., and Stevens, M.A., Eds., Marcel Dekker, New York, 63, 1986.

[43] Nash, J.F., Equilibrium points in N-person games, *Proc. Nat. Acad. Sci.*, 36, 48, 1950.

[44] Nash, J.F., The bargaining problem, *Econometrica*, 18, 155, 1950.

[45] Osborne, M.J., and Rubenstein, A., *A Course in Game Theory*, MIT Press, Cambridge, MA, 1994.

[46] Owen, G., *Game Theory*, W.B. Saunders Company, Philadelphia, PA, 1968.

[47] Panjer, H.H., Recursive evaluation of a family of compound distributions, *ASTIN Bull.*, 12, 21, 1981.

[48] Panjer, H.H., and Willmot, G.E., *Insurance Risk Models*, USA Society of Actuaries, Schaumburg, IL, 1992.

[49] Pawitan, Y., *In All Likelihood: Statistical Modelling and Inference Using Likelihood*, Oxford Science Publications, Oxford, 2001.

[50] Pollack, S., Reinsurance is an easy game, *The Actuary*, London, April 24, 2006.

[51] Renn, D., *Life, Death and Money – Actuaries and the Creation of Financial Security*, Blackwell Publishers, Oxford, 1998.

[52] Renshaw, A.E., and Verrall, R.J., A stochastic model underlying the chain-ladder technique, *Brit. Act. Journ.*, 4, 903, 1998.

[53] Ross, S., *Stochastic Processes*, 2nd ed., Wiley, New York, 1996.

[54] Seal, H.L., Risk theory, in *Encyclopedia of Statistical Sciences*, Kotz, S. and Johnson, N., Eds., John Wiley and Sons, Toronto, 8, 152, 1982.

[55] Straub, E., *Non Life Mathematics*, Lecture Notes, Univeristy of Berne, 1979.

[56] Sundt, B., *An Introduction to Non-Life Insurance Mathematics*, Veröoffentlichungen des Institus für Versicherungswissenschaft der Universität Mannheim, Ban 28, WWW Karlsruhe, 1984.

[57] Venables, W.N., and Ripley, B.D., *Modern Applied Statistics with S*, 4th ed., Springer, New York, 2002.

[58] von Neumann, J., and Morgenstern, O., *The Theory of Games and Economic Behavior*, Princeton University Press, Princeton, 1944.

[59] Waters, H.R., *Credibility Theory*, Heriot-Watt University, Edinburgh, 1993.

[60] Whitney, A., The theory of experience rating, *Proc. Cas. Act. Soc.*, 4, 274, 1918.

Appendix A

Basic Probability Distributions

In the following list of basic probability distributions for a random variable X, the density or mass function will be denoted by $f_X(x)$. The cumulative distribution function is $F_X(x) = P(X \leq x)$, and the survival function is $\bar{F}_X(x) = 1 - F_X(x)$. The moment generating function will be denoted by $M_X(t)$, and the mean and variance are given by $E(X)$ and $Var(X)$, respectively.

- **Binomial**: $X \sim B(n, p)$. X is the number of successes (with success probability $p = 1 - q$) in a sequence of n Bernoulli trials.

$$f_X(x) = \binom{n}{x} p^x q^{n-x}, \quad x = 0, 1, \ldots, n.$$
$$M_X(t) = \left(q + pe^t\right)^n$$
$$E(X) = np, \quad Var(X) = npq$$

- **Poisson**: $X \sim$ Poisson (λ). X is the number of events occurring where the rate parameter for events is λ.

$$f_X(x) = \frac{\lambda^x}{x!} e^{-\lambda}, \quad x = 0, 1, \ldots, .$$
$$M_X(t) = e^{\lambda(e^t - 1)}$$
$$E(X) = \lambda, \quad Var(X) = \lambda$$

- **Negative binomial**: $X \sim NB(k, p)$. X is the number of failures (with failure probability q, or success probability $p = 1 - q$) until the kth success in a sequence of Bernoulli trials.

$$f_X(x) = \binom{x + k - 1}{k - 1} p^k q^x, \quad x = 0, 1, \ldots, .$$
$$M_X(t) = \left(\frac{p}{1 - q e^t}\right)^k$$
$$E(X) = kq/p, \quad Var(X) = kq/p^2$$

- **Normal:** $X \sim N(\mu, \sigma^2)$.

$$f_X(x) = \frac{1}{\sigma \sqrt{2\pi}} exp\left\{ -\frac{1}{2}\left(\frac{x-\mu}{\sigma}\right)^2 \right\}, \quad \text{for } -\infty < x < \infty$$

$$M_X(t) = exp\left\{ \mu t + \frac{1}{2}\sigma^2 t^2 \right\}$$

$$E(X) = \mu, \quad Var(X) = \sigma^2$$

- **Gamma:** $X \sim \Gamma(\alpha, \lambda)$. When α is a positive integer, X may be viewed as the time to the αth event in a homogeneous Poisson process with rate parameter λ.

$$f_X(x) = \frac{\lambda^\alpha}{\Gamma(\alpha)} x^{\alpha-1} e^{-\lambda x}, \quad \text{for } x > 0$$

$$M_X(t) = \left(\frac{\lambda}{\lambda - t}\right)^\alpha \quad \text{for } t < \lambda$$

$$E(X) = \alpha/\lambda, \quad Var(X) = \alpha/\lambda^2$$

- **Exponential:** $X \sim \Gamma(1, \lambda)$. X may be viewed as the time to the first event in a homogeneous Poisson process with rate parameter λ.

$$f_X(x) = \lambda e^{-\lambda x}, \quad \text{for } x > 0$$

$$M_X(t) = \frac{\lambda}{\lambda - t} \quad \text{for } t < \lambda$$

$$E(X) = 1/\lambda, \quad Var(X) = 1/\lambda^2$$

- **Chi-square:** $X \sim \Gamma(k/2, 1/2)$. X is said to have the chi-square distribution with k degrees of freedom.

$$M_X(t) = \left(\frac{1/2}{1/2 - t}\right)^{k/2} = \left(\frac{1}{1 - 2t}\right)^{k/2} \quad \text{for } t < 1/2$$

$$E(X) = k, \quad Var(X) = 2k$$

- **Pareto:** $X \sim \text{Pareto}\,(\alpha, \lambda)$.

$$f_X(x) = \frac{\alpha \lambda^\alpha}{(\lambda + x)^{\alpha+1}} \quad \text{and} \quad \bar{F}_X(x) = \left(\frac{\lambda}{\lambda + x}\right)^\alpha \quad \text{for } x > 0$$

$$E(X) = \lambda/(\alpha - 1), \quad \text{for } \alpha > 1$$

$$Var(X) = \alpha \lambda^2 / \{(\alpha - 1)^2 (\alpha - 2)\}, \quad \text{for } \alpha > 2$$

- **Generalized Pareto:** $X \sim \text{Generalized Pareto}\,(\alpha, \lambda, k)$.

$$f_X(x) = \frac{\Gamma(\alpha + k)\,\lambda^\alpha\, x^{k-1}}{\Gamma(\alpha)\Gamma(k)\,(\lambda + x)^{k+\alpha}}, \quad \text{for } x > 0$$

$$E(X) = \lambda k/(\alpha - 1), \quad Var(X) = \lambda^2 k(k + \alpha - 1)/\{(\alpha - 1)^2(\alpha - 2)\}$$

- **Weibull**: $X \sim W(c, \gamma)$.
$$f_X(x) = c\gamma x^{\gamma-1} e^{-cx^{\gamma}} \text{ and } \bar{F}_X(x) = e^{-cx^{\gamma}}, \text{ for } x > 0$$
$$E(X) = \frac{1}{c^{1/\gamma}} \Gamma\left(1 + \frac{1}{\gamma}\right)$$
$$Var(X) = \frac{1}{c^{2/\gamma}} \Gamma\left(1 + \frac{2}{\gamma}\right) - \left[\frac{1}{c^{1/\gamma}} \Gamma\left(1 + \frac{1}{\gamma}\right)\right]^2$$

- **Lognormal**: X is lognormal if the log of X is normal. That is, $Y = \log X \sim N(\mu, \sigma^2)$ for some μ and $\sigma > 0$.
$$f_X(x) = \left[\frac{1}{\sqrt{2\pi}\sigma} e^{-(\log x - \mu)^2/2\sigma^2}\right] \frac{1}{x}, \text{ for } x > 0$$
$$E(X) = e^{\mu + \sigma^2/2},$$
$$Var(X) = e^{2\mu + \sigma^2}\left[e^{\sigma^2} - 1\right]$$

- **Uniform [a,b]**: $X \sim U[a, b]$. X is uniform on the interval $[a, b]$ if it is equally likely to take any value in that interval.
$$f_X(x) = \frac{1}{b - a} \text{ and } F_X(x) = \frac{x - a}{b - a}, \text{ for } a < x < b$$
$$M_X(t) = \frac{e^{bt} - e^{at}}{(b - a)t}$$
$$E(X) = (a + b)/2, \quad Var(X) = (b - a)^2/12$$

- **Beta**(α, β) : $X \sim Beta(\alpha, \beta)$.
$$f_X(x) = \frac{\Gamma(\alpha + \beta)}{\Gamma(\alpha)\Gamma(\beta)} x^{\alpha-1}(1 - x)^{\beta-1}, \text{ for } 0 < x < 1$$
$$E(X) = \frac{\alpha}{\alpha + \beta}, \quad Var(X) = \frac{\alpha\beta}{(\alpha + \beta)^2(\alpha + \beta + 1)}$$

- **Burr**$(\alpha, \lambda, \gamma)$: $X \sim Burr(\alpha, \lambda, \gamma)$.
$$f_X(x) = \alpha\gamma\lambda^{\alpha} x^{\gamma-1}(\lambda + x^{\gamma})^{-\alpha-1}, \text{ for } x > 0$$
$$E(X) = \lambda^{\frac{1}{\gamma}} \Gamma\left(\alpha - \frac{1}{\gamma}\right) \Gamma\left(1 + \frac{1}{\gamma}\right) / \Gamma(\alpha)$$
$$Var(X) = \lambda^{\frac{2}{\gamma}} \Gamma\left(\alpha - \frac{2}{\gamma}\right) \Gamma\left(1 + \frac{2}{\gamma}\right) / \Gamma(\alpha)$$
$$- \left[\lambda^{\frac{1}{\gamma}} \Gamma\left(\alpha - \frac{1}{\gamma}\right) \Gamma\left(1 + \frac{1}{\gamma}\right) / \Gamma(\alpha)\right]^2$$

Appendix B

Some Basic Tools in Probability and Statistics

In this appendix, we review some basic tools in probability and statistics that are of particular importance in this book. We begin with a brief review of moment generating functions and follow this with a few examples of convolutions of random variables. The concept of conditional probability is key to many of the topics studied in this book, and a review is given with particular emphasis being placed on the conditional random variables $E(X \mid Y)$ and $V(X \mid Y)$. The *double expectation theorem*

$$E(X) = E(E(X \mid Y))$$

is particularly useful in the study of loss distributions, as well as in risk, ruin and credibility theory. A brief mention of the method of maximum likelihood is given at the end.

B.1   Moment generating functions

If X is a random variable, then the *moment generating function $M_X(t)$ of X* at the point t is (when it exists) given by

$$M_X(t) = E(e^{tX}) = \begin{cases} \sum_x e^{tx} f_X(x) & \text{if X is discrete} \\ \int e^{tx} f_X(x)\, dx & \text{if X is continuous,} \end{cases}$$

where $f_X(t)$ is, respectively, the mass or density function of X. The moment generating function for most random variables exists for all t in a neighborhood of 0, and it can be a useful tool in both obtaining moments of and characterizing (through its uniqueness property) a random variable. For example, if the moment generating function $M_X(t)$ of X is continuously differentiable of order n at 0, then the nth moment m_n of X can be expressed as

$$m_n \equiv E(X^n) = \frac{d^n}{dt^n} M_X(t) \mid_{t=0} = M_X^{(n)}(0).$$

The uniqueness property of the moment generating function allows one to conclude that if X_1 and X_2 are two random variables with respective moment generating functions $M_{X_1}(t)$ and $M_{X_2}(t)$, then X_1 and X_2 have the same probability distribution if and only if $M_{X_1}(t) = M_{X_2}(t)$ for all t *near* 0. Some other interesting properties of moment generating functions include the following:

- $M_X(t) = E(e^{tX}) = E(\sum_{n=0}^{\infty} t^n X^n / n!) = \sum_{n=0}^{\infty} E(X^n) t^n / n!$.

- Let $Y = aX + b$. Then $M_Y(t) = E(e^{tY}) = E(e^{t(aX+b)}) = e^{tb} M_X(at)$.

- If Y_1 and Y_2 are independent, then

$$M_{Y_1+Y_2}(t) = E(e^{t(Y_1+Y_2)}) = E(e^{tY_1} e^{tY_2}) = E(e^{tY_1})E(e^{tY_2})$$
$$= M_{Y_1}(t) \, M_{Y_2}(t).$$

If X is a continuous random variable with density function $f_X(x)$ and finite mean $E(X)$, then using integration by parts

$$E(X) = \int_{-\infty}^{+\infty} x \, f_X(x) \, dx = \int_0^{\infty} (1 - F_X(x)) \, dx - \int_{-\infty}^0 F_X(x) \, dx.$$

Therefore when X is nonnegative (for example, representing a claim size or a survival time), then

$$E(X) = \int_0^{\infty} \bar{F}_X(x) \, dx,$$

and the second moment $E(X^2) = m_2$ (when it exists) is

$$E(X^2) = \int_0^{\infty} x^2 \, f_X(x) \, dx = \int_0^{\infty} 2 \, x \, \bar{F}_X(x) \, dx.$$

For example, suppose X is a gamma random variable $(X \sim \Gamma(n, \lambda))$ with parameters n and $\lambda > 0$, and density function

$$f_X(x) = \frac{\lambda^n x^{n-1}}{\Gamma(n)} e^{-\lambda x} \quad \text{for } x > 0.$$

Then the moment generating function of X exists for all $t < \lambda$, and is given by

$$M_X(t) = E(e^{tX}) = \int_0^{\infty} e^{tx} \frac{\lambda^n x^{n-1}}{\Gamma(n)} e^{-\lambda x} \, dx = \left(\frac{\lambda}{\lambda - t} \right)^n.$$

Hence if a random variable Y has moment generating function given by $M_Y(t) = (6/(6 - 3t))^4$, then Y is a gamma random variable with mean $4/2 = 2$ and variance $4/2^2 = 1$.

There are other *generating* functions associated with a random variable X which are also useful tools. The *cumulant moment generating function* $C_X(t)$

of the random variable X is defined by $C_X(t) = \log M_X(t)$, and is useful in calculating *central* moments of a random variable X. The nth central μ_n moment of X is given by $\mu_n = E(X - E(X))^n$, and note that for any random variable, $\mu_1 = 0$ and $\mu_2 = \sigma^2$. The nth cumulant κ_n of X is defined by

$$\kappa_n = C_X^{(n)}(t) \mid_{t=0}$$

and hence one may write

$$C_X(t) = \sum_{n=0}^{\infty} \frac{\kappa_n t^n}{n!}.$$

One can easily demonstrate that

$$\kappa_1 = m_1 = E(X), \quad \kappa_2 = \sigma^2 = Var(X) \text{ and } \kappa_3 = E(X - m_1)^3.$$

If $\mu_4 = E(X - E(X))^4$ is the 4th central moment of X, then $\mu_4 = \kappa_4 + 3\kappa_2^2$. The *skewness* γ_1 of the random variable X is given by $\gamma_1 = \kappa_3/\sigma^3$, while the *kurtosis* is given by $\gamma_2 = \kappa_4/\sigma^4 = \mu_4/\sigma^4 - 3$. While the *mean* $m_1 = E(X)$ is a measure of location and the *variance* $\sigma^2 = E(X - E(X))^2$ is a measure of spread for the random variable X, the skewness is a (standardized) measure of departure from symmetry for X. In particular, a symmetric distribution has 0 skewness, but on the other hand, a 0 skewness does not imply the distribution is in fact symmetric. The kurtosis of a random variable X is a (standardized) measure of peakedness, relative to the normal distribution. For a normal distribution $N(\mu, \sigma^2)$, the 4th central moment $\mu_4 = 3\sigma^4 = 3\kappa_2^2$ and hence the kurtosis is 0.

A positive value of γ_2 usually indicates that the distribution is more peaked and has fatter tails than the normal distribution. A negative value γ_2 usually means that the distribution has a flatter peak and thinner tails than the normal distribution. Historically (see [31]), distributions with 0 kurtosis are called *mesokurtic*. Those distributions where $\gamma_2 > 0$ are often called *leptokurtic*, while those where $\gamma_2 < 0$ are called *platykurtic*. Many of the classic distributions with positive kurtosis have, relatively speaking, more mass in one or both tails and are more peaked, while those with negative kurtosis have small tail areas and are less peaked (flatter-topped – like a platypus?).

Example B.1
Let X be normal ($X \sim N(\mu, \sigma^2)$) with mean μ and variance σ^2. Then the moment generating function of X is given by

$$M_X(t) = \int_{-\infty}^{\infty} \frac{1}{\sqrt{2\pi}\sigma} e^{tx} e^{-\frac{(x-\mu)^2}{2\sigma^2}} dx = e^{t\mu + \frac{t^2\sigma^2}{2}}.$$

Hence $C_X(t) = \mu t + \sigma^2 t^2/2$, from which it follows that $\kappa_1 = \mu$, $\kappa_2 = \sigma^2$ and $\kappa_n = 0$ for all $n \geq 3$. $\quad\square$

Example B.2
If X is a Poisson random variable with parameter λ, then

$$M_X(t) = \sum_{x=0}^{\infty} \frac{e^{tx}\lambda^x e^{-\lambda}}{x!} = e^{\lambda(e^t-1)}.$$

Therefore $C_X(t) = \lambda(e^t - 1)$, and it follows that all of the cumulants are equal to λ. Hence the skewness of X is $\gamma_1 = 1/\sqrt{\lambda}$, and its kurtosis is $\gamma_2 = 1/\lambda$. When λ is large, X has skewness and kurtosis both approximately 0, but this is not too surprising since in this case the Poisson distribution is well approximated by a normal distribution.   ▯

Example B.3
If X is exponential with parameter λ (mean $1/\lambda$), then

$$C_X(t) = \log \frac{\lambda}{\lambda - t} \quad \Rightarrow \quad \kappa_1 = \frac{1}{\lambda}, \quad \kappa_2 = \frac{1}{\lambda^2}, \quad \kappa_3 = \frac{2}{\lambda^3} \text{ and } \quad \kappa_4 = \frac{6}{\lambda^4}.$$

Note that an exponential distribution has skewness 2 and kurtosis 6 independent of its parameter λ.   ▯

The characteristic function $\phi_X(t)$ of the random variable X is defined by

$$\phi_X(t) = E(e^{itX})$$

for any t where $i = \sqrt{-1}$. A mathematical advantage of the characteristic function $\phi_X(t)$ over the moment generating function $M_X(t)$ is that it is well defined for any X and any t. The characteristic function possesses the uniqueness property of the moment generating function, and because it always exists is often used as a tool in proving limiting properties of random variables (like central limit theorems).

B.2   Convolutions of random variables

Let X and Y be independent random variables, and $S = X + Y$ be their sum or *convolution*. If X and Y are nonnegative random variables with respective density functions given by f_X and f_Y, then the distribution function of S is given by

$$F_S(s) = P(X + Y \le s) = \int_0^s F_X(s - y)\, dF_Y(y) = \int_0^s F_X(s - y)\, f_Y(y)\, dy.$$

Differentiating with respect to s, one obtains

$$f_S(s) = F'_S(s) = \int_0^s f_X(s-y)\, f_Y(y)\, dy.$$

Example B.4

Let X and Y be independent random variables both uniformly distributed on the interval $[0,1]$, and let $S = X + Y$. The range of S is $[0,2]$. If $0 \le s \le 1$, then

$$f_S(s) = \int_0^s f_X(s-y) f_Y(y)\, dy = \int_0^s 1 \cdot 1\, dy = s,$$

while if $1 \le s \le 2$ then

$$f_S(s) = \int_0^s f_X(s-y) f_Y(y)\, dy = \int_{s-1}^1 1 \cdot 1\, dy = 2 - s.$$

$\square$

Example B.5

Let X and Y be independent $N(\mu, \sigma^2)$ random variables. Then

$$f_{X+Y}(s) = \int_{-\infty}^{+\infty} \frac{1}{\sqrt{2\pi}\sigma} e^{-\frac{1}{2}(\frac{s-y-\mu}{\sigma})^2} \frac{1}{\sqrt{2\pi}\sigma} e^{-\frac{1}{2}(\frac{y-\mu}{\sigma})^2}\, dy$$

$$= \frac{1}{\sqrt{2\pi}\sigma\sqrt{2}} \int_{-\infty}^{+\infty} \frac{1}{\sqrt{2\pi}\sigma/\sqrt{2}} e^{-\frac{[2(y-\frac{s}{2})^2 + 2(\mu-\frac{s}{2})^2]}{2\sigma^2}}\, dy$$

$$= \frac{1}{\sqrt{2\pi}\sigma\sqrt{2}} e^{-\frac{(s-2\mu)^2}{2(\sigma^2 2)}} \int_{-\infty}^{+\infty} \frac{1}{\sqrt{2\pi}\sigma/\sqrt{2}} e^{-\frac{(y-s/2)^2}{2(\sigma/\sqrt{2})^2}}\, dy$$

$$= \frac{1}{\sqrt{2\pi}\sigma\sqrt{2}} e^{-\frac{(s-2\mu)^2}{2(\sigma^2 2)}},$$

and therefore $X + Y \sim N(2\mu, 2\sigma^2)$. Note, however, that this may be proved more directly using moment generating functions, since if X and Y are independent

$$M_{X+Y}(t) = M_X(t) M_Y(t) = (e^{\mu t + \sigma^2 t^2/2})^2 = e^{2\mu t + (\sqrt{2}\sigma)^2 t^2/2},$$

and hence using the uniqueness property of moment generating functions it follows that $X + Y$ must be normal with mean 2μ and variance $2\sigma^2$. $\square$

B.3 Conditional probability and distributions

Conditional distributions are of immense importance in applied probability, and in particular in areas of risk analysis (like insurance) where we are interested in one variable (for example, a death, an accident or a default on a loan), conditioned on another variable (such as being of a certain age, gender, health status or educational background).

Let X and Y be jointly continuous (or discrete) random variables with joint density (mass) function given by $f_{X,Y}(x,y)$. Then the *conditional* density (mass) function of X given Y is given by

$$f_{X|Y}(x \mid y) = \frac{f_{X,Y}(x,y)}{f_Y(y)} \quad \text{if } f_Y(y) > 0.$$

Of course, if X and Y are independent, then the conditional distribution of X given Y is the same as the distribution of X – that is, knowing the value of the random variable Y has no influence on the value that X might take.

We define the conditional expectation $E(X \mid Y = y)$ of the random variable X given that $Y = y$ by

$$E(X \mid Y = y) = \begin{cases} \sum_x x \, f_{X|Y}(x \mid y) & \text{if } X \text{ is discrete} \\ \int x \, f_{X|Y}(x \mid y) \, dx & \text{if } X \text{ is continuous.} \end{cases}$$

$E(X \mid Y = y)$ is essentially the average value of the random variable X given knowledge that $Y = y$. Of course, if the random variables X and Y are independent, then knowing the value of the random variable Y tells us nothing about X and thus $E(X \mid Y = y) = E(X)$ for any value y of Y.

Example B.6
Let us suppose that for each randomly selected claimant in an insurance company, we record the age A, the gender G and the claim amount C. Then $E(C \mid A = 25)$ is the average claim size for an individual of age 25 years. Similarly, we can define the expectation of one random variable conditioned on several others – for example, $E(C \mid A = 19$ and $G = F)$ is the average claim size for a female of age 19. ⬚

Example B.7
Employees in a large company have been offered group rates on both life and health insurance by an insurance company. If X is the proportion of employees who will opt for the life scheme and Y is the proportion of people who go for the health scheme, then experience suggests that an appropriate model for the joint distribution of X and Y is given by

$$f_{X,Y}(x,y) = \begin{cases} \frac{2}{5}(x + 4y) & 0 < x < 1, \ 0 < y < 1 \\ 0 & \text{otherwise.} \end{cases}$$

Since

$$f_X(x) = \int_0^1 \frac{2}{5}(x + 4y)dy = \frac{2}{5}(x + 2) \quad \text{for } 0 < x < 1,$$

it follows that the mean acceptance rate for life insurance is

$$E(X) = \int_0^1 x \frac{2}{5}(x+2)\, dx = \frac{8}{15}.$$

Similarly,

$$f_Y(y) = \frac{2}{5}\left(\frac{1}{2} + 4y\right) \text{ and hence } E(Y) = \frac{19}{30}.$$

The probability of at least a 30% acceptance to the life scheme is given by

$$P[X > 0.3] = \int_{0.3}^1 \frac{2}{5}(x+2)\, dx = 0.742.$$

We now determine the probability that, at most, 50% will opt for the health insurance given there is a 20% acceptance rate for the life scheme. Now

$$f_{Y|X=0.2}(y\mid 0.2) = \frac{(2/5)(x+4y)}{(2/5)(x+2)}\Big|_{x=0.2} = \frac{0.2+4y}{2.2}.$$

Therefore

$$P(Y \le 0.50 \mid X = 0.2) = \int_0^{0.5} \frac{0.2+4y}{2.2}\, dy = 0.27 \text{ and hence}$$

$$E(Y \mid X = 0.2) = \int_0^1 y\, \frac{0.2+4y}{2.2}\, dy = 0.65.$$

⬚

B.3.1 The double expectation theorem and $E(X)$

Given two random variables X and Y, the *random variable* $E(X \mid Y)$ is defined on the sample space of events Ω by

$$E(X \mid Y) : \omega = E(X \mid Y = Y(\omega)).$$

It is important to realize that $E(X \mid Y)$ is actually a random variable, and not a real number (like, for example, $E(X \mid Y = y)$). Once the value of the random variable Y is known (for example, $Y = y$), then the value of $E(X \mid Y)$ is determined. For instance in Example B.6, $E(C \mid A)$ is the random variable that associates with any person the average claim size for all individuals with the same age as that person. Given that $E(X \mid Y)$ is a random variable, we can determine its mean and variance. The following theorem is a classic result in probability theory, and is often referred to as the *double expectation theorem*.

THEOREM B.1
Given any random variables X and Y with finite means, we have that

$$E(X) = E_Y(E(X \mid Y)) \equiv E(E(X \mid Y)).$$

PROOF   (for the discrete case)

$$E_Y(E(X \mid Y)) = \sum_y E(X \mid Y = y)P(Y = y)$$

$$= \sum_y \left[\sum_x x\,P(X = x \mid Y = y) \right] P(Y = y)$$

$$= \sum_y \sum_x x\,\frac{P(X = x, Y = y)}{P(Y = y)}\,P(Y = y)$$

$$= \sum_x \sum_y x\,P(X = x, Y = y)$$

$$= \sum_x x\,P(X = x)$$

$$= E(X).$$

∎

The subscript Y in the above expression $E_Y(E(X \mid Y))$ indicates that expectation is take with respect to Y. Remember that once Y is known, $E(X \mid Y)$ is determined. In some cases, we will find it easier to find $E(X)$ by conditioning on some other random variable Y than by calculating it directly.

Example B.8

A study of smoking habits and general health was undertaken in a certain community. The smoking habit S was rated 0 (nonsmoker), 1 (occasional smoker), 2 (mild smoker) or 3 (heavy smoker). General health H was rated 0 (poor), 1 (average) or 2 (very healthy) for each individual. The results are summarized in Table B.1, which gives the joint distribution of (S, H) for this community.

TABLE B.1
General health (H) and smoking status (S).

		Health (H)			
		0	1	2	
Smoking	0	0.05	0.10	0.20	0.35
status	1	0.10	0.10	0.05	0.25
(S)	2	0.10	0.05	0.05	0.20
	3	0.10	0.10	0.00	0.20
		0.35	0.35	0.30	**1.00**

Suppose we select at random a person from this community. Let us consider the random variable $E(H \mid S)$, which is the expected health rating conditioned on smoking status. If the selected individual is a nonsmoker ($S=0$), then the value of this random variable is given by

$$E(H \mid S = 0) = 0 \cdot P(H = 0 \mid S = 0) + 1 \cdot P(H = 1 \mid S = 0)$$
$$+ 2 \cdot P(H = 2 \mid S = 0)$$
$$= 0 \cdot (0.05/0.35) + 1 \cdot (0.10/0.35) + 2 \cdot (0.20/0.35)$$
$$= 50/35.$$

Hence if we know that a randomly selected person from this community is a nonsmoker, then the expected health rating for this person is $50/35$. Similarly, one may calculate $E(H \mid S = i)$ for $i = 1, 2, 3$. In doing so, one observes that the range of the random variable $E(H \mid S)$ is $(50/35, 20/25, 15/20, 10/20)$, with probability distribution as given in Table B.2.

TABLE B.2
Probability distribution of $E(H \mid S)$.

$S = i$	Value of $E(H \mid S)$	Probability
0	$50/35 = 1.43$	0.35
1	$20/25 = 0.80$	0.25
2	$15/20 = 0.75$	0.20
3	$10/20 = 0.50$	0.20

Using the double expectation theorem (conditioning on the random variable S), we find

$$E(H) = (50/35)(0.35) + (20/25)(0.25) + (15/20)(0.20) + (10/20)(0.20)$$
$$= 0.95.$$

One way of interpreting this is that the average health rating $E(H)$ is a weighted average of the average health ratings of those who are nonsmokers, occasional smokers, mild smokers and heavy smokers – with the weights being the proportions of these groups in the community. Of course, in this example we can calculate $E(H)$ more directly since we know the marginal distribution of H, and hence $E(H) = 0(0.35) + 1(0.35) + 2(0.30) = 0.95$.

From Table B.2, one may observe that the average health rating for a randomly selected individual depends on their smoking habit, and that the average health rating decreases as the degree of smoking increases. This can also be seen from calculating the correlation between S and H. Since

$$E(SH) = 1 \, (0.10) + 2 \, (0.10) + 3 \, (0.10) + 4 \, (0.05) = 0.80,$$

$E(S) = 1.25$, $Var(S) = 1.2875$, $E(H) = 0.95$ and $Var(H) = 0.6475$, it follows that $corr(S, H) = -0.4244$. $\Box$

B.3.2   The random variable $V(X \mid Y)$

Another useful tool similar to the double expectation theorem allows us to determine the variance of a random variable X by conditioning on any other random variable Y. In some cases, it may be easier to calculate the variance of X by conditioning than by doing it more directly. We define the random variable $V(X \mid Y)$ by

$$V(X \mid Y) : s = Var(X \mid Y = Y(s)).$$

Again, once the value of the random variable Y is known (say $Y = y$), then the value of the random variable $V(X \mid Y)$ is determined. For instance, using the terminology in Example B.6, $V(C \mid A)$ is the random variable which assigns to any individual s the variance of the claim sizes for all individuals with the same age as s. The following result establishes a useful link between $Var(X)$ (the variance of X) and both the mean of the random variable $V(X \mid Y)$ and the variance of the random variable $E(X \mid Y)$.

THEOREM B.2

For any random variables X and Y,

$$Var(X) = E(V(X \mid Y)) + Var(E(X \mid Y)) \tag{B.1}$$

PROOF   Making use of the double expectation theorem, the following argument gives a proof for the case when X and Y are discrete.

$$
\begin{aligned}
Var(X) &= E(X^2) - E_Y[E^2(X \mid Y)] + E_Y[E^2(X \mid Y)] - E^2(X) \\
&= E[E(X^2 \mid Y)] - E[E^2(X \mid Y)] \\
&\quad + E[E^2(X \mid Y)] - E^2[E(X \mid Y)] \\
&= \sum_y E(X^2 \mid Y = y) f_Y(y) - \sum_y E^2(X \mid Y = y) f_Y(y) \\
&\quad + Var(E(X \mid Y)) \\
&= \sum_y \{E(X^2 \mid Y = y) - E^2(X \mid Y = y)\}\, f_Y(y) + Var(E(X \mid Y)) \\
&= \sum_y Var(X \mid Y = y)\, f_Y(y) + Var(E(X \mid Y)) \\
&= E(V(X \mid Y)) + Var(E(X \mid Y)).
\end{aligned}
$$

∎

In Example B.6, $V(H \mid S)$ assigns to any individual the variance of the health ratings of all other individuals with the same smoking status. For example, suppose that we condition on a nonsmoker $(S = 0)$. Then

$$Var(H \mid S = 0) = E(H^2 \mid S = 0) - E^2(H \mid S = 0)$$

$$= \left[0^2 \cdot \frac{0.05}{0.35} + 1^2 \cdot \frac{0.10}{0.35} + 2^2 \cdot \frac{0.20}{0.35}\right] - \left(\frac{50}{35}\right)^2$$

$$= \frac{650}{35^2} = 0.5306.$$

Table B.3 gives the probability distribution of the random variable $V(H \mid S)$ based on the joint distribution of H and S (Table B.1). Therefore it follows

TABLE B.3
Probability distribution of $V(H \mid S)$.

$S = i$	Value of $V(H \mid S)$	Probability
0	$650/1225 = 0.5306$	0.35
1	$350/625 = 0.5600$	0.25
2	$275/400 = 0.6875$	0.20
3	$100/400 = 0.2500$	0.20

that

$$E(V(H \mid S)) = \frac{650}{1225}(0.35) + \frac{350}{625}(0.25) + \frac{275}{400}(0.20) + \frac{100}{400}(0.20) = 0.5132$$

and

$$Var(E(H \mid S)) = E[E^2(H \mid S)] - E^2[E(H \mid S)]$$

$$= \left(\frac{50}{35}\right)^2 (0.35) + \left(\frac{20}{25}\right)^2 (0.25)$$

$$+ \left(\frac{15}{20}\right)^2 (0.2) + \left(\frac{10}{20}\right)^2 (0.2) - (0.95)^2$$

$$= 0.1343.$$

Thus

$$Var(H) = E(V(H \mid S)) + Var(E(H \mid S)) = 0.5132 + 0.1343 = 0.6475.$$

Example B.9
Let $\{X_i\}_{i=1}^{\infty}$ be a sequence of independent and identically distributed random variables, and assume that N is a nonnegative integer valued random variable which is also independent of this sequence. The random variable defined by

$$S = X_1 + \cdots + X_N$$

is said to have a compound distribution (note that the number of terms in the sum is itself a random variable). By conditioning on N, one is able to use the double expectation theorem to establish compact formulae for the mean, variance and moment generating function of a compound random variable as follows:

$$E(S) = E(X)\,E(N)$$
$$Var(S) = E^2(X)\,Var(N) + Var(X)\,E(N), \quad \text{and}$$
$$M_S(t) = M_N(\log M_X(t)).$$

$\square$

B.4   Maximum likelihood estimation

If $F_\theta : \theta \in \Theta$ is a one-parameter family of distributions, then the maximum likelihood estimator $\hat{\theta}$ of θ based on a sample $\mathbf{x}$ of size n has very desirable asymptotic properties. In particular, one has the asymptotic property that

$$\hat{\theta} \,\dot{\sim}\, N(\theta, 1/nI(\theta)),$$

where $\dot{\sim}$ means *approximately distributed as*, and $I(\theta)$ is the Fisher information given by

$$I(\theta) = E\left[\frac{\partial}{\partial\theta} \log f(X \mid \theta)\right]^2 = -E\left[\frac{\partial^2}{\partial\theta^2} \log f(X \mid \theta)\right].$$

In particular, note that the asymptotic variance of $\hat{\theta}$ is the Cramér–Rao lower bound for unbiased estimators of θ.

Suppose now that we have a random sample of observations from a two-parameter family of distributions, where the parameters are denoted by (θ_1, θ_2). The maximum likelihood estimates $(\hat{\theta}_1, \hat{\theta}_2)$ are those values of (θ_1, θ_2) which maximize the likelihood function $L_\mathbf{x}$ for $\mathbf{x} = (x_1, x_2, \ldots, x_n)$. That is, using f to denote the density function for the random variable of interest X,

$$max\, L_\mathbf{x}\,(\theta_1, \theta_2) = max_{(\theta_1, \theta_2)} \left[\prod_{i=1}^{n} f(x_i \mid \theta_1, \theta_2)\right] = \prod_{i=1}^{n} f(x_i \mid \hat{\theta}_1, \hat{\theta}_2)$$
$$= L(\hat{\theta}_1, \hat{\theta}_2).$$

These estimates may, in some cases, be found by differentiating the log-likelihood function $l(\theta_1, \theta_2) = \log L(\theta_1, \theta_2)$ with respect to θ_1 and θ_2 and, setting the resulting equations, equal to 0. Again, one of the main reasons for using the maximum likelihood estimators is their asymptotic properties. The 2×2 information matrix $I(\theta_1, \theta_2)$ is given by

$$I(\theta_1, \theta_2) = \begin{pmatrix} -E\left[\frac{\partial^2}{\partial \theta_1^2} \log f(X \mid \theta_1, \theta_2)\right] & -E\left[\frac{\partial^2}{\partial \theta_1 \partial \theta_2} \log f(X \mid \theta_1, \theta_2)\right] \\ -E\left[\frac{\partial^2}{\partial \theta_1 \partial \theta_2} \log f(X \mid \theta_1, \theta_2)\right] & -E\left[\frac{\partial^2}{\partial \theta_2^2} \log f(X \mid \theta_1, \theta_2)\right] \end{pmatrix},$$

and hence if the sample size n is large,

$$\begin{pmatrix} \hat{\theta}_1 \\ \hat{\theta}_2 \end{pmatrix} \overset{\cdot}{\sim} N\left(\begin{bmatrix} \theta_1 \\ \theta_2 \end{bmatrix}, [n \cdot I(\theta_1, \theta_2)]^{-1}\right).$$

Example B.10
Let $X \sim N(\mu, \sigma^2)$. This is a two-parameter family of distributions with parameters μ and σ, and the information matrix $I(\mu, \sigma)$ takes the form

$$I(\mu, \sigma) = \begin{pmatrix} 1/\sigma^2 & 0 \\ 0 & 1/2\sigma^4 \end{pmatrix}.$$

In particular, it follows that the maximum likelihood estimators $\hat{\mu} = \bar{x}$ and $\hat{\sigma} = \sqrt{\sum(x_i - \bar{x})^2/n} = s\sqrt{(n-1)/n}$ are asymptotically independent and unbiased. Of course, a classic result of Gosset and Fisher states that $\bar{x}$ and s^2 (equivalently, $\bar{x}$ and s) are independent when sampling from a normal distribution, and hence we know that in this situation $\hat{\mu}$ and $\hat{\sigma}$ are independent for any n. ▯

Appendix C

An Introduction to Bayesian Statistics

If θ is an unknown characteristic or parameter related to a population or random variable, then the frequentist statistician will attempt to make inferences about θ on the basis of the sample information $\mathbf{x}$. The Bayesian statistician, however, would always believe that there is additional prior information available about θ that should be combined with the sample information $\mathbf{x}$ in order to make inferences about θ.

C.1 Bayesian statistics

The Bayesian (statistician) would express prior knowledge about an unknown parameter θ by means of a *prior* probability distribution with density or mass function $f_\Theta(\Theta)$ and corresponding distribution function $F_\Theta(\theta)$. Any sample information $\mathbf{x}$ is observed conditional on the unknown true parameter θ, and we use $f_{\mathbf{X}|\theta}$ to represent this conditional density (or mass function). The prior and sampling distributions give rise to a joint distribution for θ and $\mathbf{X}$ given by

$$f_{\Theta,\mathbf{X}}(\theta, \mathbf{x}) = f_{\mathbf{X}|\Theta}(\mathbf{x} \mid \theta) f_\Theta(\theta).$$

The marginal distribution for the sample $\mathbf{X}$ is essentially an average of the conditional distributions $\mathbf{X} \mid \theta$ with respect to our prior belief about θ, and is given by

$$f_{\mathbf{X}}(\mathbf{x}) = \int f_{\mathbf{X}|\Theta}(\mathbf{x} \mid \theta) f_\Theta(\theta) \, d\theta = \int f_{\Theta,\mathbf{X}}(\theta, \mathbf{x}) \, d\theta.$$

However, in practice it is our feeling or belief about θ after observing $\mathbf{x}$ which is used for inference about θ. This is called the posterior distribution for θ given $\mathbf{X} = \mathbf{x}$, given by

$$f_{\Theta|\mathbf{X}}(\theta \mid \mathbf{x}) = \frac{f_{\Theta,\mathbf{X}}(\theta, \mathbf{x})}{f_{\mathbf{X}}(\mathbf{x})} = \frac{f_{\mathbf{X}|\Theta}(\mathbf{x} \mid \theta) f_\Theta(\theta)}{f_{\mathbf{X}}(\mathbf{x})}.$$

Calculation of the posterior distribution can often be tedious and difficult; however, in some situations simulation methods may be useful in generating samples from such a distribution. It is also important to note that

$$f_{\Theta|\mathbf{X}}(\theta \mid \mathbf{x}) \propto f_{\Theta,\mathbf{X}}(\theta, \mathbf{x}) = f_{\mathbf{X}|\Theta}(\mathbf{x} \mid \theta) f_\Theta(\theta), \tag{C.1}$$

where $\propto$ indicates "is proportional to" as a function of θ. Hence $1/f_{\mathbf{X}}(\mathbf{x})$ is simply a factor ensuring that as a function of θ, $f_{\Theta|\mathbf{X}}(\theta \mid \mathbf{x})$ integrates to 1. Therefore in many cases one need not actually calculate the marginal distribution $f_{\mathbf{X}}(\mathbf{x})$ in order to get the general form of the posterior.

C.1.1   Conjugate families

In some cases, the posterior distribution for θ will have the same functional form as the prior distribution for a given sampling distribution. When this occurs, we say the prior and posterior families of distributions are conjugate with respect to the sampling distribution. The following example shows that the family of beta distributions is conjugate for samples from a binomial distribution. In Chapter 5 on credibility theory, it is shown that the normal family of distributions is conjugate for samples from a normal distribution, and the gamma family of distributions is conjugate for samples from a Poisson distribution. Mathematically, it is attractive to work with conjugate families of distributions, but is important to note that conjugacy is an exception, and one often has to resort to simulation to find the posterior distribution in practice. Markov Chain Monte Carlo (MCMC) methods are a commonly used tool for this purpose.

Example C.1 Binomial | Beta model
Assume that q is the probability of an event of interest, for example, the probability of a person dying in a given year (mortality rate) or of a defective being found when an item is randomly selected from a production line (quality control). The beta family of distributions is a rich family, and often one may select one such distribution $Beta(\alpha, \beta)$ to represent prior feeling about the unknown value of q. Such a distribution has mean μ, variance σ^2 and density function $f(q)$ given, respectively, by

$$\mu = \frac{\alpha}{\alpha + \beta}, \ \sigma^2 = \frac{\alpha\beta}{(\alpha + \beta)^2(\alpha + \beta + 1)} \text{ and } f(q) = \frac{\Gamma(\alpha + \beta)}{\Gamma(\alpha)\Gamma(\beta)} q^{\alpha-1}(1 - q)^{\beta-1}$$

(C.2)

for $0 < q < 1$, and some positive parameters α and β.

Assume now that we may also observe some sample information $\mathbf{x}$ related to q, in particular, that we have a binomial observation that is the number of successes $\sum x_j$ in a sequence of n independent trials with success probability q. This may represent the number of deaths from a sample of n individuals, or the number of defects observed in a random sample of n items from a production line.

From Equation (C.1) it follows that the posterior for q is of the form

$$f(q \mid \mathbf{x}) \propto \binom{n}{\sum x_j} q^{\sum x_j}(1 - q)^{n - \sum x_j} \frac{\Gamma(\alpha + \beta)}{\Gamma(\alpha)\Gamma(\beta)} q^{\alpha-1}(1 - q)^{\beta-1}$$

$$\propto q^{\alpha + \sum x_j - 1}(1 - q)^{\beta + n - \sum x_j - 1}.$$

Therefore when the prior for q is $Beta(\alpha, \beta)$ and the sampling distribution is $B(n, q)$, then the posterior for q is the $Beta(\alpha + \sum x_j, \beta + n - \sum x_j)$ distribution. This situation is therefore referred to as the binomial | beta model. Note in particular that the mean of the posterior distribution for q can be written in the form

$$\frac{\alpha + \sum x_j}{\alpha + \beta + n} = \frac{n}{\alpha + \beta + n} \frac{\sum x_j}{n} + \frac{\alpha + \beta}{\alpha + \beta + n} \frac{\alpha}{\alpha + \beta}$$

$$= \mathbf{Z}\,\bar{x} + (1 - \mathbf{Z})\,\mu_0,$$

where $\mu_0 = \alpha/(\alpha + \beta)$ is the prior mean. Hence the posterior mean can be written as a *credibility estimate* for q with *credibility factor* $\mathbf{Z} = n/(\alpha + \beta + n)$.

Let us consider the example of a life insurance portfolio consisting of 400 policies, each independent with respect to mortality. Assume that the probability of death q will be the same for all policyholders next year, and that prior information about q may be expressed by a beta distribution with mean 0.04 and variance 0.000191. Using the expressions for the mean and variance of a beta distribution given in Equation (C.2), we have that

$$\alpha/(\alpha + \beta) = 0.04 \implies \alpha = 4\beta/96, \quad \text{and}$$

$$0.000191 = \alpha\beta/[(\alpha + \beta)^2\,(\alpha + \beta + 1)] = 4(96)/[100^2(100\beta/96 + 1)],$$

yielding $\beta = 192.045$ and $\alpha = 8.0018$. If 36 policyholders die within the coming year, then the posterior mean for q may be written in the form

$$\frac{\alpha + 36}{\alpha + \beta + 400} = \frac{400}{400 + \alpha + \beta} \frac{36}{400} + \frac{\alpha + \beta}{400 + \alpha + \beta} \frac{\alpha}{\alpha + \beta}$$

$$= 0.6666\,(0.09) + 0.3334\,(0.04) = 0.0733,$$

where the *weight* put on the sample mean of 0.09 is $\mathbf{Z} = 0.6666$.   ▯

C.1.2   Loss functions and Bayesian inference

Beginning with prior knowledge of θ given by $f_\Theta(\theta)$, the posterior distribution summarizes our current belief about θ, having observed the sample $\mathbf{x}$. How might we use the posterior distribution to make inferences about θ? In many cases, we might consider giving some type of point estimate $\dot{\theta}$ of θ based on the posterior distribution. Given a prior distribution for θ, the posterior distribution for θ is determined by the sample information $\mathbf{x}$, hence any reasonable estimate of θ will be a function of $\mathbf{x}$ which we write in the form $\dot{\theta} = g(\mathbf{x})$. We may view this as a decision problem in which there is a loss or penalty $l(\theta, \dot{\theta}) = l(\theta, g(\mathbf{x}))$ to pay as a result of estimating θ by $g(\mathbf{x})$. There are many possible loss functions, but the following are the most common:

1. quadratic or squared error loss where $l_1(\theta, g(\mathbf{x})) = [\theta - g(\mathbf{x})]^2$

2. absolute value loss where $l_2(\theta, g(\mathbf{x})) = |\theta - g(\mathbf{x})|$

3. zero-one (or all or nothing) loss where

$$l_3(\theta, g(\mathbf{x})) = \begin{cases} 1 & \text{if } \theta \neq g(\mathbf{x}) \\ 0 & \text{if } \theta = g(\mathbf{x}). \end{cases}$$

Given a particular loss function l, we define the *Bayesian estimator* of θ to be that value $\dot{\theta} = g(\mathbf{x})$ which minimizes the expected loss with respect to our posterior belief about θ.

For any random variable Y, it is well known that the value of g which minimizes $L_{Y,1}(g) \equiv E(l_1(Y, g)) = E(Y - g)^2$ is $g = E(Y)$.

Letting $L_{Y,2}(g) = E(|Y - g|)$, then (when Y has continuous density $f_Y(y)$) one has

$$L_{Y,2}(g) = E(|Y - g|) = \int_{-\infty}^{+\infty} |y - g| f_Y(y) \, dy$$

$$= \int_{-\infty}^{g} (g - y) f_Y(y) \, dy + \int_{g}^{+\infty} (y - g) f_Y(y) \, dy$$

$$= g F_Y(g) - \int_{-\infty}^{g} y f_Y(y) \, dy - g[1 - F_Y(g)] + \int_{g}^{+\infty} y f_Y(y) \, dy$$

$$= 2g F_Y(g) - g + E(Y) - 2 \int_{-\infty}^{g} y f_Y(y) \, dy.$$

Therefore setting $L'_{Y,2}(g) = 2F_Y(g) - 1 = 0$, one notes that the value of g minimizing $E(|Y - g|)$ is the median $g = F_Y^{-1}(1/2)$.

Another approach is to find the mean of $|Y - g|$ by integrating its survival function $\bar{F}_{|Y-g|}$ and minimizing $E(|Y - g|) = \int_0^\infty \bar{F}_{|Y-g|}(t) \, dt$. Now

$$\bar{F}_{|Y-g|}(t) = \begin{cases} \bar{F}_Y(g + t) & \text{if } g \leq t \\ \bar{F}_Y(g + t) + F_Y(g - t) & \text{if } g > t. \end{cases}$$

Therefore

$$E(|Y - g|) = \int_0^\infty \bar{F}_{|Y-g|}(t) \, dt$$

$$= \int_0^\infty \bar{F}_Y(g + t) \, dt + \int_0^g F_Y(g - t) \, dt$$

$$= \int_g^\infty \bar{F}_Y(y) \, dy + \int_0^g F_Y(w) \, dw.$$

Letting $G(g) = E(|Y - g|)$, then $G(0) = E(Y)$ and $G(\infty) = \infty$. Moreover, $G'(g) = -\bar{F}_Y(g) + F_Y(g) = 2 F_Y(g) - 1$, which is 0 if $g = F_Y^{-1}(1/2)$. Note that $G''(g) = 2 f_Y(g) \geq 0$, showing that the median is a minimum for $E(|Y - g|)$.

Finally, consider the zero-one loss function and $L_{Y,3}(g) \equiv E(l_3(Y, g))$. If Y is discrete with mass function $f_Y(y)$, then $E(l_3(Y, g)) = 1 - f_Y(g)$ is clearly minimized where $f_Y(g)$ is maximized (at the modal value for Y). In the continuous case $E(l_3(Y, g)) = 1$ for all g, and hence there is no optimization problem here strictly speaking. However, we may motivate the use of the zero-one loss function in this situation by considering it as the limit as $\epsilon \to 0$ of loss functions $l_3^\epsilon(\theta, g(\mathbf{x}))$ defined by

$$l_3^\epsilon(\theta, g(\mathbf{x})) = \begin{cases} 1 & \text{if } |\theta - g(\mathbf{x})| > \epsilon \\ 0 & \text{if otherwise,} \end{cases}$$

where $\epsilon > 0$. Now

$$E(l_3^\epsilon(Y, g)) = 1 - \int_{g-\epsilon}^{g+\epsilon} f_Y(y) \, dy \doteq 1 - \int_{g-\epsilon}^{g+\epsilon} f_Y(g) \, dy = 1 - 2\epsilon f_Y(g).$$

For any ϵ this is minimized where $f_Y(g)$ is maximized, that is, at the modal value for Y, and

$$E(l_3(Y, g)) = \lim_{\epsilon \to 0} E(l_3^\epsilon(Y, g)).$$

Returning to the Bayesian context and our interest in estimating the parameter θ after observing the sample information $\mathbf{X} = \mathbf{x}$, the random variable (or posterior distribution) of interest is that of $Y = [\Theta \mid \mathbf{X} = \mathbf{x}]$. It has been demonstrated that the Bayesian estimator for θ having observed $\mathbf{X} = \mathbf{x}$ is therefore, respectively, the mean, median or mode of the posterior distribution $[\Theta \mid \mathbf{X} = \mathbf{x}]$ when using the quadratic, absolute value or zero-one loss function.

Example C.2
The number of monthly claims in a small portfolio of household theft policies is to be modeled by a Poisson distribution with parameter λ, where λ is initially assumed to be one of the values $(3, 4, 5)$ with prior distribution $f_\Lambda(3, 4, 5) = (0.3, 0.4, 0.3)$. If at the end of two months a total of 11 claims have been observed (6 in month 1 and 5 in month 2), what would you estimate λ to be?

Given the sample information $\mathbf{x} = (6, 5)$, the posterior distribution takes the form

$$f_{\Lambda|\mathbf{x}}(\lambda \mid \mathbf{x}) \propto \prod_{i=1}^{2} [\lambda^{x_i} e^{-\lambda}] \, f_\Lambda(\lambda)$$

$$\propto \lambda^{11} e^{-2\lambda} f_\Lambda(\lambda)$$

$$\propto (131.731, \ 562.813, \ 665.038)$$

$$\propto (0.097, \quad 0.414, \quad 0.489),$$

giving Table C.1 of the prior and posterior distributions for λ. Therefore the estimate might be the posterior mean (4.392), median (4) or mode (5),

TABLE C.1

Distributions for λ in household thefts.

parameter λ		3	4	5
prior	$f_\Lambda(\lambda)$	0.3	0.4	0.3
posterior	$f_{\Lambda|\mathbf{X}}(\lambda \mid \mathbf{x} = (6,5))$	0.097	0.414	0.489

depending on the loss function that one is using (respectively, quadratic, absolute value or zero-one).   ⬚

We refer to any interval $[c_l, c_u]$ containing $(1-\alpha)$ of the posterior probability of θ as a $100(1-\alpha)\%$ Bayesian belief interval for θ. It expresses a subjective belief about where we think the value of θ lies, and its interpretation is not to be confused with the meaning of a frequentist $100(1-\alpha)\%$ confidence interval for θ. Such frequentist confidence intervals are constructed on the basis of sample information alone, and the *degree of confidence* in such an interval refers to *how often the procedure used generates an interval containing the unknown θ.*

Example C.3 Burr | Gamma model
The claim size arising from a certain group policy is modeled by a Burr distribution with density function given by $f(x|\theta) = 2\theta x/[1 + x^2]^{1+\theta}$ for $x > 0$, where θ is an unknown parameter. Prior information on θ suggests that a gamma $\Gamma(21, 15)$ distribution is appropriate. A sample of size $n = 100$ will be taken in order to make inferences about θ. We will determine the form of the Bayesian estimators for θ using the three loss functions l_1, l_2 and l_3, as well as a 95% Bayesian belief interval for θ.
    From Equation (C.1) we see that the posterior density for θ given $\mathbf{x}$ will be of the form

$$f_{\Theta|\mathbf{X}}(\theta \mid \mathbf{x}) \propto \frac{15^{21} \theta^{20} e^{-15\theta}}{\Gamma(21)} (2\theta)^{100} \frac{\prod x_j}{\prod[1 + x_j^2]^{1+\theta}}$$

$$\propto \theta^{120} e^{-\theta(15+\sum \log[1+x_j^2])},$$

and hence the posterior distribution is $\Gamma(121, 15 + \sum \log[1 + x_j^2])$.
    For the gamma distribution $\Gamma(\alpha, \lambda)$, the mean is α/λ and the mode is $(\alpha - 1)/\lambda$ (unless $\alpha \leq 1$, in which case the gamma density is decreasing). Therefore the usual Bayesian estimator (for a quadratic loss function) in this case is the posterior mean, which can be written in the form

$$\frac{100 + 21}{15 + \sum \log[1 + x_j^2]} = \mathbf{Z} \frac{100}{\sum \log[1 + x_j^2]} + (1 - \mathbf{Z}) \frac{21}{15},$$

where the weight (credibility factor) put on the sample statistic (in this case, the maximum likelihood estimator of θ) $\hat{\theta} = 100/\sum \log[1 + x_j^2]$ is given by

$\mathbf{Z} = \sum \log[1 + x_j^2]/(15 + \sum \log[1 + x_j^2])$. When using a zero-one loss function, the Bayesian estimator would be $120/(15 + \sum \log[1 + x_j^2])$. For the absolute value loss function, one would usually refer to a statistical package to determine the median or 50th percentile of the $\Gamma(121, 15 + \sum \log[1 + x_j^2])$ distribution. The gamma distribution $\Gamma(121, 15 + \sum \log[1 + x_j^2])$ can usually be well approximated by the normal distribution with the same mean and variance, and hence an approximate 95% Bayesian belief interval for θ when using quadratic loss is given by

$$\left[\frac{121}{15 + \sum \log[1 + x_j^2]} \pm 1.96 \frac{\sqrt{121}}{15 + \sum \log[1 + x_j^2]} \right].$$

▯

Appendix D

Answers to Selected Problems

D.1   Claims reserving and pricing with run-off triangles

1. An additional amount of 142,441 should be reserved for paying ultimate claims arising from the origin year 2006.

3. Necessary reserves would be about 2,835.53 ($000's).

5. Approximately $1,420,000.

7. Reserves of approximately 7,350 are needed.

9. Reserves of approximately 4076.83 would be needed.

11. Reserves of 1562.27 should be set aside using the average cost per claim method, while the estimate would be 1571.00 if we ignored the information on claim numbers.

13. Reserves of 984,340 should be set aside at the end of 2006.

D.2   Loss distributions

1. (a) $W \sim$ Pareto$(\alpha, 60{,}000)$, (b) $\hat{\alpha} = 4.6931$ and sd$(\hat{\alpha}) \doteq 2.09887$. (c) Letting W^* denote the excess over 40,000 in 2006, $E(W^*) = 16{,}462.64$.

3. $\hat{\alpha} = 7/(0.807430 + 0.640722) = 4.833747$.

5. Next year, the expected average amount paid will be $E(Y) = 394.31$.

7. If Y represents the random amount paid this year and Y^* next year, then (a) $E(Y) = 300 - 64.8 = 235.2$ (with a mean reduction of 64.8) and (b) $E(Y^*) = 250.34$.

9. Using $\bar{x} = 272.675$ and $s = 461.1389$, one obtains $\tilde{\alpha} = 3.075243$ and $\tilde{\lambda} = 565.8669$. This gives observed and expected values for the five intervals

of $(9, 9, 10, 4, 8)$ and $(8, 8, 8, 6, 10)$, respectively. The resulting χ^2 test statistic is 1.8167, and since $\chi^2_{0.95} = 5.991$, we have no reason to reject the fit of this Pareto distribution.

11. (a) For the exponential fit, $E_8 = 6, E_9 = 18$ and $E_{10} = 12$. The chi-square test statistic for the exponential fit is $\chi^2 = 41.64$ with a p-value less than 0.001, and hence the exponential fit is unacceptable.
(b) For the the Weibull fitted distribution, $E_8 = 6.21, E_9 = 5.44$ and $E_{10} = 0.02$. One should probably combine the last two intervals whereby the resulting expected number of observations would be 5.468414. The resulting chi-square test statistic would be $\chi^2 = 5.665817$ with a p-value of $1 - pchisq(5.665817, df = 7) = 0.580$, suggesting a much better fit! If one further combines the intervals 1 and 2 to form one interval, then the resulting χ^2 test statistic on 5 degrees of freedom is 5.03836 with a p-value of 0.539.

13. Using R, one obtains 173 and 369.5 as the first and third quartiles, respectively. With the method of percentiles, one finds that $\hat{\gamma} = 2.072234$ and $\hat{c} = 6.624574e - 06$. This gives rise to expected values of

$$(5.43, \; 6.30, \; 6.82, \; 6.53, \; 4.92)$$

in the respective intervals, and a chi-square test statistic of $\chi^2 = 3.455929$ with a p-value of 0.178.

15. Let $X \sim N(\mu_N, \sigma_N^2)$ be the normal random variable modeling claim size, and Y be the lognormal model where $\log Y \sim N(\mu_{LN}, \sigma_{LN}^2)$. Using the method of moments, one finds $\tilde{\mu}_N = 800, \tilde{\sigma}_N^2 = 350^2$ and $\tilde{\mu}_{LN} = 6.597045, \tilde{\sigma}_{LN}^2 = 0.1751343$. The 95th percentiles of X and Y are, respectively, $w_N = 1375.699$ and $w_{LN} = 1458.845$. If the distribution of X fits the data well, we would expect about 81 observations less than 125, while if that of Y does then we would expect none.

17. (a) One maximizes $\prod_1^n [e^{-(y_i - \mu)^2 / 2\sigma^2}]/\sigma$ as a function of μ and σ^2.
(b) The average size of a claim is $= 340.36$ and $P(X > 400) \doteq 0.27$.
(c) $\log W = \log k + \log X$.

19. $E(Y) = 629.3333$, while next year it would be $E(Y^*) = 645.2003$.

21. Using the method of moments, $\tilde{\mu} = 5.921898$ and $\tilde{\sigma}^2 = 0.329753$. Hence
(a) $P(X < 200) = 0.138757$ and (b) $P(X > 1100) = 0.029865$. (c) If Z is the amount of a claim X paid by the reinsurer, then $E(Z) = 9.10$ and the average amount for those involving the reinsurer is 304.71.

23. $E(X) = 0.2$ and $Var(X) = 0.22$.

D.3   Risk theory

1. $Var(S) = 6\lambda/\beta^2$ and $skew(S) = 4/\sqrt{6\lambda}$.

3. $E(S) = 100$, $Var(S) = 49{,}950$, $skew(S) = 2.2327$ and $P(S > 600) = 0.0174575$.

5. $E(S) = 22/3$, and $Var(S) = 16.55556$. $P(S = 9) = 0.0949$.

7.
$$P(N = n) = \left(q + \frac{(k-1)q}{n}\right) P(N = n - 1).$$

9. Letting S^* denote aggregate claims if business increases by a factor of k and θ^* be the necessary security factor, then
$$\theta^* = z_{1-\alpha}[\sqrt{Var(S^*)}/E(S^*)] = z_{1-\alpha}[\sqrt{Var(kS)}/E(kS)] = \theta/\sqrt{k}.$$

11. Using $E(S) = 4$, $Var(S) = 32$ and $skew(S) = 6/\sqrt{8}$, one finds $\alpha = 8/9$, $\delta = 1/6$ and $\tau = -4/3$. $P(S > 6) = 0.2522069$ and $P(S > 8) = 0.1825848$. Letting $S_{norm} \sim N(4, 32)$, then $P(S_{norm} > 6) = 0.3618368$ and $P(S_{norm} > 8) = 0.2397501$. With $S_{TG} = -4/3 + \Gamma(8/9, 1/6)$, $P(S_{TG} > 6) = 0.2525313$ and $P(S_{TG} > 8) = 0.1777525$.

13. The necessary reserves would be $U_A = 7248.58$, $U_B = 9907.89$ and $U_{AB} = 9483.11$. Note that $U_{AB} < U_B$.

15. $E(X) = 300$, $E(X^2) = 306{,}000$ and necessary reserves are $U = 15{,}724.06$. If $\theta = 0.3$, then reserves of $U = -6775.94$ would do. With Y, one finds that $P(Y > 400) = 0.2230367$, and that the same amount of reserves would do.

17. $\alpha = 0.9029363$, and the probability that the net premiums of the reinsurer will meet its claims is 0.9331928.

19. $P(S_I^M + P_R^M > 2{,}500{,}000) = 0.05083777$ at $M = 100{,}000$.

21. $M^* = 269.6938$. $M = 300 \Rightarrow EAP = 1440$, $M = 800 \Rightarrow EAP = 16{,}560$ and $EAP = 28{,}406.25 \Rightarrow M = 2000$.

25. (a) $E(S) = 91{,}000$, $Var(S) = 176{,}500{,}000$ and $skew(S) = 0.15299416$.
    (b) $U_0 = 17{,}256.3$, and (c) $P(S_1 > 50{,}000) = 0.02530859$.

D.4   Ruin theory

1. The probability that the first process is nonnegative for the first two years is 0.567147, while it is 0.507855 for the second.

3. $R = 0.014645$ and the Lundberg upper bound is 0.556668 for the first process, while $R = 0.029289$ with a corresponding Lundberg upper bound of 0.002857 for the second.

5. $R = 0.116204$, and $\psi(25) \leq 0.054743$.

7. Using a, b and c for the three processes, respectively, $R_a = 0.050364$ and $\psi_a(150) \leq 0.000524$. For b, $R_b = 1/60$ and $\psi_b(150) = 0.068404$. For c, $R_c = 0.012250$ and $\psi_c(150) \leq 0.159210$.

9. $R_0 = 0.0000697$ and $R = 5.776542e - 05$ with an upper bound on the ruin probability of 0.003099 when reserves are 100,000.

11.

$$ _2\psi(U) = \frac{e^{-\alpha U}}{2+\theta}\left[1 + \frac{\alpha U}{(2+\theta)} + \frac{1+\theta}{(2+\theta)^2}\right]. $$

13. (a) $\alpha \geq 0.25$, $R = 0.004076$ and $\psi(450) \leq 0.159735$.

15. $R(\alpha) \geq R(1) = 1/300 \Leftrightarrow \alpha \geq 6/13$.

17. (a) $\alpha \geq 0.2$ and $R = 0.009231$ with $\psi(300) \leq 0.062710$.

19. (a) $R \leq 2\theta(a + 2b)/(2a + 6b)$ and (b) $R = 0.152147$ with $\psi(50) \leq 0.000497$.

D.5   Credibility theory

1. To estimate $E(S) = \lambda E(X)$ with the desired precision, $r = 9$ years of data would suffice, while only 3 would be needed to estimate λ alone.

3. The numbers of lives needed are, respectively, 75,293, 76,193 and 76,193.

5. The partial credibility Z given to 1,200 claims would be 0.5772.

7. The posterior has mean 3.6529, median 4 and mode 3.

9. The Bayes estimate is the posterior mean $242/(20+\sum \log(1+x_i^2))$, with a 90% Bayesian belief interval for θ (using a normal approximation to the posterior) of the form

$$\frac{242}{20 + \sum \log(1 + x_i^2)} \pm 1.645 \frac{\sqrt{242}}{20 + \sum \log(1 + x_i^2)}.$$

11. (a) $\alpha = 2$ and $r = 100$. (b) The posterior for λ is $\Gamma(548, 10)$. A 95% Bayesian interval for λ would be of the form $(50.212, 59.388)$.

13. (a) The prior mean for θ is 625. (b) The posterior for θ is given by the density $f(\theta \mid \mathbf{x}) = 15(800)^{15}/\theta^{16}$ for $\theta \geq 800$, and has mean 857.1429.

15. The posterior mean and variance for α are, respectively, $(\alpha_1 + n)/(\beta_1 - \sum \log y_i)$ and $(\alpha_1 + n)/(\beta_1 - \sum \log y_i)^2$.

17. The pure premium would be

$$E(S_5 \mid \mathbf{s}) = (0.2725)\,\bar{s} + (0.7275)\,E[m(\Theta)]$$
$$= (0.2725)\,1125 + (0.7275)\,665 = 790.34,$$

where the credibility factor is $\mathbf{Z} = 4/(4+K) = 0.2725$ and $K = 10.6797$. $\mathbf{Z}$ is small and K is quite large due to the fact that the expected value of the process variance is much higher than the variance of the hypothetical means.

19. The posterior mean

$$E(\lambda \mid N = k) = \mathbf{Z}\,\frac{k}{n} + (1 - \mathbf{Z})\,\frac{\alpha}{\beta},$$

where $\mathbf{Z} = n/(n + \beta)$ is an increasing function of the sample size n and a decreasing function of β. It does not depend on α. We would estimate the number of claims next year to be 73.33.

21. (a) A *reasonable* prior for μ is $\mu \sim N(\mu_0 = 150{,}000, \sigma_0^2 = 10{,}204.08^2)$. (b) The posterior is $N(142{,}166.67, 4749.77^2)$, and a 95% Bayesian belief interval for μ is $(132{,}857.1, 151{,}476.20)$. (c) The classical (frequentist) 95% confidence interval would be $(129{,}481.53, 150{,}518.46)$.

23. The credibility premiums for regions A, B, C and D are, respectively, (in millions of $) 183.115, 249.137, 90.310 and 206.939.

25. For Model 1, the credibility (pure) premiums for the three risks are, respectively, 4,476.36, 3,952.50 and 4,746.14. Using Model 2, the credibility premiums (per unit risk) for risks 2 and 3 are, respectively, 7.56 and 6.28.

D.6   No claim discounting in motor insurance

1. He would have paid $400 in 2002 with the soft rule and $640 with the severe rule.

3.
$$\mathbf{P}^8 = \begin{pmatrix} 0.1863592 & 0.2440922 & 0.5695486 \\ 0.1859750 & 0.2444764 & 0.5695486 \\ 0.1859750 & 0.2440922 & 0.5699327 \end{pmatrix}.$$

5. $\pi = (0.2028, 0.1711, 0.1402, 0.1458, 0.1020, 0.2381)$.

8. (a) $E_0 \rightarrow 400, E_1 \rightarrow 560, E_2 \rightarrow 160$. (b) The respective rates for making claims are $(0.2298, 0.1779, 0.2875)$. (c) $\pi = (0.0450, 0.2195, 0.7355)$. (d) The long-run expected premium is $9,062,578.

9. Given that an individual has a (first) loss, the chances a claim will be made are, respectively, $(0.7985, 0.6977, 0.8353, 0.8353)$ for those in discount levels $E_{0\%}, E_{20\%}, E_{40\%}$ and $E_{50\%}$.

11. Thresholds for making claims are $E_0 \rightarrow 160, E_{20} \rightarrow 240, E_{40} \rightarrow 80$ and the limiting distribution is $\pi = (0.0068475, 0.0860274, 0.9071251)$.

13. (a) The thresholds for making claims are $(250, 350, 100)$, with corresponding probabilities of making a claim $(0.1136094, 0.0853332, 0.192)$, for the different discount levels. (b) The limiting distribution is found to be $\pi = (0.0164279, 0.1706438, 0.8129282)$, and in the limit, expected numbers in the respective levels are given by $(329, 3413, 16, 258)$.
(c) Expected premiums for next year would be in the region of $7,340,828, and in the long run they would be about $5,423,427.

15. (a) Threshold values for the discount classes are, respectively, 300, 420 and 120.
(b) First claim incidence rates are $(0.2440776, 0.2308395, 0.2499405)$.
(c) The expected number getting full discount next year is 3932.
(d) The total expected premium is $4,027,781.

17. $E(C_N) = 0.1647114$.

D.7   Generalized linear models

1. (a) 57.7%, (b) age ≤ 19.22, (c) a maximum of 84.7% for a 17-year-old Dublin student on 640 points, and a minimum of 51.2% for a 20-year-old student from outside Ireland.

3. Consider y_i as being fixed and we show that the function $g(x)$ given by

$$g(x) = [y_i \left(\log y_i - \log x\right) - (y_i - x)]$$

is nonnegative. Now g has only one critical point at $x = y_i$ since $g'(x) = (-y_i/x) + 1$. Since $g'' > 0$, this is a minimum for g and $g(y_i) = 0$.

5. The predicted probability of success for an equity-based fund with a \$2 million promotional budget would be $\hat{p} = 0.53644$. A budget level of $x = 32.96078$ (in \$ million) would make the property and bond products equally likely to be successful.

7. (a) 0.16696, (b) 0.68386 at age 20.

9. $\mathbf{X}^T\mathbf{y} = (1127.2,\ 117{,}648.6,\ 13{,}133{,}185.7)$, and $\hat{\sigma}^2 = 9.686355$. A significant change seems to have occurred in year 20 (see Figure D.1), and this should be taken into account in modeling the future.

12.

$$l(\theta) = A\,\frac{2\,y\,\theta/2 - \gamma(2\,\theta/2)}{\phi} + \tau(y, \phi/A)$$
$$= A^*\frac{y\,\theta^* - \gamma^*(\theta^*)}{\phi} + \tau^*(y, \phi/A^*).$$

13. Y can be expressed in exponential form where $\theta = \log(1-p)$, $A = \phi = 1$, $\gamma(\theta) = -k\log(1 - e^\theta)$, and $\tau(y, \phi) = \log \binom{y+k-1}{k-1}$.

15. Figure D.2 gives a plot of the predicted accident rates for companies A and B.

D.8 Decision and game theory

1. (a) is not really a game, while one could argue that (b), (c) and (d) are.

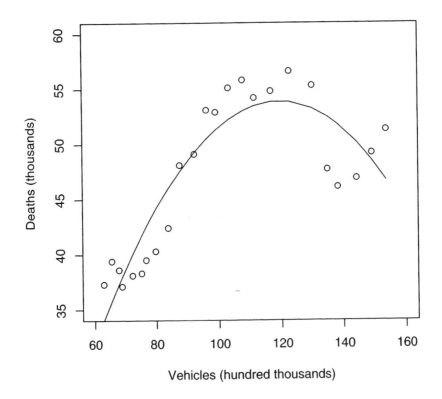

FIGURE D.1
Automobile deaths and vehicle registrations.

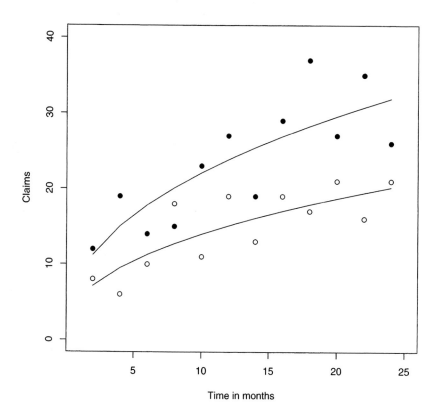

FIGURE D.2

Predicted accident rates for manufacturing companies A and B using a Poisson model.

3. The thief should visit warehouse 1 with probability 0.2, and the value of the game is 20. The optimal strategy for the security agent is to guard the more valuable warehouse (W1) 80% of the time.

5. There are two saddle points here and the game has a value of 5.

7. If $X = 7$ and $Y = 1$, then the value of the game is 7. When $X = 9$ and $Y = 6$, the optimal strategy for Ann is to play II with probability $p = 2/9$ and III otherwise, resulting in a game with value $74/9$.

9. The optimal strategy for Richie is to pick I and IV with equal probability $1/2$. Mort would choose strategy 2 with probability $1/4$ and 3 otherwise. The value of the game is 3.

11. $(Rugby, Rugby)$ and $(Football, Football)$ are both points of Nash equilibrium.

13. Decision function d_1 (where $d_1(0) = 0$ and $d_1(1) = 1$) has the minimum Bayes risk of $1/3$.

15. Both the minimax (risk) and the Bayes decision rules are d_2, and the Bayes risk using d_2 is $7/16$.

17. The probability a triangle can be formed is $1/4$. The Bayes risk associated with taking action Yes is $3/4$ while that for action No is $2/4$, and hence the optimal Bayes decision is to say No.

19. The risk function for d_1 takes the values $(4/81, 5/48, 13/81)$ while that for d_2 is $(7/81, 1/6, 7/81)$. Hence d_1 is both the minimax and Bayes decision function.

21. $P_1 = 138.63$, $P_2 = 115.07$ and $P_{1+2} = 253.70$. $P_1 + P_2 = P_{1+2}$ because X_1 and X_2 are independent.

Index

accident year, 2
Ace-Six Flats die
    example, 281–283
adjustment coefficient, 134
    adjustment equation, 135
    upper bound for, 137
aggregate claims, 78
    premiums and reserves, 99–107
    reinsurance for, 107–119
    setting aside reserves for, 103–107
AIC, *see* Akaike information criterion
Akaike information criterion, 58, 251
analysis of covariance - ANCOVA, 226
analysis of variance - ANOVA, 226
    deviance and, 245
Anderson–Darling test, 53
Arrow's theorem, 293–295
average cost per claim method, ix, 11–14

basic probability distributions, 309–311
basic tools in probability and statistics, 313–325
    conditional probability distributions, 317–319
    convolutions of random variables, 316–317
    double expectation theorem, 313, 319–324
    $E(X \mid Y)$, 319
    $V(X \mid Y)$, 322
    maximum likelihood, 324–325
    moment generating functions, 313–316

    cumulant, 314–315
Bayesian credibility, 160, 164–170
    Normal - Normal model, 165
    Poisson - Gamma model, 167
Bayesian statistics, vii, x, 327–333
    an introduction, 327
    and loss functions, 329–333
    Bayesian belief interval, 332
    Bayesian estimator, 330, 331
    conjugate families, 328–329
    Binomial | Beta model, 328
    Burr | Gamma model, 332
Bonus–Malus systems, x, 192–193
Bornhuetter–Ferguson method, ix, 4, 8, 14–19
Bühlmann credibility
    estimators, 174
    parameter, 160

censored losses, 63
central limit theorem, 97
chain ladder method, 3–14
    average cost per claim, 11–14
    basic, 5–8
    inflation-adjusted, 8–10
characteristic functions, 316
chi-square goodness-of-fit test, 54–56
claim severity, 24
claims
    cumulative, 27
    IBNER, 1
    IBNR, 1, 14
    incremental, 8, 9
    incurred, 1–2, 14, 24
    open, 1
    outstanding, 1, 2, 24, 28, 30
    paid, 5, 14, 15